Temporary Works
Part Two

emerald
PUBLISHING
ice Publishing

Temporary Works Part Two

Further principles of design and construction

Edited by
Ray Filip

Published by Emerald Publishing Limited, Floor 5,
Northspring, 21-23 Wellington Street, Leeds LS1 4DL.

ICE Publishing is an imprint of Emerald Publishing Limited

Other ICE Publishing titles:
Temporary Works: Principles of design and construction, Second edition
Peter F Pallett and Ray Filip. ISBN 978-0-7277-6338-9
Structural Design of Buildings: Fundamentals in Design, Management and Sustainability
Edited by Feng Fu and David Richardson.
ISBN 978-1-8354-9577-3
ICE Manual of Bridge Engineering, Third edition
Edited by Gerard Parke and Nigel Hewson. ISBN 978-0-7277-6305-1

A catalogue record for this book is available from the British Library

ISBN 978-1-8360-8525-6

Cover photo: Courtesy of MACE

Commissioning Editor: Michael Fenton
Content Development Editor: Cathy Sellers
Production Editor: Benn Linfield

Typeset by KnowledgeWorks Global Ltd.
Index created by David Gaskell
Printed and bound by CPI Group (UK) Ltd, Croydon, CR0 4YY

Dedication

This book is dedicated to the memory of Peter Pallett and Stuart Marchand, both of whom contributed so much to temporary works over the last five decades and have left us too soon.

Peter was my mentor, colleague and friend. He was Technical Manager for Rapid Metal Developments for many years, a member of the British Standards Institute technical committee responsible for BS 5975 from 1977 onwards, a founding member of Temporary Works Forum, Fellow of the Institution of Civil Engineers and Fellow of the Concrete Society. Over the last 30 years, he trained numerous temporary works coordinators in many countries. Initially I did a temporary works training 'double-act' with Peter, whereby he taught 'above-ground' and I taught 'below-ground'. Peter was involved in many technical publications and was a Law Society expert witness. I collaborated with Peter on many publications and co-edited part 1 of this publication. His attention to detail was second to none and he taught me so much.

Stuart was founder and Managing Director of Wentworth House Partnership, also a long standing member of the BS 5975 technical committee and a founding member of Temporary Works Forum. He was a mentor to many young engineers and contributed to numerous industry publications, including previous versions of this publication. Stuart was my friend for 20 years and over the years I jointly worked on many publications with him, benefiting from his knowledge and experience.

Peter and Stuart were both 'go to' people for complex engineering problems and questions.

The editor, chapter authors and whole temporary works community would like to commemorate the memory of our much-respected colleagues Peter and Stuart and thank them for their contributions. They were pioneers and made us all proud to call ourselves 'temporary works engineers'. They are truly irreplaceable and will be missed.

Contents

Foreword

Temporary works have always been, and remain very much, a key stage in the safe delivery of construction projects, whether they be a simple shallow excavation in the ground or a complex multi-location project like HS2. How much people are aware of this, or provide due regard and early consideration of it, is still up for debate amongst many professionals. Comments heard over the years, like 'temporary works are something the contractor has to consider', 'our insurance dosesn't cover temporary works', or the long-lived 'we've always done it like that in the past', are not appropriate in a modern industry. Neither were they appropriate in the past, but they have all too often been inappropriately used.

Our industry has improved for the better. The large-scale temporary works collapses and failures of the 1960s and 1970s are now a relatively rare occurrence in the UK construction sector. The continuing development of BS 5975 has led to improvements in the way temporary works are designed, built, inspected, used and removed. That is not to say that near misses and close calls haven't occurred, and we should all learn from them.

Those involved daily in the field of temporary works often mutter phrases such as 'why has the permanent works designer done that?', 'surely the contractor knew this had to be carried out in a directed sequence?' and 'well, it was obvious that would happen, wasn't it?'. It's at those in the wider industry that the shared experience and knowledge contained within this book is aimed, as well as practising temporary works professionals. The intent and purpose of this book align very much with the objectives of the Temporary Works Forum which are as follows:

- Give authoritative guidance and, when required, professional leadership to the industry.
- Consider aspects of permanent works and interfaces between permanent works and temporary works, as are relevant.
- Consider both current practice and likely development.
- Be aware of trends and innovations in design, construction and use.

Continuing professional development should be something that we all strive for, not just for our own benefit but for the benefit of those working on projects with us, the members of the public impacted by our works during construction and those after completion. This is very much aligned with one of the principles of the Health and Safety at Work etc. Act and very much with the broader principles of the Construction (Design and Management) Regulations ('CDM'). This revision and update builds on the original chapters while adding further areas of knowledge and shared wisdom, not just to enhance our engineering knowledge but to share key and important issues with the wider construction community.

This publication is not just aimed at temporary works specialists, but anyone involved in the temporary works process who would like further information. It should be read in conjuction with *ICE Temporary Works: Principles of Design and Construction*, 2nd edition. The definition of temporary works covers a wide scope of items and the information provided is relevant to all sectors of the construction industry (civil engineering, building, demolition, utilities etc.).

My thanks go to all those involved, directly and indirectly, in sharing valuable experience.

Robert Millard
Managing Director, De Construction Solutions;
chair, Temporary Works Forum, UK

Preface

Temporary works is a wide subject which covers all aspects of the construction industry. It has developed and continues to develop with an increasing awareness of its importance. I have been involved in temporary works now for over 35 years and have seen many changes. Perhaps the most satisfying is the positive change of status of temporary works for the engineers and other professionals involved. Temporary works is now recognised as a separate specialist subject which requires a set of unique skills, perhaps above all 'practicality' – knowing what solution 'will work' effectively and safely. I have been fortunate to have worked for, collaborated with and known some outstanding engineers, from whom I have learnt so much.

The inquiry into the Grenfell Tower disaster highlighted many failings across the construction industry and regulatory framework. Competence of organisations and individuals was identified as a significant issue. Knowing what you are competent in and being able to prove and maintain / develop that competence are vitally important. Considering this, there is some concern amongst senior members of the temporary works community due to changes to the attributes of professionally qualified membership (Chartered Engineers), whereby 'designers' no longer require 'site experience'. For example, the site experience requirement was removed by the Institution of Structural Engineers (IStructE) and Institution of Civil Engineers (ICE) a few years ago.

The IStructE criteria for the 'Construction' core objective are (from IStructE website):

The candidate must demonstrate that they have gained experience of:

- construction techniques, construction plant and machinery
- designing, specifying or reviewing temporary systems
- reviewing construction programmes and construction sequencing
- reviewing or producing fabrication or shop drawings
- erection methodologies
- identification and correction of potential errors
- how their observations of construction / fabrication have informed their approach to design / buildability.

ICE criteria include (from ICE website):

- Understanding and Practical Application of Engineering.
- Maintain and extend knowledge of engineering theory and practice, and how technology assists its application.
- Identify, review and select techniques, procedures and methods to undertake engineering tasks.

This would appear to be at odds with:

- requirements for continuing professional development
- Construction (Design and Management) Regulations 2015 - Part 3 - Section 9 - Duties of designers
- Management Regulations - Schedule 1 - General Principles of Prevention
- ICE attribute: for all professional grades to have an understanding and practical application of engineering and to exercise sound engineering judgement.

If a designer has not had first-hand experience of the site environment, site risks and the construction process, they are likely to lack the awareness of how to minimise their design risks and how to produce a buildable scheme. This could directly result in impractical, inappropriate, unsafe and poor value designs. A design which properly addresses design and construction risks is one that is both practical and safe and can be considered as 'buildable'. Site is where design buildability is tested. Without site experience one of the two components of competence (experience and knowledge) is compromised. This therefore compromises the ability to produce a design that is both practical and safe.

Anyone who considers themselves to be a competent, well-rounded designer should have direct hands-on experience of site working practices and have an insight into practical issues and conditions prevalent in the construction environment. Furthermore, site experience is considered by many to be an essential attribute for a temporary works designer and needs to be maintained throughout their career.

The editor, in conjunction with Temporary Works Forum (TWf), has been carrying out research into temporary works accidents and near misses (see chapter 18) and capturing / recording senior temporary works professionals' experience. One of the legacy questions we asked was:

> What advice would you give to a young engineer wanting a career in temporary works?

The most common type of answer that was given was:

> Get site experience first to understand how things are built and the risks involved in working on site.

Finally, section 6 of the Bragg Report states:

> the effectiveness of training depends on its integration with experience on site… The Institution should accept that an essential qualification for corporate membership should be proven awareness of the importance of safety and safe working.

Ray Filip BEng (Hons) MSc DIC CEng FICE
Managing Director, RKF Consult Ltd, UK

Acknowledgments

I would like to thank all those authors who contributed and shared their experience in the production of this book. I would also like to thank all those who contributed to my research in chapter 18 and the Temporary Works Forum legacy project (this work will continue after this book is completed and will be made publicly available).

Finally, I extend my thanks to: Chris Bennion, Godfrey Bowring, Ben White, Bill Hewlett and Ben Beaumont for their advice and help at various stages during the production of this book.

About the editor

Ray Filip BEng (Hons) MSc DIC CEng FICE
Managing Director, RKF Consult Ltd, UK

Ray is a Chartered Civil Engineer and Fellow of the Institution of Civil Engineers, with over 35 years of experience in temporary works. Educated at University of Surrey and Imperial College, London, he spent the first 18 years of his career working for major contractors and specialist subcontractors in the UK and Southern Africa. He was involved in various aspects of temporary works, including design, design checking, pricing, construction and inspection. In 2007 he formed his own specialist temporary works consultancy, RKF Consult Ltd. He has had the roles of Designated Individual, Temporary Works Coordinator and Temporary Works Supervisor at various points.

Ray has designed numerous items of temporary works and advised many UK and international organisations on temporary works procedures, design management and risk management. He also carries out site audits and inspections and is regularly consulted by the Health and Safety Executive on temporary works issues. He has provided a wide variety of temporary works and other technical training courses to organisations in the UK, a number of European countries and Australia.

Ray has been involved in producing technical publications for Construction Industry Research and Information Association (CIRIA), The Concrete Society, Contractors Plant-Hire Association, Deep Foundations Institute and Institution of Structural Engineers. Since 2016 he has been a member of the committee responsible for the British Standard on temporary works (BS 5975).

About the contributors

Adrian Bywaters BEng (Hons) CEng MICE
Adrian is a Chartered Civil Engineer with 35 years' experience in the construction industry. He is an Operations Director for Engineering at Mace and has wide-ranging knowledge of all stages of the construction process, from tender, through delivery and to completion. He is recognised for expert knowledge, with particular respect to technical management, temporary works, safety and quality management and project assurance. He is also passionate about the mentoring and professional development of our young civil engineers.

Nick Cook
Formerly with Laing, Laing O'Rourke and Balfour Beatty (now retired), Nick has been designing and involved with temporary works in heavy civil engineering for 49 years. His work has included temporary steel structures, such as specialist lifting equipment, slipform shutters, loading and access platforms, excavation supports, façade retention, lifting and moving systems. Projects have included British Sugar silos, Sizewell B, Cardiff Principality Stadium, Second Severn and Oresund Crossings, Heathrow Terminal 5 and Crossrail, to mention a few.

Professor Colin Eddie FREng BSc (Hons) CEng FICE
Colin is an internationally recognised expert in tunnel design and construction with over 40 years of experience. He has a deep understanding of various tunnelling methods and ground improvement techniques. As the former Engineering Director at a major UK tunnelling contractor, he has contributed significantly to the sector by developing numerous innovations in tunnelling and underground space which have been credited with improving the safety, efficiency and sustainability of many recent major projects. Today he and his organisation (CECL-Global) provide expert support and advice to an international client, and he manages an active research and development programme.

Matthew Hall BA (Hons) MA PhD
Matthew is Director of Technology at Richter, heading up the development of their in-house autonomous design generation tools. He is dedicated to a career of innovation within temporary works and is driven by a mission to reduce harm on site while maintaining fully engineered and optimised solutions. Matthew spent much of his career at Costain as a temporary works designer and found his passion for driving

such initiatives while on site as a temporary works supervisor. Since graduating with a PhD from Swansea University, he hopes to further innovate the developed streamlined processes with the Richter in-house tools by incorporating finite element analysis with a rapid modelling interface, creating new suites of tools targeted directly at the swift delivery of onsite temporary works design support.

Sam Hurst MEng CEng CEnv MICE
Sam has over a decade of experience as a Chartered Civil Engineer and Chartered Environmentalist. He has a proven track record in leading sustainability initiatives and implementing carbon reduction strategies on large-scale projects, particularly within the infrastructure sector. Sam's expertise extends to temporary works design and construction, where he champions the integration of sustainable practices. He is a recognised authority in his field, contributing to writing multiple guidance notes and 'how-to guides' on low-carbon temporary works and construction practices. He is a Senior Sustainability Engineer.

Jan de Klerk BSc MSc CEng MIMechE
Jan started his engineering career in 2011 at TWD in the Netherlands. In 2019 he moved to the UK to help set up the UK branch of TWD and is currently the technical lead there. Over the years he gained extensive experience in designing working methods, equipment and temporary works for the installation of a wide variety of projects in the marine civils and offshore industry.

EUR ING Ronan O'Driscoll BEng (Hons) MSc CEng IntPE(UK) MICE MIMechE MIStructE LCGI DIS MNucl
Ronan is a Chartered Engineer with over 25 years of experience in construction engineering, specialising in temporary works. He is the temporary works lead on the Hinkley Point C construction project, ensuring compliance with BS 5975 standards. Ronan has worked on a variety of projects, both managing and delivering temporary works on site as well as designing temporary works and management off site. Ronan regularly shares valuable industry insights on LinkedIn. He authored *Construction Engineering – A Guide to Temporary Works in Construction & BS 5975:2019*, which he has shared freely to benefit the industry since December 2022.

Jeff Mahony BEng (Hons) CEng EUR ING FIEI FICE
Jeff is Head of Temporary Works for the HS2 (S1/S2) project which, when it was started, was considered the largest infrastructural project in Europe. With over 27 years of experience in major civil engineering projects throughout the UK and Ireland, his experience extends from major road and rail schemes, major pipelines, significant demolition works in the centre of London to time in the military as an engineer officer. The last 12 years of his career have been completely dedicated to the temporary works industry, most of which have been with Costain. Passionate about innovation, he has led various improvement projects in temporary works within the HS2 programme, along with intercompany reviews and improvements at Costain. He is also heavily involved with the Temporary Works Forum on a regular basis and has recently chaired a working group for the 'Design and Construction of Granular Working Platforms for Construction Plant'.

Robert Millard BEng (Hons) MSc CEng MICE MIDE
Rob is Managing Director of De Construction Solutions and the current Chair of the Temporary Works Forum (since June 2022), having previously been a director of the board. With 26 years of experience across many sectors of construction, he has spent the last ten years providing specialist engineering support to the demolition, enabling and construction fields. In addition, through the development and delivery of temporary works courses, Rob is able to provide relevant and up-to-date learning opportunities in the field of temporary works. More recently his SKEB has been put to further service to clients, providing expert witness support involving demolition engineering, temporary works and construction practices.

Dr Henry Pairaudeau MA MEng PhD CEng MICE
Henry has over 14 years of experience in engineering design, management and research for major infrastructure and tunnelling projects. His major project contributions include the Lee Tunnel, Thames Tideway West and the UK's Geological Disposal Facility for radioactive waste. Recently, he has completed a PhD at the University of Warwick, researching novel materials for tunnel linings.

Martin Pike MSt MEng CEng MICE
Martin is a Chartered Civil Engineer with over 20 years of experience in the construction industry. With expertise in technical engineering, prefabrication projects, and health,

safety and quality management, Martin has contributed to the successful delivery of high-profile projects, including airports, data centres and residential developments. As an Associate Director at Mace UK, he has led the development of innovative construction methodologies, particularly in off-site manufacturing, ensuring value and efficiency. His practical problem-solving approach and knowledge of temporary works and concrete structures have been key to delivering successful outcomes.

Malachy Ryan BEng MSc CEng MICE
Malachy is a Chartered Civil Engineer and serves as the Strategy Director at Alan White Design in Scotland. With a wealth of experience, Malachy has held the role of ICE Supervising Civil Engineer and previously served as a Director of the Temporary Works Forum. Malachy's expertise extends to his past involvement as a committee member of BSI B/525/9, focusing on the structural use of aluminium. He holds a master's degree in concrete technology from the University of Dundee, further enhancing his knowledge in the field. Over the span of 15 years, Malachy has made significant contributions to the temporary works industry.

Simon R Smith CEng FICE FIStructE DipEM FFB
Simon is Managing Director of Contractors Design (CDS) Ltd, which he established in 2000. CDS has now become a well-established business offering a wide range of engineering design services in the fields of demolition, geotechnics, building refurbishment and basements, together with all facets of temporary works. Simon has over 40 years of experience in the construction industry, most of which have been acquired in the management and design of temporary works.

David Thomas MSc BSc (Eng) AKC CEng FICE CFIOSH MInstRE
David is a Chartered Civil Engineer and a Chartered Safety and Health Professional. Largely retired now, his career has included engineering and management consultancy, enforcement, contracting and promoting industry best practice. He has provided support and advice on work at height and the management of temporary works. David was a Director and Company Secretary, Temporary Works Forum from 2014 to 2023. He worked at the Health and Safety Executive (1997 to 2007) and was latterly an HM Principal Specialist Inspector of Health and Safety (Construction). He was SHE Director

at William Hare Ltd (2007 to 2011), a global leader in engineered steel solutions. In 2023 he was awarded the Garth Watson Medal by the Institution of Civil Engineers (ICE) for dedicated and valuable service in the field of temporary works and the whole of the civil engineering industry, and in 2024 was awarded the Darren Saffill Lifetime Achievement Memorial Award by the Northwest Construction Safety Group. He is the Director of Heightsayfe Ltd. He is also trustee of No Falls Foundation.

Mike Webster BEng MSc PhD DIC CEng FICE FIStructE
Mike is a chartered engineer with over 30 years of experience and the Director of MPW R&R Ltd. He specialises in CDM, construction and structural safety and provides consulting, forensic and expert witness services in those areas. He is a member of the Temporary Works Forum, the Institution of Structural Engineers' Health and Safety Panel, Structural-Safety and CROSS (the panel for Collaborative Reporting for Safer Structures). Mike has led independent evaluations of CDM 1994 and CDM 2007 and, in 2021, was appointed by HSE to lead a review of the CDM 2015 Principal Designer role. Mike frequently acts as an expert witness following collapses and other problems on construction sites (including temporary works) and has been instructed by both prosecution and defence teams.

Steve Williams BEng (Hons) CEng FIStructE FICE MPWI
Steve is a Buildings and Civils Engineering Expert for the Technical Authority in Network Rail. He has played a pivotal role in the design, execution and assurance of numerous projects for over 35 years in various sectors of the construction industry. His specialisms include the structural engineering of building and civil engineering structures, and the management of temporary works interfaces to ensure efficient design and construction in the railway sector. He leads Network Rail's Health & Safety by Design Buildings and Civils Working Group. He is actively involved in many forums, panels and initiatives, including the Temporary Works Forum, Collaborative Reporting for Safer Structures UK (panel expert), Institution of Structural Engineers' Design Practice and Structural Safety Committee (panel member) and the Get It Right Initiative.

Kit Yardley
Kit has a career in the design of temporary works spanning five decades, with George Wimpey, Laing and Laing O'Rourke.

He has designed a large variety of temporary works elements on projects such as National Bank of Dubai, Houses of Parliament Victoria Tower refurbishment, Sizewell B NPS, Second Severn Crossing, Cardiff Principality Stadium, A55 North Wales and Edinburgh Airport Control Tower. He is now semi-retired and but still delivers temporary works training.

Ray Filip
ISBN 978-1-83608-525-6
https://doi.org/10.1108/978-1-83608-524-920251001

Chapter 1
Introduction, understanding temporary works duties and recent updates

Ray Filip
RKF Consult Ltd, UK

Robert Millard
De Construction Solutions, UK; Chairman, Temporary Works Forum, UK

David Thomas
Retired Former Secretary, Temporary Works Forum; Formerly HM Specialist Inspector Health and Safety Executive UK

Dr Mike Webster
MPW R&R Ltd, UK

1.1. Introduction

Temporary Works: Principles of Design and Construction, 2nd edition (Pallett and Filip, 2019) is now six years old. Most of the information provided in the second edition is still relevant and valid today. This publication should be read in conjunction with the second edition, as the aim is to provide relevant updates and expand the information by providing 18 new chapters, covering subjects which were not included in the second edition. It further highlights the wide variety of items and subjects under the heading of 'temporary works'. This book summarises and signposts existing guidance and information available into a single document for easy use and reference. In addition, some of the existing guidance and information is now several decades old so, where appropriate, technical updates and personal experience have been added. This publication also provides some pertinent learning from Temporary Works Forum (TWf), recent major projects such as Hinkley Point C and High Speed 2, as well as many smaller projects.

The construction industry continues to evolve, and this book considers important trends which can affect temporary works, such as modern construction methods and the increasing use of digital technology and artificial intelligence. Finally, across society, there is an increasing awareness and concern regarding the detrimental effect that humans are having on our planet and this publication highlights how we should consider sustainability and the environment for temporary works purposes.

1.2. Continuing industry changes

It is now 50 years since the publication of the Health and Safety at Work etc. Act (HMG, 1974) and almost 50 years since the publication of the Bragg Report (Bragg, 1975) (see 2nd edition, section 1.2). The last half century has seen significant changes in the construction industry and both these documents have contributed to significant improvements in overall health and safety performance. However, temporary works failures are still commonplace and often catastrophic. Chapter 18 of this publication explores some of the common reasons for temporary works incidents, near misses and failures. It includes a summary of issues obtained from the Health and Safety Executive records, personal archives and responses to a questionnaire received from eminent temporary works engineers, who between them have extensive temporary works experience and knowledge.

Section 1.2 of Pallett and Filip (2019) highlighted some industry changes that have occurred since the early 1970s. The following may also be added as of result of further evolution in recent years (including the evolution of legislation):

- Temporary works has traditionally been seen as the domain of principal / main contractors and some specialist subcontractors (concrete frame subcontractors, groundworkers etc.). There is an increasing realisation that many other specialist sectors are involved in temporary works, such as mechanical / electrical contractors, process engineering (oil / gas sectors), utilities and building maintenance contractors. There continues to be an increasing level of awareness of the important roles and responsibilities for those involved in the management of temporary works (see section 1.3).
- There is a continued drive in the industry towards offsite manufacture, prefabrication, standardisation of components and automation, with elements being assembled on site, rather than traditional construction techniques using traditional construction skills.
- The traditional role of main / principal contractor, utilising their own in-house equipment, plant, labour and temporary works design skills, has continued to diminish. These contractors seek to divert risk and duty downwards and manage package subcontractors with their own specialist skills (including specialist temporary works design and construction). The common name for this is management contracting.
- The development and utilisation of specialist technology (software, digitalisation, automated equipment etc.) in the construction industry and its use for temporary works purposes. Technology is rapidly developing and evolving at an ever-increasing rate. It could be argued that the construction industry is struggling to keep up with and embrace some of these developments, especially when compared to other industries, such as manufacturing or pharmaceuticals.
- Relatively large independent temporary works design consultants now offer a wide variety of professional services to multiple contractors. However, it should be noted that relevant professional indemnity insurance is becoming increasingly costly and comes with significant exclusions (CROSS, 2021 and TWf, 2022a).
- There is some continued development in awareness of clients, principal designers (TWf, 2017 and Webster, 2023) and permanent works designers of their duties relating to temporary works (it could be argued that progress is relatively slow – see section 1.4 and chapter 2). Also, increasingly permanent works designers are recognising the opportunity

to develop their business model by offering a temporary works design service, thereby creating a 'one stop shop'. However, as mentioned in section 1.2(e), there can be issues with professional indemnity insurance.

- The construction industry is taking steps towards a more sustainable, reusable, considerate and environmentally aware future. Temporary works cannot be excluded from this ambition, but the overall rate of progress has been relatively slow and further positive action is becoming increasingly critical.
- The construction industry is international, with UK companies having to compete against multinational companies. In the UK domestic market there is an increasing choice of international organisations offering a wide range of specialist services, products, equipment and labour for temporary works purposes.
- Since the worldwide Covid pandemic of the early 2020s, working practices are changing. Flexible and remote working are becoming increasingly commonplace, along with the use of remote meetings and training. The construction industry is starting to embrace these opportunities but must also be aware of the limitations (reduced mentoring / supervision and less hands-on experience).

1.3. Awareness of temporary works and management roles

There is an increasing awareness of the importance of temporary works to the safe, efficient and successful completion of a project. BS 5975 Parts 1 and 2 (BSI, 2024a, 2024b) are widely recognised in the UK and referred to around the world – for example, Eire, Australia and New Zealand. Many organisations around the world now follow the general principles of BS 5975.

Temporary Works Forum (TWf) has now been in existence for 15 years and membership in the UK continues to increase. Members are now not only from Tier 1 contractors but include a wide range of organisations with an interest in temporary works (e.g. clients, equipment suppliers, specialist subcontractors, designers, academia). TWf has been recognised by the Institution of Civil Engineers (ICE) as a 'specialist knowledge society' and continues to produce specific guidance and provide a focal point for all matters relating to temporary works. Also, there are now several active TWf branches around the world (the original group started in London), including Australia, New Zealand and Hong Kong. TWf has links with international organisations such as the American Society of Civil Engineers (ASCE) to share knowledge and experience.

The importance of the specific management roles recommended by BS 5975 (BSI, 2024a, 2024b) have been recognised. There is now a series of widely recognised training courses (for temporary works coordinators and temporary works supervisors) administered by Construction Industry Training Board (CITB), with a transferable qualification, granted following the successful completion of an examination. However, it should be stressed that the qualification alone does not infer competence of an individual and should be part of a wider overall assessment of competence, which includes specific experience and knowledge of temporary works and personal attributes. TWf has also started to produce a series of e-learning modules, which are available for all. Some training providers also provide temporary works design courses.

In 1975 the Bragg Report (Bragg, 1975) made a recommendation (number 20) to include 'falsework' in university civil engineering design courses; over the years the uptake on this has been slow (this recommendation is also included in BS 5975-1 (BSI, 2024a, clause 4.2). Temporary works subjects are increasingly being taught at universities in the UK (such as Swansea University), offering specific temporary works modules or carrying out research into temporary works issues. Individual companies and TWf provide funding for research and offer sponsorship to courses and student scholarship.

1.4. Understanding temporary works duties and responsibilities

It is accepted that a suitable temporary works management procedure plus trained and experienced personnel are a good starting point for demonstrating legal compliance. However, across the construction industry there continues to be some confusion regarding these legal obligations and duties relating to temporary works. The term 'temporary works' is rarely explicitly mentioned in law or general regulations or contracts; however, they can be implied. It is important to differentiate between criminal law and contract law, whereby criminal law (such as health and safety laws and duties) take precedent over civil law (such as contracts and contractual duties – i.e. client requirements). A client cannot attempt to contract out of their legal duties or require an individual or organisation they employ to break the law. Figure 1.1 is an overview of the hierarchy in the UK.

Pallett and Filip (2019) section 1.7 provides further information on contractual obligations.

The Health and Safety Executive (HSE) is tasked with enforcement (Office of Rail and Road undertakes this role for rail construction projects and Office for Nuclear Regulation undertakes this role for nuclear construction projects) and their inspectors focus attention on ensuring that appropriate temporary works management procedures are in place, commensurate with the scale and complexity of the project and the construction risk involved. HSE inspectors can issue an improvement notice, a prohibition notice, seize, render harmless or destroy dangerous items and can ultimately prosecute. If there is a material breach the HSE will charge a 'fee for intervention' at an hourly rate.

It is essential that roles, duties and responsibility are clearly identified, clarified, assigned and communicated to avoid ambiguity and confusion and to improve operational efficiency. A spreadsheet (matrix) can be created covering the headings of: responsible, accountable, consulted and informed (RACI) (see TWf, 2019a), with the various stakeholders listed on an organisational project-wide and individual level.

Note on definitions:

- Duty – an obligation or commitment that someone must fulfil in accordance with laws, regulations or conventions and is performed for the 'greater good' rather than for 'self-interest'. Duty is often an integral part of any role and cannot be transferred to other parties.
- Responsibility – being accountable, the act of accepting and acting on an assigned task to achieve a desired result; these actions may be voluntary.

Figure 1.1 Hierarchy of control documentation (editor's own)

Acts – primary legislation (the law) enacted by Parliament (can be wide-ranging but without specific detail).

↓

Regulations – secondary legislation approved by a Minister. Contain specific details.

↓

Approved Code of Practice – practical examples of good practice (measuring tool for what is reasonably practicable) with special legal status. A court can find against you if the relevant provisions are not followed.

↓

Guidance – produced by organisations such as the HSE with the purpose to interpret the law. Following the guidance normally ensures compliance with the law.

↓

Standards – international, European and British standards are written by a committee of subject-matter experts for the industry and provide specific recommendations on good practice.

↓

Industry guidance – industry bodies produce good practice guides and advice in their own field of expertise.

↓

Company guidance – produced either internally or externally at the request of senior managment and may include policies, procedures, lessons learnt and guidance notes, setting out how the company operates.

1.4.1 Legal obligations and requirements (Acts)

If in breach of part of an Act, then the individual or organisation can be prosecuted (legal criminal action). Table 1.1 provides some examples of Acts which can directly or indirectly relate to temporary works.

1.4.2 Regulations

Regulations usually accompany HSWA but can include those accompanying European Directives (now part of UK law). If in breach of part of a regulation, then the individual or organisation can be prosecuted (legal criminal action). Table 1.2 provides some examples of regulations which can directly or indirectly relate to temporary works.

1.4.3 Standards / codes of practice / guidance

Table 1.3 contains examples of standards and organisations that have an influence or interest in temporary works. They produce good practice recommendations and specific guidance. Following these recommendations is a reasonable course of action under law. However, it is not obligatory to follow these recommendations (free to take other actions) but, if something goes wrong, they may be used in evidence against you in court and you would have to justify

Table 1.1 Some Acts which can relate to temporary works

(HSWA) Health and Safety at Work etc. Act 1974 (HMG, 1974)	HSWA does not specifically mention temporary works (temporary works is implied) but it gives general duties relating to health and safety: ▪ duties on employers in relation to employees and the public ▪ duties on suppliers in relation to their services and products ▪ duties on employees not to risk themselves or others (i.e. employers must manage their work safely and hence by implication they must manage temporary works safely). Most of the duties are qualified by the principle of 'so far as is reasonably practicable' whereby there is balance between feasibility and cost.
Corporate Manslaughter and Corporate Homicide Act 2007 (HMG, 2007)	This Act was created to overcome limitations of the common law offence of 'gross negligence manslaughter' as applied to companies. It clarifies the criminal liabilities of an organisation (corporate liability) where the way in which its activities (including temporary works) are managed or organised cause a person's death and amounts to a gross breach of a relevant duty of care owed by the organisation to the deceased. Individuals (e.g. directors and site managers) could still be prosecuted for other health and safety offences (e.g. gross negligence manslaughter).
Health and Safety Offences Act 2008 (HMG, 2008)	See Pallett and Filip (2019) section 1.6 for summary.
Building Safety Act 2022 (HMG, 2022a)	See section 1.5.1

Table 1.2 Some regulations which can relate to temporary works (continued on next page)

Lifting Operations and Lifting Equipment Regulations 1998 (HMG, 1998a)	Place duties on organisations and individuals who own, operate or have control over lifting equipment. Such equipment is often used for temporary works purposes.
Provision and Use of Work Equipment Regulations 1998 (HMG, 1998b)	Requires that any equipment provided for work (including temporary works equipment) is suitable for its intended purpose, safe, well maintained, inspected to ensure it does not deteriorate with use and persons using the equipment are provided with adequate training and information.

Table 1.2 Continued

Management of Health and Safety at Work Regulations 1999, amended 2003 and 2006 (HMG, 1999)	More explicit guidance given to employers on management of health and safety (under HSWA) and requires employers to make suitable and sufficient assessment of risks to employees and those not in their employ who may affected by their work and to record the significant findings and implement the risk control measures identified. General principles of prevention are to be applied.
Work at Height Regulations 2005 (HMG, 2005)	To adequately plan any necessary work at height and control risks to prevent death and injury caused by falls from height. It is common for temporary works to provide protection from falls from height (e.g. temporary edge protection systems). See Pallett and Filip (2019) section 1.5 for summary.
Reporting of Injuries, Diseases and Dangerous Occurrences Regulations 2013 (HMG, 2013a)	Accidents and certain dangerous occurrences relating to temporary works activities may occur and may need to be reported; records of such accidents are to be kept.
Construction (Design and Management) Regulations 2015 (HMG, 2015)	Requires that health and safety is managed through all stages of a project from conception, design, construction, maintenance to demolition. Introduced a series of specific duty holder roles. Contractors must plan, manage and monitor construction work (this includes permanent and temporary works). Principle contractors are also required to coordinate the construction work. See Pallett and Filip (2019) section 1.4 for summary and chapter 2 in this publication.

Note: for secondary legislation (regulations) the duties may be qualified by 'so far as is reasonably practicable' or 'practicable'.

Table 1.3 Some standards and organisations which produce guidance relating to temporary works (continued on next page)

Health and Safety Executive (HSE)	National regulator responsible for preventing workplace deaths, injuries and illness. Provides specific advice, guidance and information to influence workplaces and raises awareness of issues. Information helps interpret the law, helps with compliance and provides technical advice. Undertakes targeted inspections and investigations and prosecutes when necessary. Includes a specialist construction division.

Table 1.3 Continued

British Standards BS 5975-1:2024 and BS 5975-2:2024 (BSI, 2024a, 2024b)	Provides overall good practice recommendations on how to manage and design temporary works (see section 1.5.2).
Construction Industry Research and Information Association (CIRIA)	Public bodies whose activities are defined by statutory instruments and many others set up as charitable organisations.
Building Research Establishment (BRE)	Undertake research and publish guidance and warnings on specific subjects or products, many of which will be directly related to temporary works.
Temporary Works Forum (TWf)	Some members provide training courses.
The Concrete Society (CS)	
Steel Construction Institute (SCI)	
The Concrete Centre (TCC)	
Office for Nuclear Regulation (ONR)	Set up under the Energy Act (HMG, 2013b) as the UK's independent nuclear regulator with legal authority to regulate security and safety and to ensure high standards during construction and operation of nuclear facilities and environmental protection of nuclear waste.
Office for Rail and Road (ORR)	Non-ministerial government department responsible for the economic and safe regulation of Britain's railway network. ORR regulates Network Rail's (NR) activities and funding, regulates rail network access and licenses operators of railway assets. ORR regulates competition on the railway network and enforces consumer protection law.
Institution of Civil Engineers (ICE) Institution of Structural Engineers (IStructE)	The ICE, IStructE and other professional bodies are centres of excellence and provide industry voices which maintain relevant professional standards and award professional qualifications.

that what you were doing is equal to or better than the recommendations (the onus of proof would be on you to prove compliance with the law by other means).

An Approved Code of Practice (ACoP) is approved by HSE with endorsement from the Secretary of State and has special legal status. An ACoP will describe preferred or recommended methodology or standards to be met to comply with regulations and duties imposed by HSWA.

It has been suggested that guidance, either from the government or HSE, is not legally binding, it is just 'guidance'. So, on the face of it, it is designed to afford flexibility to employers. However, the underlying reality is that the guidance is much more prescriptive in nature. Employers have to prove they have done all that is reasonably practicable, and the courts tend to judge that according (in part) to the regulatory or industry guidance as to acceptable practice. This imposes a very substantial burden on any employer to justify departure from generic guidance such as this, and in most cases makes it effectively mandatory, unless there is a state-of-the-art alternative.

1.5. Some specific legislative and regulatory updates

1.5.1 Building Safety Act 2022

The Building Safety Act (BSA) (HMG, 2022a) was developed in response to the 2017 Grenfell fire. The BSA introduces new requirements through secondary legislation and reforms existing legislation, such as the Defective Premises Act (HMG, 1972) and the Building Act (HMG, 1984). It also introduces new duties relating to fire and structural safety. Temporary works are not considered explicitly in the BSA as it focuses on compliance with the Building Regulations (HMG, 2010) – this section is provided for information. However, solutions specified for Building Regulations compliance may impact on the constructability of a structure and should not be considered in isolation, but as part of the overall design, construction, maintenance and demolition process. The BSA applies to all buildings and includes additional requirements for higher-risk buildings (HRBs). HRBs are currently defined as buildings that are 18 m or more in height or have seven or more storeys and contain at least two residential units. The BSA applies to residential buildings (such as blocks of flats), care homes and hospitals during design and construction and building works during refurbishment. During occupation, it covers only residential buildings (and not care homes and hospitals). The BSA names the HSE as the new Building Safety Regulator (BSR).

The primary aims of the BSA are to manage risks to ensure the safety of people in or about buildings, quality of buildings, transparency of the building process and accountability of professionals involved in decision making. A key definition is that of 'building work' which can include 'erection or extension of a building' and 'work involving underpinning a building'. It is possible that some temporary works designers may be asked to sign off a Building Regulations Compliance Declaration in relation to items such as underpinning. However, most temporary works would have been removed (or incorporated into the permanent works) by the time 'building work' has been completed so it is likely this requirement would not apply. There may also be further duties under BSA (as opposed to Construction (Design and Management) Regulations (HMG, 2015) (CDM 2015)) on temporary works engineers (both designers and contractors) involved in work that can affect structural safety of a building during its occupancy phase, but at time of writing this requirement is unclear.

There are several separate strands within the BSA with separate audiences for each. One strand refers to all new buildings, with HRBs having additional duties and costs. Another refers to existing residential buildings. Many of these buildings are now over 50 years old with few 'as built' drawings and questionable maintenance regimes to improve robustness. These could be a significant area for temporary works as items such as scaffolding or propping will be required if investigations reveal structural problems.

The BSA introduces new duties on individuals and organisations over and above those contained in CDM 2015 (HMG, 2015). These duties are to ensure that clients, principal designers, principal contractors, designers and contractors comply with the Building Regulations. As with CDM 2015, individuals must demonstrate the appropriate skills, knowledge, experience (and, additionally, behaviours), and organisations must demonstrate organisational capability. There is a duty on anyone that appoints an individual or organisation to ensure that they meet the competency requirements. PAS 8671 (BSI, 2022a) provides a framework for the competence of principal designers, while PAS 8672 (BSI, 2022b) provides a framework for the competence of principal contractors. At the time of writing, specific competency standards for occupations including designers, managers, supervisors and installers were still being developed.

From 1 October 2023, the Building Safety Regulator (BSR) oversaw the building control approval process for all higher-risk buildings. These buildings are subject to three gateway reviews.

- Gateway 1: fire safety is considered as part of the planning permission process.
- Gateway 2: building work on higher-risk buildings will only be able to start after the developer has approval from the BSR. The BSR will also carry out regular inspections during construction (it is likely that these inspections will focus on compliance with the BSA and Building Regulations; enforcement of construction activities including temporary works will remain with HSE).
- Gateway 3: HRBs will only be allowed to be occupied after the BSR has checked that the building work is compliant and has issued a building control certificate.

Existing duty holders under CDM 2015 (HMG, 2015) (e.g. clients, designers, contractors etc.) have additional duties to plan, manage and monitor their activities in relation to building regulations.

At the time of writing, the industry, government and BSR guidance supporting the BSA, as well as secondary legislation, are still evolving – some has yet to be published, some is being modified. In addition, the final report of the Grenfell Inquiry (Moore-Bick *et al.*, 2024) has now been published and the recommendations from this report may impact on guidance and practice. As such, the reader will need to keep abreast of those developments.

Further guidance on the BSA is available on the UK government website (HMG, 2022b) and the HSE website (HSE, 2022a). Build UK (2023) has also published a summary document which is regularly updated.

1.5.2 BS 5975:2024 – Parts 1 and 2

The law (HSWA, HMG, 1974) requires that any work / project must be managed safely and therefore by implication temporary works must be managed safely. How can an organisation demonstrate compliance? One way is to follow the good practice recommendations of the relevant British standard BS 5975 (BSI, 2024a, 2024b) which is written by a committee of temporary works experts with input from the wider industry by way of a public consultation process.

BS 5975 (BSI, 2024a, 2024b) applies to all organisations involved in temporary works, such as clients, designers, contractors, sub-contractors and equipment suppliers. As set out in section 1.2(a), BS 5975 good practice recommendations can be applied to many sectors (not just civil engineering and building) and these sectors are increasingly realising that they need to manage their own involvement in temporary works. Increasingly, various organisations (not just Tier 1 principal / main contractors but clients, subcontractors, equipment suppliers) are producing their own bespoke temporary works procedures which generally follow the recommendations of BS 5975. Those involved in specialist sectors should produce a procedure which follows the general intent of BS 5975 but with allowance made for their own specialist sector risks and ways of working. For example, a specialist contractor working in the gas supply or distribution sector should have a bespoke temporary works procedure which covers the specific risks of working with or near a gas supply and will include specialist sector training (i.e. the risks of working with gas are different from the risks of working with wet concrete or steel frame construction).

BS 5975 (BSI, 2024a, 2024b) has been split into two parts. Part 1 is now a stand-alone procedural guide; the text has not been substantially changed from the 2019 revision (BSI, 2019). Part 2 has been significantly revised and updated and is now a design and implementation guide. There has been significant recent discussion relating to environmental loading (particularly wind loading), which not only applies to falsework but can include other items of temporary works, such as formwork, hoardings, site cabins, façade retention and so on. Designers should familiarise themselves with the recommendations of BS 5975-2 (BSI, 2024b), clause 6.4.

TWf have published *Temporary Works Sample Procedure* (TWf, 2019b), which is a template intended to help small to medium-sized contractors utilising temporary works on their projects. It contains sample text and forms that may be modified to suit an individual company's requirements. Although written for an earlier version of BS 5975, the contents are still largely valid.

Various organisations have produced simplified temporary works procedures flowcharts, such as a free download from Andun Ltd (www.andun.co.uk/temporary-works-flowchart).

1.5.3 HSG168 *Fire Safety in Construction* (HSE, 2022b)

In 2022, HSE updated its guidance on fire safety. This update supersedes the 2010 version and is compatible with the requirements of CDM 2015 (HMG, 2015). HSG168 (HSE, 2022b) contains three parts covering:

- Part 1 – fire risk assessment
- Part 2 – detailed guidance on fire risk assessment and fire precautions
- Part 3 – legal and enforcement responsibilities.

Appendix 4 clarifies what each of the five CDM 2015 duty holders is required to do during the preconstruction (design / planning) and construction phases in relation to managing construction fire risks. It provides an overview of the fire safety roles and responsibilities of each of the CDM 2015 duty holders.

1.6. Some specific updates from Pallett and Filip (2019)

1.6.1 Chapter 5 – site roads and working platforms

- TWf has published *Design of Granular Working Platforms for Construction Plant* (TWf, 2019c). This guide supplements current guidance and provides improved guidance for temporary works designers on the design of working platforms. It suggests suitable methods for the application of Eurocodes and considers ways of achieving greater economy and sustainability. There is also further development in improving the efficiency of working platform design by using a 'giant shear box'. A team from HS2, along with Wentworth House Partnership and City University London, presented their initial findings at the TWf 55th quarterly meeting. The aim is to further refine, enhance and revise the TWf document *Design of Granular Working Platforms for Construction Plant*.

- European Federation of Foundation Contractors (EFFC) and Deep Foundations Institute (DFI) have published the *Guide to Working Platforms* (EFFC and DFI, 2019). This guide highlights the importance of stability of tall machinery, summarises the issues involved and describes good practice. The guide starts with determination of track loading and continues through design, installation, testing, inspection, maintenance, responsibility and training of those involved. It is available from the EFFC website.
- Steel Piling Group (SPG) has published *Guidance Note on Working Platform Requirements for Tracked Excavator Piling Rigs* (SPG, 2023). This guide enforces current good practice for the requirements to have a properly designed, constructed, maintained and tested working platform for tracked plant.
- TWf 57th quarterly meeting: Dr Andrew Lees presented a new performance-based design method for construction of haul roads. An associated paper has been published (Robinson *et al*., 2023) and a recording is available to members from the TWf website.
- Although not directly related to tracked machines, TWf has published *Assessment and Management of Outrigger Loading* (TWf, 2022b). This guide considers the loading applied from cranes, concrete pumps, lorry loaders and other machinery outriggers (or stabilisers). It sets out good practice for the assessment and management of loading and provides clarity on the selection of a suitable spreader mat arrangement to adequately spread the applied load.
- Further useful information is available from Selmanpakoglu (2024) on crane outrigger load distribution through HDPE mats. This research highlights some of the advantages and notes that caution should be exercised when relying on information from HDPE mat suppliers' websites.

1.6.2 Chapter 6 – control of groundwater

- Cashman and Preene (2020) is an updated edition of *Groundwater Lowering in Construction*.
- Roberts *et al*. (2023) have published a paper which discusses how traditional groundwater strategies are evolving in response to ever increasing environmental regulations and the changing nature of construction projects. It discusses various techniques and associated risks.
- The Geological Society has a series of seven online training modules – *Temporary Works Groundwater Control for Construction* – available from their website.

1.6.3 Chapter 11 – sheet piling

- Steel Piling Group (SPG) published an extensive suite of guidance notes between 2018 and 2024 on a wide variety of subjects relating to sheet piling. Subjects covered include typical project information for selection of steel sheet piles, noise during sheet pile installation and corrosion performance of sheet piles.

1.6.4 Chapter 21 – access and proprietary scaffolds

- TWF has published *Effective Management of Scaffolding to BS 5975:2019* (TWf, 2020a). This guide applies to both tube and fitting and proprietary systems and provides advice on temporary works management procedures to comply with legislation, regulations and industry best practice. It explains when a scaffold requires a design and when a 'standard detail' could be appropriate. It also highlights the duties and responsibilities of those involved in scaffolding, from concept, design, assembly, inspection, and alteration to final removal. The guide remains valid for BS 5975-1:2024 (BSI, 2024a). It is accompanied by a free e-learning programme which is available via the TWf website.
- National Access and Scaffolding Confederation (NASC) has published TG20:21: *Good Practice Guidance for Tube and Fitting Scaffolding* (NASC, 2021). This guide provides guidance for scaffolding constructed with tube and fitting in the UK. There are four elements: *Operational Guide*, *Design Guide*, *User Guide* and *eGuide* software (all available from NASC). Some revisions, alterations and additions have been made to the range of TG20-compliant scaffolds, for which an updated compliance sheet (providing all information to construct the specified scaffold and further information on loading for foundations and ties) can be produced. NASC has also published TG30:24: *Comprehensive Guidance Suite for System Scaffolding* (NASC, 2024), which is a comprehensive guidance suite designed for system scaffolding. It includes *eGuide* software, *Design Guide* and *Operational Guide* and will complement TG20:21. Further additions to TG30:24 are due in 2025.
- National Federation of Demolition Contractors (NFDC, 2020) has produced a safety video on scaffolding in the demolition industry.

1.6.5 Chapter 22 – falsework

- TWf has published *Lateral Stability: Top-Restrained Falsework (or not)?* (TWf, 2023). This guide considers industry concerns relating to the use of top-restrained falsework on site. These systems do not provide their own lateral stability but rely on external support. This concept has been misunderstood, consequently leading to issues. BS 5975-2:2024 (BSI, 2024b), clause 7 continues to highlight the important issue of lateral stability in top-restrained falsework.
- BS EN 17293 Temporary Works Equipment (BSI, 2020) describes the manufacturing requirements for temporary works equipment (components) manufactured from steel, aluminium and timber and applies to falsework, formwork and scaffolding.

1.6.6 Chapter 23 – formwork

Embodied carbon is a major consideration in the concrete industry, especially with the use of cement. Low carbon concrete is increasingly being used and this can take longer to gain strength. The effect on concrete pressure and curing times for slabs and foundations (plus associated issues with striking falsework and back-propping) needs to be considered.

1.6.7 Chapter 31 – digital project delivery

- TWf has published *Yearbook 2023/24* (TWf, 2024a). There are two articles on digitising temporary works.

1.6.8 Chapter 32 – rebar stability

TWf has published two documents on rebar stability because of continuing industry issues and a general lack of understanding of the behaviour of cages in the temporary condition and wire ties:

- *Temporary Condition of Reinforcement Cages Prior to Concreting: Part 1 – Management Guidance* (TWf, 2020b) highlights the challenges and risks associated with reinforcement cages which are built in situ or prefabricated and lifted into position. It provides guidance on the division of duties of various parties involved and highlights some practical measures to prevent incidents occurring during any stage of the lifecycle of a cage, prior to it being encased in concrete.
- *Temporary Condition of Reinforcement Cages Prior to Concreting: Part 2 – Technical Guidance* (TWf, 2021) provides a detailed engineering explanation of potential modes of failure and specific design guidance for solutions. The process of joining bars together using thin wire by steel fixers (tying) was identified as a significant variable and issue (little research and testing on the ability of individual ties to carry a load) for the overall stability of cages. The guide introduces two types of ties for holding individual bars in place or maintaining overall stability and safety:
  - positional ties – where the tie is only required to hold a bar in position long enough for it to be encased in concrete. This type of tie can be a single wire slash tie and does not require a formal design as it is not intended to carry any significant loading. Standard tying patterns can be used
  - robust structural ties – where ties are required to carry a significant load or may be safety critical (e.g. lifting a cage), a design is necessary and double wire ties are to be used. Two further approaches are then suggested:
    - (i) method A – where a project cannot justify tie testing (due to time or cost constraints) and a conservative safe working load for ties of 0.35 kN/tie is used (for 1.6 mm black annealed wire and 1.2 mm stainless steel wire). This load is established from extensive tie testing commissioned by TWf and applies a significant factor of safety on failure to allow for variations in tie quality and to limit overall cage displacement
    - (ii) method B – this method requires ties 'to do more' than in method A and some tie stretch is assumed. It is likely to be suited to larger projects where the amount of rebar to be tied is large. Site testing of ties is recommended to establish an 'actual' failure load with a lower factor of safety (when compared with method A) and can be applied with additional quality control measures. The guide provides a specification for carrying out testing.
- TWf *Yearbook 2023/24* (TWf, 2024b) provides a further summary and shows the tie testing apparatus used by TWf.

1.6.9 Chapter 28 – heavy moves

- TWf has published *Self-Propelled Modular Transporters (SPMTs): A Brief Guide* (TWf, 2022c). SPMTs are used in some specialist temporary works operations; this guide summarises key elements of operation and provides a basic understanding of capabilities, site preparation and project planning.

REFERENCES

Bragg SL (1975) *Final Report of the Advisory Committee on Falsework*. Department of Employment and Department of the Environment, The Stationery Office, London, UK.

BSI (2019) BS 5975: Code of practice for temporary works procedures and the permissible stress design of falsework. BSI, London, UK.

BSI (2020) BS EN 17293: Temporary works equipment – Execution – Requirements for manufacturing. BSI, London, UK.

BSI (2022a) PAS 8671: Framework for competence of individual principal designers. BSI, London, UK.

BSI (2022b) PAS 8672: Framework for competence of individual principal contractors. BSI, London, UK.

BSI (2024a) BS 5975-1:2024: Management procedures for the control of temporary works. Code of practice. BSI, London, UK.

BSI (2024b) BS 5975-2:2024: Temporary works – Falsework: design and implementation. Code of practice. BSI, London, UK.

Build UK (2023) Building Safety Regime. https://builduk.org/wp-content/uploads/2023/01/Building-Safety-Regime.pdf (accessed 28/07/2024).

Cashman PM and Preene M (2020) *Groundwater Lowering in Construction*, 3rd edn. CRC Press, London, UK.

CROSS (Collaborative Reporting for Safer Structures) (2021) *Safety Report 13, August 2021, Report ID 1022*. CROSS-UK, IStructE, London, UK.

EFFC (European Federation of Foundation Contractors) and DFI (Deep Foundations Institute) (2019) *Guide to Working Platforms*. EFFC, Faversham, UK.

HMG (Her Majesty's Government) (1972) Defective Premises Act. The Stationery Office, London, UK.

HMG (1974) Health and Safety at Work etc. Act. The Stationery Office, London, UK.

HMG (1984) Building Act. The Stationery Office, London, UK.

HMG (1998a) Lifting Operations and Lifting Equipment Regulations. The Stationery Office, London, UK.

HMG (1998b) Provision and Use of Work Equipment Regulations. The Stationery Office, London, UK.

HMG (1999) Management of Health and Safety at Work Regulations 1999 as amended in 2003 and 2006. The Stationery Office, London, UK.

HMG (2005) Work at Height Regulations. The Stationery Office, London, UK.

HMG (2007) Corporate Manslaughter and Corporate Homicide Act. The Stationery Office, London, UK.

HMG (2008) Health and Safety Offences Act. The Stationery Office, London, UK.

HMG (2010) Building Regulations. The Stationery Office, London, UK.

HMG (2013a) Reporting of Injuries, Diseases and Dangerous Occurrences Regulations. The Stationery Office, London, UK.

HMG (2013b) Energy Act. The Stationery Office, London, UK.

HMG (2015) Construction (Design and Management) Regulations. The Stationery Office, London, UK.

HMG (2022a) Building Safety Act. The Stationery Office, London, UK.

HMG (2022b) https://www.gov.uk/guidance/the-building-safety-act (accessed 28/07/2024).

HSE (Health and Safety Executive) (2022a) https://www.hse.gov.uk/building-safety/ (accessed 28/07/2024).

HSE (2022b) *HSG168: Fire Safety in Construction*. The Stationery Office, London, UK.

Moore-Bick M, Akbor A and Istephan T (2024) *Grenfell Tower Inquiry: Phase Two Report*. The Stationery Office, London, UK.

NASC (National Access and Scaffolding Confederation) (2021) *TG20:21: Good Practice Guidance for Tube and Fitting Scaffolding*, NASC, London, UK.

NASC (2024) *TG30:24: Comprehensive Guidance Suite for System Scaffolding*. NASC, London, UK.

NFDC (National Federation of Demolition Contractors) (2020) Safety video on scaffolding in the demolition industry. https://www.Youtube.com/watch?v=gr8KIjI (accessed 28/07/2024).

Pallett PF and Filip RK (2019) *Temporary Works: Principles of Design and Construction*, 2nd edn. ICE Publishing, London, UK.

Roberts T, Preene M and Powrie W (2023) The changing nature of groundwater control for temporary works. *Proceedings of the Institution of Civil Engineers – Civil Engineering* **176(5)**: 11–20.

Robinson WJ, Wayne MH, Lees A and Han J (2023) An investigation of tyre contact area on unbound aggregate. *76th Canadian Geotechnical Conference Journal, Saskatoon, Canada*.

Selmanpakoglu O (2024) *Investigation of the Load Distribution of HDPE Mats Under Crane Outriggers*. Ozgur Engineering Ltd, London, UK.

SPG (Steel Piling Group) (2023) *Guidance Note 06: Working Platform Requirements for Tracked Excavator Piling Rigs*. SPG, Ascot, UK.

TWf (Temporary Works Form) (2017) Information Sheet 3: The Construction (Design and Management) Regulations 2015; Principal Designer: Guidance on Temporary Works. https://www.twforum.org.uk/viewdocument/twf-information-sheet-no-3-the-c (accessed 07/01/2025).

TWf (2019a) TWf2019:01: Precast Concrete: Good Practice Guide and Common Issues in Temporary Works. https://www.twforum.org.uk/viewdocument/precast-concrete-good-practice-and (accessed 07/01/2025).

TWf (2019b) TWf 2019:03: Temporary Works Procedure: Sample (revised May 2023). https://www.twforum.org.uk/viewdocument/twf201903-temporary-works-procedu (accessed 07/01/2025).

TWf (2019c) TWf2019:02: Working Platforms: Design of Granular Working Platforms for Construction Plant: A Guide to Good Practice (updated 2025). https://www.twforum.org.uk/viewdocument/working-platforms-design-of-granu (accessed 07/01/2025).

TWf (2020a) TWf2020:01: Effective Management of Scaffolding to BS 5975:2019. https://www.twforum.org.uk/viewdocument/effective-management-of-scaffolding (accessed 07/01/2025).

TWf (2020b) TWf 2020:03: Temporary Condition of Reinforcement Cages Prior to Concreting: Part 1 – Management Guidance. https://www.twforum.org.uk/viewdocument/rebar-guide-part-one-management (accessed 07/01/2025).

TWf (2021) TWf 2021:01: Temporary Condition of Reinforcement Cages Prior to Concreting: Part 2 – Technical Guidance. https://www.twforum.org.uk/viewdocument/rebar-guide-part-two-technical (accessed 07/01/2025).

TWf (2022a) *Yearbook 2021/22*, p. 19. https://www.library.myebook.com/Hub/temporary-works-forum-2021-22-yearbook/3327 (accessed 28/07/2024).

TWf (2022b) TWf2022:02: Assessment and Management of Outrigger Loading. https://www.twforum.org.uk/viewdocument/assessment-and-management-of-outrig-3 (accessed 07/01/2025).

TWf (2022c) TWf2022:01: Self-Propelled Modular Transporters (SPMTs): A Basic Guide. https://www.twforum.org.uk/blogs/david-thomas/2022/03/02/self-propelled-modular-transporters (accessed 07/01/2025).

TWf (2023) TWf23.067: Information Sheet 17: Lateral Stability: Top-Restrained Falsework (or not)? https://www.twforum.org.uk/blogs/david-thomas/2023/05/09/lateral-stability-top-restrained-falsework (accessed 07/01/2025).

TWf (2024a) Digitising temporary works. *Yearbook 2023/24*, pp. 23–24 and p. 68. https://www.library.myebook.com/Hub/temporary-works-forum-2023-24-yearbook/4906 (accessed 28/07/2024).

TWf (2024b) Equipment for testing ties. *Yearbook 2023/24*, pp. 15–18. https://www.library.myebook.com/Hub/temporary-works-forum-2023-24-yearbook/4906 (accessed 28/07/2024).

Webster M (2023) *Report RR1198: Implementation of the Principal Designers Role within CDM 2015*. HSE (Health and Safety Executive), The Stationery Office, London, UK.

USEFUL WEBSITES

BRE (Building Research Establishment) – www.bregroup.com
BSI (British Standards Institution) – www.bsigroup.com
CIRIA (Construction Industry Research and Information Association) – www.ciria.org
CITB (Construction Industry Training Board) – www.citb.co.uk
CPA (Construction Plant Hire Association) – www.cpa.uk.net
CROSS (Collaborative Reporting for Safer Structures) – www.cross-safety.og
CS (Concrete Society) – www.concrete.org.uk
DFI (Deep Foundations Institute) – www.dfi.org
EFFC (European Federation of Foundation Contractors) – www.effc.og
GSL (Geological Society of London) – www.geolsoc.org.uk
HSE (Health and Safety Executive) – www.hse.gov.uk
IStructE (Institution of Structural Engineers) – www.istructe.org
NASC (National Access and Scaffolding Confederation) – www.nasc.org.uk
NFDC (National Federation of Demolition Contractors) – www.demolition-nfdc.com
SCI (Steel Construction Institute) – www.steel-sci.com
SPG (Steel Piling Group) – www.steelpilinggroup.org
TCC (The Concrete Centre) – www.concretecentre.com
TWf (Temporary Works Forum) – www.twforum.org.uk

FURTHER READING

O'Driscoll R (2019) *Construction Engineering: A Guide to Temporary Works in Construction and BS 5975:2019*. www.constructionengineering.uk (accessed 07/01/2025).

Underwood J (2019) *Temporary Works Toolkit Part 16: Legal Requirements in the UK*. https://shop.istructe.org/journal/volumes/volume-95-(2017)/issue-10/temporary-works-toolkit-part-16-legal-requirements/ (accessed 07/01/2025).

Ray Filip
ISBN 978-1-83608-525-6
https://doi.org/10.1108/978-1-83608-524-920251002

Chapter 2
Considering temporary works in permanent works design

Ray Filip
RKF Consult Ltd, UK

Robert Millard
De Construction Solutions, UK; Chairman, Temporary Works Forum, UK

David Thomas
Retired Former Secretary, Temporary Works Forum; Formerly HM Specialist Inspector
Health and Safety Executive UK

Dr Mike Webster
MPW R&R Ltd, UK

2.1. Introduction

Temporary works (including the temporary condition of the permanent works) is an essential part of the construction process and needs to be considered for the safe and effective completion of a project (see BS 5975-1 (BSI, 2024a), clause 4.1.1.1, for the wide definition of temporary works). Traditionally, temporary works and construction methodology were solely the domain of contractors, and temporary works designers (TWDs) were considered as 'poor relations' to other designers, such as permanent works designers (PWDs). It is now recognised that temporary works needs to be considered as early as possible by organisations and professionals, such as clients (who should check that any designers and contractors they appoint have necessary temporary works experience), architects, planners and cost consultants. It is unlikely and not expected that these professionals have expert temporary works knowledge or are able to design temporary works. Consequently, early involvement of specialists should be encouraged (early contractor involvement has been established as an example of this initiative). Some clients have developed their own internal guidance – for example, Network Rail's *Early Focus on Constructability and Temporary Works* (Network Rail, 2023).

BS 5975-1:2024 (BSI, 2024a) builds on the requirements of the Construction (Design and Management) Regulations (CDM 2015) (HMG, 2015) which apply to all stages in the construction process, from concept through to construction, maintenance and eventual decommissioning. There has been significant recent discussion regarding the CDM role of 'designer' (see definition below), particularly PWDs and the principal designer (PD), in relation to temporary works. Issues can stem from a client's scope of work to the PD failing to consider constructability / sequencing / temporary states and associated temporary works. The main concern is that not all PWDs / PDs are fully aware of their duties in relation to these issues and the range of guidance and information available (see chapter 1 and References and Further reading sections at the end of this chapter); this chapter aims to help in this regard.

The definitions of 'designer' and 'design' in CDM 2015 (HMG, 2015) are wide.

- Designer – any person who in the course or furtherance of a business:
  - prepares or modifies a design
  - arranges for, or instructs, any person under their control to do so, relating to a structure or to a product or mechanical or electrical system intended for a particular structure. A person is deemed to prepare a design where a design is prepared by a person under their control. This can apply equally to PWDs and TWDs in addition to other professionals, such as architects, surveyors and possibly clients if they become actively involved in design.
- Design – includes drawings, design details, specifications and bills of quantities (including specification of articles or substances) relating to a structure, and calculations prepared for the purpose of a design.

2.2. Permanent works designer's duties

Some PWDs are exclusively involved in the design of the permanent works and may not recognise their duties in relation to temporary works. They may feel that contractual arrangements prohibit their involvement in temporary works and that clients will not pay them for this service or may consider themselves to be insufficiently resourced or inexperienced (not competent) in temporary works. Some may not have appropriate professional indemnity insurance (see section 2.4) as they consider temporary works to be a minor part of their business activity, and the potential income would not cover the additional costs. However, PWDs have increasingly identified a business opportunity to offer a 'one stop shop' by also offering a temporary works design service. They should ensure they are competent to offer this service (see section 2.4) and contractually they are permitted to do so and that there is no conflict of interest.

This section attempts to clarify the duties of PWDs relating to temporary works, regardless of whether they are involved or not involved directly in designing temporary works.

CDM 2015 regulation 9 (HMG, 2015) requires the PWD to (can apply to any 'designer': see definitions in section 2.1)

> eliminate, reduce or control the risks through design. Where it is not possible to eliminate the risks, they should provide information on significant residual risks that could not be designed out to the PD, contractors and other designers who could be affected by those residual risks. As part of this, permanent works designers are expected to understand how their structure is to be constructed and the requirement for temporary works.

In practice, regulation 9 means three principles.

- Designers should address health and safety risks from the start of the design stage, as decisions they take will have an impact on the construction phase. Designs shall be safe to construct, maintain, operate and eventually decommission – and designers can assume competent persons will undertake these activities. Most designs result in 'standard' methods of work with well-known risks. It is the non-standard risks, those that cannot be 'designed out' (e.g. difficult access or temporary instability of the structure) that require additional site management measures, and these should be highlighted (from preconstruction stage).

- Designs shall comply with contemporary industry good practice with regard to health and safety, unless there is good reason for not complying. Guidance is available from organisations such as Health and Safety Executive (HSE), British Standards Institution (BSI) and Temporary Works Forum (TWf) (see section 1.4 and Further reading section at the end of this chapter).
- Information on significant and relevant residual risks pertinent to the design shall be effectively communicated to those who need to know. It is not normally necessary to communicate obvious risks which should be known to a competent contractor or person, but those which are unusual or likely to be difficult to manage.

BS 5975-1:2024 (BSI, 2024a), section 7.3 gives procedural recommendations for PWDs in relation to temporary works.

- Address the buildability of permanent works and interfaces with temporary works – the permanent works cannot be 'wished in place' and they should demonstrate how their design can be built, operated and decommissioned safely. PWDs need to give particular attention to preconstruction information (unless obvious) on buildability, access / logistics, project restrictions, working space, temporary stability and deflection criteria, any assumptions they have made, any unusual processes or where additional specialist design is to be carried out by others. A contractor can propose alternative methodology but may need to demonstrate their methodology does not have a detrimental effect on the permanent works.
- Clearly state loads (in an appropriate manner for the target audience – see section 2.6) for which the structure has been designed.
- Identify and make provision for any temporary works and temporary conditions required by their design and assumed method of construction. They should give particular attention to the stability of any existing or partially constructed / demolished structures and, where this is not obvious, provide information on how temporary stability could be achieved.
- Liaise with the PD to provide all necessary information for any necessary temporary works or temporary conditions to the principal contractor (PC).

BS 5975-1:2024 (BSI, 2024a), clause 5.1.1.2 and clause 7 recommend that PWDs should have a bespoke and appropriate management procedure for the type and complexity of work they are involved in. As a minimum it should describe how the organisation will fulfil its CDM 2015 duties to temporary works. If they are involved in the design of temporary works, the procedure should follow the recommendations of BS 5975-1:2024 (BSI, 2024a). Examples of items that should be included in a procedure for PWDs involved in designing temporary works include the following.

- Appoint a suitable senior person (ideally a board director) to be the designated individual (DI) with overall responsibility and authority for establishing and maintaining a procedure for controlling the aspects of temporary works (and associated risks) that the organisation has responsibility for or which they can constrain or influence. The procedure should be approved by the main board, regularly audited and periodically updated as necessary.

- Ensure individual designers have appropriate and sufficient skills, knowledge, experience and behaviours (SKEB) and provide regular and relevant temporary works risks, design and practical awareness training.
- Identify whether they can provide a complete solution or whether parts will be subcontracted to other specialist designers. Ensure any subcontractors the organisation employs to design or detail temporary works have sufficient SKEB and are adequately resourced to carry out the work. This should include adequate procedures and appropriate professional indemnity insurance (see section 2.4). Often some design and / or detailing (e.g. scheduling reinforcement) may be carried out abroad – these checks should still apply and an effective communication and checking system should be established.
- Fulfil CDM 2015 (HMG, 2015) duties as described in section 2.2 by identifying the need for temporary works as early as possible (for all stages in the life of the structure), following the general principles of prevention. Consider stability and regularly carry out constructability / buildability reviews (TWf, 2020a). Clearly identify any significant residual risks (those that are pertinent to the design).
- At construction stage, request that the names of the principal contractor's temporary works coordinator / temporary works coordinator (PC's TWC / TWC) are confirmed as they are the main point of contact for all correspondence relating to temporary works.
- Do not start any meaningful design work unless an appropriate and adequate design brief is provided (otherwise design work is carried out at own risk) by the client, PD contractor or others.
- Clearly identify any temporary works design assumptions and how these are to be confirmed on site.
- Identify any critical components, critical stages in construction sequences or special (unusual) materials (these should be added to site inspection checklists).
- Identify any specific testing or monitoring that may be required during construction and any remedial measures that may be necessary. Testing and monitoring specifications should be provided, including performance criteria that should be achieved.
- Ensure temporary works designs are appropriately checked in accordance with BS 5975-1:2024 (BSI, 2024a), table 2. Identify how the design check category will be determined. High-risk (implementation risk class from BS5975-1:2024 (BSI, 2024a), table 1), very complex or novel schemes, where the consequences of failure could be catastrophic, should also be peer-reviewed with particular attention given to external stakeholders, whose approval may also be required (e.g. if working adjacent to Network Rail).
- Keep regular communication with site so that any site changes are clearly communicated to the designer for appropriate action. This necessitates close liaison with the PC's TWC and / or TWC.
- Advise that any sacrificial items of temporary works are recorded in the health and safety file for the client.
- Procedures should be reviewed and updated on a regular basis and lessons learnt from each project.

PWDs need to appreciate that decisions they take can affect the 'implementation risk class' (see BS 5975-1:2024, BSI, 2024a, table 1) of any subsequent temporary works scheme.

- Implementation risk class is a function of
  - execution risk – risks associated with the construction, use and removal of the temporary works (e.g. workmanship, materials, experience of workers and manner of use)
  - consequence of failure of the temporary works – risks associated with the location of the temporary works and what might be affected by its failure.
- The implementation risk for the same temporary works item erected in different locations can vary significantly (e.g. scaffold erected in a field can be a different risk from the same scaffold erected adjacent to a railway line).

Neither CDM 2015 (HMG, 2015) nor the Health and Safety Executive's guidance on CDM (L153, HSE, 2015) explicitly explain how a PWD can determine whether their risk management is compliant. However, dangerous site situations resulting from poor design decisions must be avoided. HSE (2015) has less mention of temporary works than the document it superseded (L144, HSE, 2007). This was an Approved Code of Practice (ACoP) whereas L153 is not. It could be argued this is a backward step, but ultimately this will make no difference in court as L153 is used in the same way as L144 to illustrate how a duty could be fulfilled.

2.3. Principal designer's duties

Principal designers (PDs) are required to work with clients and all designers during the preconstruction phase (in CDM 2015 regulation 2, the definition of preconstruction also includes 'preparatory work') to produce efficient, buildable designs that can also be used, maintained and decommissioned safely. They must ensure that all designers comply with their duties and take reasonable steps to ensure cooperation. Many different organisations have identified a business opportunity to offer their services as PD, but they do not have to be a designer on the project. (They have to be a designer – this is where the issue of implicit or explicit designers becomes relevant. A quantity surveyor can claim that because their decisions attract designer duties, they are a designer and can thus be PD.) Many of these organisations advertise as experts in CDM but omit their expertise in construction. The Institution of Structural Engineers has useful information (Banks, 2016).

CDM 2015 regulation 11 (HMG, 2015, p.8) requires the PD to

> plan, manage, monitor and coordinate health and safety in the preconstruction (design) phase of a project to ensure that, as far as reasonably practicable, the project is carried out without risks to health and safety.

This includes working with all designers to identify foreseeable risks, follow the general principles of prevention and communicate residual risks (see section 2.2). Their duties do not require them to undertake, check or approve designs or undertake risk assessments to inform the design. These are the responsibility of specific designers.

The PD should ensure (so far as is reasonably practicable) that the PWD has

- considered overall project restrictions from the client or adjacent stakeholders and sought relevant permissions and approvals as necessary
- ensured that the principles of prevention have been followed
- considered buildability, such as: adequate working space, sequences, temporary conditions and access for the construction process
- provided adequate preconstruction information relating to temporary works to enable TWDs to undertake their design
- proposed and communicated a safe method and sequence of construction (which does not adversely affect the permanent works), with attention given to 'non-standard', new or unusual processes or materials
- identified where a detailed design is to be carried out by others
- stated the loads for which the structure has been designed in a clear and appropriate manner (see section 2.6)
- considered that temporary works are to be removed unless they become sacrificial or part of the permanent works
- retained any appropriate information relating to temporary works for the health and safety file
- ensured that the preconstruction information is developed as planning and design progresses.

With temporary works the 'preconstruction phase' continues during the 'construction phase' as 'design' (which includes the design of temporary works) is unlikely to stop becoming a consideration on contract award to the PC. The role of the PD therefore continues from the preconstruction phase into the construction phase.

The PD should be familiar with the recommendations of BS 5975-1:2024 (BSI, 2024a), clause 7.5. This may include ensuring that a temporary works register for key items has been developed at preconstruction stage and included in the tender documentation. The Health and Safety Executive has carried out a review (Webster, 2023) of the PD role within CDM 2015 (HMG, 2015) which highlights that compliance by PDs with the functions set out in CDM 2015 with regards to temporary works is 'variable'. Temporary Works Forum (TWf, 2017) has useful guidance on the role of PD in respect of temporary works.

2.4. Designer competency

As mentioned in the editor's preface, competence has been highlighted as a significant issue by the Grenfell Tower Inquiry (Moore-Bick *et al.*, 2024). Any organisation employing another organisation to design or check the design of temporary works should ensure they are competent to carry out the role. Also, any organisation wishing to offer their services as a TWD (or design checker) should ensure they have the appropriate SKEB. Some organisations have appropriate SKEB in many aspects of temporary works design, but some may specialise (e.g. scaffolding or excavations).

The design and / or design checking can be performed by

- principal contractor's in-house temporary works design service, who will mostly have generalised knowledge
- subcontractor's in-house temporary works design service, who will mostly have specialist knowledge

- permanent works designers, who also offer a temporary works design service
- equipment hire companies, who have expertise in their own equipment but will exclude others' equipment or interfaces with permanent works
- temporary works design consultants; some will have skills in multiple disciplines, whereas others will specialise (e.g. dewatering or demolition).

There is no single 'correct choice' and, on a typical project, various elements of temporary works or the whole scheme can be designed, or design checked, by organisations, providing they have appropriate SKEB.

To justify appropriate SKEB in temporary works, the PWD should:

- be up to date with construction methodology and common temporary works risks and solutions
- have some previous and successful experience of designing or checking the design of temporary works; this includes understanding the similarities and differences between temporary and permanent works design (see section 2.7)
- have relevant professional qualifications and appropriate, up-to-date training in temporary works risks, design and practical awareness; keep up to date with relevant continuing professional development
- have adequate policies and procedures (in accordance with BS 5975-1:2024 (BSI, 2024a) – see section 2.2) to manage design and design checking (see BS 5975-1:2024, BSI, 2024a, table 2)
- be adequately resourced to complete the tasks within the agreed timescale
- ensure they are contractually permitted to offer the service and there is no conflict of interest
- have adequate and appropriately customised professional indemnity insurance (see chapter 1.2.e). Checks should be carried out on the wording of policies to ensure adequate temporary works design cover is provided, with particular attention paid to exclusions (some policies specifically exclude temporary works). Certain projects may require a minimum level of insurance cover which may be more than provided by the policy. An exclusion which is becoming more common in many policies relates to fire. This can be an issue when designing temporary fire escape routes (scaffolding and staircases) which should be fire-rated. Designers are often not permitted to specify fire-rated materials unless they have specialist expertise
- be familiar with BS 5975-2 (BSI, 2024b) which has extensive temporary works design guidance.

Temporary Works Forum (TWf, 2021c) has useful information on pre-qualification.

2.5. Economics – are temporary works required?

Engineers are problem solvers, but how often do they consider why the problem exists, or how to eliminate the problem at source? The Institution of Structural Engineers' *Temporary Works Toolkit Part 3: Integration of Permanent and Temporary Works* (Markham, 2016) starts with the phrase: 'temporary works is waste' (p. 1). It is also often said that: 'the best temporary works are no temporary works' (p. 1). Both these phrases infer that temporary works can be costly, can be difficult to install / remove and can be time consuming.

The IStructE document highlights the following possibilities.

- Design the permanent works to eliminate the need for any temporary works – especially if temporary works cannot be easily removed – or integrate any sacrificial temporary works into the permanent works.
- Minimise temporary works by utilising either all or part of the permanent works to provide a temporary works support function.
- Simplify the permanent works so that minimal or simple temporary works are required to construct them – some discussions may be required with the client and architect.

Examples include the following.

- Ensure there is sufficient space (wherever possible) for site offices, welfare, public protection measures, cranes and storage areas, without the need for complex temporary works.
- Design a building slab or floor to accommodate temporary construction loading (e.g. construction live loading, material storage, weight of construction machinery) to eliminate the need for back-propping.
- Ensure the permanent works construction sequence considers temporary lateral stability so that permanent bracing or shear walls can be utilised to also provide the required level of temporary stability (see section 2.12).
- Prefabricate or precast components so they can be transported to site and lifted into position. If reinforced concrete elements can be precast then formwork and falsework for casting elements in situ can be eliminated (this may also improve the quality of the finished concrete – chapter 7 discusses this further).
- Use profiled steel decking (spanning between permanent steelwork) to construct a reinforced concrete slab instead of falsework. The steel decking is sacrificial but can become part of the permanent works as it can perform as a composite together with the concrete.
- Position an opening in a load-bearing brick wall (shear wall) so there is sufficient remaining brickwork either side of the opening to provide overall lateral stability, rather than adding additional temporary stability measures (this is likely to involve discussions with clients and architects).
- Position shallow excavations in locations where there is sufficient room to safely batter the sides rather than using temporary ground support such as sheet piling.
- Design precast concrete elements and steel beams so that edge protection systems can easily be connected to them.
- Enhance a reinforced concrete lift shaft base to become a tower crane foundation, as the lift shaft is a ready-made hole through the building for the crane mast.
- Consider the design and position of permanent piled foundations to accommodate temporary loads from plant such as tower cranes.
- Consider whether working platforms for tracked construction machinery (such as piling rigs or crawler cranes) or temporary site access roads can be used as part of the permanent construction beneath slabs, car parks or permanent roads.
- A temporary sheet pile wall used to support the ground for a basement excavation can become sacrificial (if too difficult or too expensive to remove). The thickness of the

steel in the sheet piles may be governed by driveability (Pallett and Filip, 2019 section 11.5.14), rather than bending capacity. By welding the sheet pile clutches to prevent water ingress, the sacrificial sheet piles can become the permanent wall (instead of reinforced concrete) if adequate consideration is given to longevity, fire protection and insulation (for habitable basement).

- Concrete capping beams on retaining walls can be enhanced to distribute significant point loads from construction machinery (outrigger loads from mobile cranes) into the retaining wall. It can be advantageous to position crane outriggers directly onto a retaining wall (compression in the wall) rather than behind the wall (bending in the wall).
- Can a precast or prefabricated element of permanent works be designed to be sufficiently stiff and robust so that it can be transported / moved / lifted / jacked into its final position? If not, then temporary works may be required to prevent excessive deformation. Temporary stresses need to be considered and lifting points included.
- Detail prefabricated or built in situ reinforcement cages so they are stable without the need for external propping (see TWf, 2020b, 2021c).
- When lifting items of permanent works, weight, item stability and centre of gravity need to be considered. A single steel bridge beam may become unstable in its final position, so temporary propping may be required. If beams can be designed to be lifted in pairs with the permanent bracing included (assuming the site crane has sufficient capacity and reach), the issue can be eliminated. Some items are easier to install / position / remove if they are lifted so they do not hang vertically. The position of lifting points in relation to the centre of gravity could be adjusted if temporary stresses are adequately considered.
- Design a roof with a higher parapet to eliminate the need for any temporary edge protection during maintenance.

The construction industry already has significant experience in eliminating or minimising temporary works and the PWD should educate themselves in the possibilities and associated challenges. Responsibility and liability should be clearly established. Temporary Works Forum (2021b) has further useful information on early consideration of temporary works and improving buildability.

2.6. Design methodology and communication

The introduction to Pallett and Filip (2019) discusses different types of design methodology. Currently in the UK, a significant proportion of temporary works designs are still carried out using permissible stress methodology, with the use of limit state design methodology increasing. However, almost all permanent works designs are carried out using limit state design methodology. It is too simplistic to claim that this is a generational distinction with the older generation preferring permissible stress design, as this is becoming increasingly blurred. It is important that designers do not mix methodologies as they are not compatible and there can be conflicts when a design is carried out using one methodology and checked using the other. Designing temporary works items to European standards is discussed in TWf (2014) and Harris and Angelino (2017).

Most items of temporary works will be reused many times on various projects (e.g. spreader mats beneath mobile cranes outriggers) or will use proprietary components (from hire companies) which will be used on site and then returned to be hirer. These items should be designed within elastic limits for their intended purpose and be sufficiently robust to avoid permanent deformation or other damage which could render them unsuitable for reuse.

Designers should consider how they communicate residual risk information. Many provide a 'residual risk schedule'. It is not acceptable for this schedule to extend to many pages for a simple project. Many of the risks are often generic (slips, trips and falls etc.) and thus the specific / unusual risks can be diluted and lost among the other mass of information. The designer should consider whether they have sufficiently carried out their CDM 2015 duty by 'eliminating risks wherever possible'. It is often more appropriate to provide residual risks information on the relevant drawing (often notated with warning triangles, but this should not be a 'scatter gun approach') as drawings are more likely to be used and understood by the site team.

Designers should provide information on any assumptions and limitations to their designs, such as maximum permitted loading they have designed the temporary works to accommodate. They should realise that most site personnel are generally not aware of the difference between permissible stress and limit state and tend to be more familiar and comfortable with safe working loads (permissible stress). These limits should be communicated in a manner that is appropriate to the target audience. It is usual for designers to use kilonewtons (kN or kN/m^2). Anecdotal evidence from the editor of this publication (based on several decades of discussions with site management and site operatives) identifies that typically only around 50% to 60% of site management and site engineers understand kilonewtons and as little as 10% to 20% of site operatives. This may become an issue as an item of temporary works could become inadvertently overloaded leading to collapse because of a basic lack of understanding of what a kilonewton is. Designers should consider converting kilonewtons into units the site team are more likely to be familiar with – for example, number of people, bags of cement or bricks or perhaps the weight / size of a piece of construction machinery. An easy conversion is that 10 kN is approximately 1 ton (1000 kg), which is approximately equivalent to the weight of a small car.

2.7. Practical temporary works design considerations

The basic engineering skills for a PWD and a TWD (buildability, eliminating risk where possible, structural analysis etc.) are similar but there can also be significant differences, which need to be recognised. Practical experience of the construction process and understanding temporary works risks are arguably the most important attributes for a PWD and a TWD.

It is often not reasonable to simply 'cut and paste' performance specifications for permanent works items and assume they can be directly applied to items of temporary works. This may lead to unrealistic requirements – for example, concrete cover, fatigue or movement, which may be difficult, time consuming and uneconomic to achieve.

Table 2.1 highlights some of the issues.

An item of temporary works should be robust.

Table 2.1 Differences and similarities between temporary works and permanent works (continued on next page)

	Temporary works	Permanent works
Design method	Often permissible stress	Mainly limit state
Preconstruction information	Often very little relevant and useful ground or existing structure information or this information is at a location away from any intended temporary works position	Desk studies, intrusive surveys, site investigations, testing etc. are commissioned and carried out
Loading	Short term (days, weeks, months, perhaps several years). Often statistical reductions can be made based on exposure time and risk Elastic shortening (e.g. in props) and tensile elongation (e.g. in ties) often need to be considered	Long term (decades) Structural components often have high stiffness
Risk	During the construction phase many items can be high risk and can be prone to damage, collapse, lack of stability and the location can increase the risk (e.g. adjacent to a railway). Could be multiple interfaces between different parties and different equipment / machinery	Once completed most structures are deemed low risk. This may change during maintenance or decommissioning Will generally have a few 'owners'
Utilisation	Live loading is often a high percentage of the total loading and under typical loading conditions the utilisation of element capacity is high (often >90%)	Live loading can be low compared to dead loading and under typical loading conditions the utilisation of the element capacity is low (often <50%)
Durability	Rarely a major consideration. Minimum concrete cover is usually satisfactory Exceptions could include: • marine cofferdams or jetties in use for several years where corrosion becomes an issue (coatings and designing for sacrificial loss of section are common measures)	A major consideration with materials being selected to minimise maintenance over the life of the structure

Table 2.1 Continued

	Temporary works	Permanent works
	• timber and plywood hoardings on major projects that are exposed to the elements for several years • site haul roads and working platforms on major projects exposed to the elements and heavy traffic for several years	
Sustainability	Becoming an increasing consideration Questions being asked about embodied carbon in temporary works Most proprietary elements are reused	Major consideration, especially embodied carbon
Construction tolerance	Considered – realistic and achievable tolerances should be considered to ensure permanent structure is built to acceptance standards	Considered to ensure overall structure performs as intended and meets client's expectations
Materials	Mainly proprietary equipment (reused many times), second-hand (steelwork) or recycled materials (piling mats from demolition crush) Confirmation of exact properties may be difficult without testing and checking condition	Generally new materials which are supplied with certification
Fatigue	Often ignored, as the number of cycles over the life of the structure will be low (measured in hundreds or thousands) There are some exceptions, such as temporary bridges or cofferdams subject to flowing water If an item is reused, it is accepted good practice that the item is inspected on site prior to use and is replaced if there is any damage or signs of fatigue	Considered, as the number of cycles over the life of a structure could be high (measures in millions)

Decomissioning	Most elements are standard components which need to be removed and reused Load must be released and transferred to permanent supports in an agreed sequence to allow the temporary works to be removed	Mostly bespoke elements which at some point need to be replaced (e.g. bridge bearings), repaired in situ or demolished
Aesthetics	Generally, not considered Exceptions could include high profile projects with significant public interfaces where the client or contractor may be concerned about image – temporary steelwork may be painted, site hoarding may be decorated and decorative sheeting added to scaffolding	Generally, an important architectural consideration (temporary works can affect the appearance of permanent works e.g. formwork affects concrete surface finish)
Fire	Needs to be considered for temporary emergency escape routes and the temporary condition of structures until permanent fire protection is installed	Major consideration
Support and stability	Often rely on external or existing elements for support and stability (e.g. scaffold ties into structures, top-restrained falsework) Can try to utilise existing foundations and use permanent works elements for lateral stability (see section 2.12)	Mostly bespoke support and stability. New foundations constructed and shear walls / bracing / cores included in the new structure
Thermal effects	Often not considered as individual elements are generally small and expansion / contraction will be negligible Exceptions could include: • large props in deep excavations • large props to support substantial structural loads • steel girder viaduct works	Major consideration as elements can be massive and expansion / contraction will be significant. Movement joints, crack control measures etc. are provided

Table 2.1 Continued

	Temporary works	Permanent works
Deflection, movement and settlement	See section 2.11	Major consideration to prevent structural damage and for aesthetic appearance
Weatherproofing	Considered for temporary site structures such as offices and workshops. Also considered during demolition / refurbishment projects	Major consideration
Timescale	Mainly quick turnaround (sometimes immediate due to an emergency) Designs are often built as soon as they are suitably checked and approved	Planning consent and design evolution can take years. A designer can wait years for their design to be built
Warranties	Clients will generally not require assurance and warranties but there are exceptions on some major projects	Clients will generally require assurance and warranties

Some specific considerations during the temporary works design process include the following.

- Identify overall design responsibility and responsibility for details (and design checking) – for example, structural and geotechnical design for temporary works may be split but an overview of compatibility is required.
- Identify whether the permanent works is constructed in the final position or needs to be transported / moved / lifted / jacked into place. Such operations induce temporary stresses and deformations into the permanent works element, which need to be considered. Section 1.2.2 mentions there is an industry drive for offsite manufacture (see chapter 7) and designers should be aware of the principles of design for manufacture and assembly (DfMA) for optimisation.
- Wherever possible, an item of temporary works should be removed and reused. Often, hired proprietary equipment is the most economic and simple solution for shorter-term use. For longer-term use, hire costs may start to become prohibitive and the contractor may buy the equipment or use bespoke solutions such as structural steel. Proprietary equipment has the benefit that it has been designed / tested so the performance and limitations are well known. Also, connection details have been developed and many different types of components are available. Components from different hire companies should not be mixed as there will be issues with compatibility and liability. Designers should become familiar with the various types, manufacturer's instructions and what can be achieved.
- Contractors may have preferred suppliers (who deliver the service they require and may offer a bulk discount on hire or purchase) or may have favoured methodologies or machinery (which they may own) and would use wherever possible. Availability and logistical constraints could play a part in selection and designers should discuss the options with the contractor. It may be prudent for the designer to provide several alternative options so the contractor can then select the most effective / efficient version for their project.
- Designers should avoid specifying multiple different section sizes and components, as this can lead to site errors and may not be economic to hire or purchase. Simplicity and standardisation should be encouraged, as site operatives will become increasingly familiar and confident in repeating the same operation.
- Often, temporary works components and equipment are reused (second-hand) and it is reasonable for designers to assume that any equipment or components are correct and in good condition (but this should be confirmed on site) and will perform as intended.
- Clear load paths and support conditions need to be established for temporary and permanent works. Lateral stability should also be assessed and achieved (see section 2.12). This is particularly important for temporary works items such as falsework and for the permanent works during structural alterations and demolition.
- How are the temporary works to be installed? Critical components or construction stages should be identified and incorporated into site-specific inspection checklists. Temporary works should be easy and quick to install: its purpose is to facilitate the construction of permanent works and not be a hinderance. Its position should be carefully considered in relation to the construction sequence to avoid clashes with permanent works.

- Construction tolerances (positional and level tolerances) need to be considered and agreed (should be specified and provided on a design brief), as well as how the temporary works can be adjusted for installation / removal, positional accuracy, verticality and level. Construction techniques, workmanship, components and materials have differing tolerances; some can be large – for example, typical pile installation is +/-75 mm in position – whereas structural steel can be manufactured and installed to within a few millimetres. Designers need to allow for construction tolerance as it can cause eccentric loading or unintended load paths. Jacks, wedges, packs, shims, slotted holes for bolts or anchors, turnbuckles for tensioning tie rods are all practical solutions for accommodating dimensional tolerance.
- Some items of temporary works may impart a load onto permanent works – for example, proprietary edge protection systems clamped or bolted to permanent steelwork or falsework supported by a suspended slab beneath. Design responsibility should be established for these interfaces. Often PWDs are best placed to confirm that the temporary works do not have a detrimental effect on the permanent works or to advise whether other measures are required. It is advisable for these issues to be identified at preconstruction stage.
- How are loads applied to the temporary works? Designers should be aware that loads can be applied progressively (pouring concrete) and not in an even manner (heaping of concrete when pouring or stacking bundles of reinforcement before fixing). Support to some localised areas may need to be enhanced.
- When and how can the temporary works be removed? If the temporary works is removed too early the permanent works could be damaged. Monitoring or testing may be required – for example, strength testing in situ reinforced concrete slabs for striking falsework. Consideration should be given to how load is transferred from the temporary works onto the permanent works supports. For larger and more complex schemes, jacking systems may need to be incorporated for load transfer. PWDs and specialists should be consulted.
- A specific removal sequence should be provided to avoid damage to the permanent works or temporary works components. There is a misconception that temporary works should be removed in the reverse of the installation sequence. Ideally the removal sequence allows the loads to be transferred through the permanent structure to the supports and foundations as the PWD intended.
- Consideration must equally be given to how the load is to be released from the temporary works elements and transferred onto the permanent works in a controlled manner. It is generally preferable for the load to be released from the temporary works by unloading jacks (manually or hydraulically) or removing wedges and the temporary works removed and reused. There may be occasions where project-specific restrictions prevent this or it becomes preferable to maintain the load, such as when preloading a foundation (to induce settlement). Designers should allow for the jack (and perhaps other structural members) to be sacrificial and built into the permanent works. Resin or grout can be pumped into proprietary 'flat jacks' which inflate, the resin / grout will harden to maintain the applied load and then can be relatively easily concealed within the permanent works.
- Designers should be aware that the critical / most onerous loading conditions on an item of temporary works could occur during removal (see Figure 2.1). These conditions

Figure 2.1 Understanding basement construction sequence to determine 'most onerous cases' for geotechnical and structural analysis (editor's own)

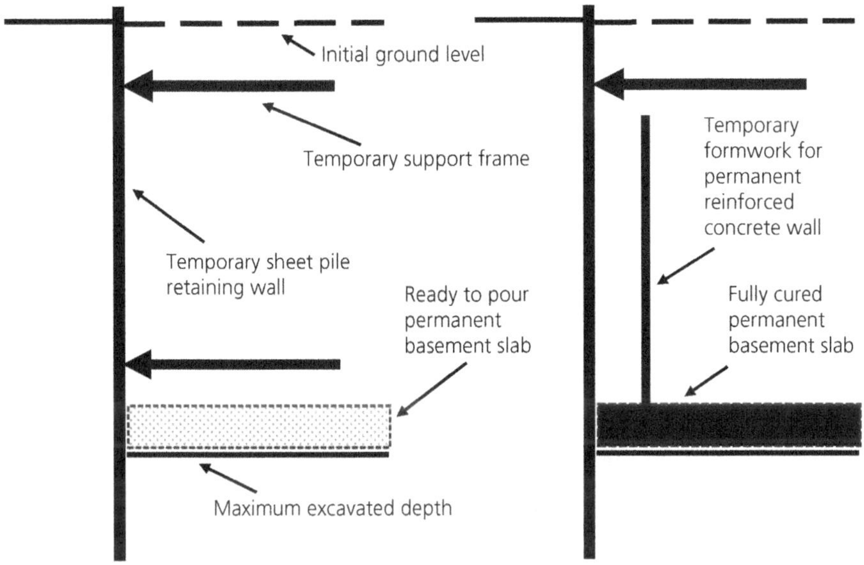

need to be identified so that an effective analysis is carried out and members suitably designed.

- If the temporary works cannot be removed, can it be incorporated into the permanent works for economic or structural benefits (see section 2.5)? This should then be highlighted in the health and safety file.

Figure 2.1. highlights a simplistic basement construction. The generalised methodology for the reinforced concrete basement construction could be as follows.

1. Install the temporary retaining wall (e.g. sheet piles).
2. Excavate to the underside of upper support frame and install the upper frame.
3. Excavate to the underside of lower support frame and install the lower frame.
4. Construct the reinforced concrete base slab and allow the concrete to cure.
5. Remove the lower support frame and install the formwork ready to construct the permanent reinforced concrete wall and then construct the wall. If de-bonded from the concrete the sheet piles can be extracted and reused.

Some designers might assume that the left-hand diagram would give the most onerous temporary condition. However, when the lower temporary support frame is removed, the span of the sheet piles increases (from basement slab to upper support frame); hence, the bending moment may increase and the loading in the upper support frame may also increase. An understanding of methodology must take priority over analysis so the most onerous conditions can be determined. For different elements this may occur at different stages and all stages should be identified.

In this example the reinforced concrete wall could also be enhanced by the PWD (designed to act as a cantilever above basement level slab rather than spanning between basement and ground floor slabs) to provide support to the ground as the upper frame would need to be removed to allow the construction of the ground floor slab. If the concrete is prevented from bonding to the sheet piles (using a membrane), then the sheet piles could be removed.

If a temporary works concept or scheme is innovative or complex, consideration should be given to carrying out a trial to accurately replicate and monitor the intended works. The methodology, components, plant, equipment and tolerances can be developed and adjusted to ensure the scheme is feasible. Such trials are regularly carried out when structures are slid or jacked into position, especially when the works is to be carried out in a safety- and time-critical environment (e.g. railway and motorway possessions).

It is prudent for designers to consider the sensitivity of their designs to site changes (e.g. ground conditions are not exactly as expected) and a 'what if' scenario. An example from the editor's experience: a 3 m-deep temporary trench support scheme at a refinery was designed for 'short-term – cohesive' ground conditions. Approximately 50 m of trench was excavated ahead of pipework delivery. The company providing the specialist pipework suddenly went into liquidation and it was estimated it could take up to 6 months to source suitable alternative pipework. There were concerns about the safety of the trench and it was decided to fence off the area and temporarily backfill the trench.

2.8. Construction phase loading

PWDs require an understanding of typical construction phase loading and loading combinations. It is common for the loads in the temporary condition to be different and occasionally more onerous than in the permanent condition. Understanding how the permanent works is to be constructed is critical to understanding the loading, constraints, temporary lateral stability and overall behaviour. Problems and extensive temporary works can be avoided if the PWD makes adequate provision for these loads in the temporary condition. However, a balance must be made between enhancing the permanent works (with respect to strength or section size, which can inadvertently affect items such as embodied carbon or transportation) and the cost and constraints of temporary works. The short-term nature of temporary works can be considered – for example, length of time of exposure to wind loading.

Typical construction phase loading that designers should consider can include:

- construction live loading – access or operatives and their tools
- environmental loading – wind, snow, flowing water
- geotechnical – soil and water pressures plus any surcharge (machinery behind wall or adjacent structures)
- materials storage – on a structure because of lack of space elsewhere on site; these could be significant localised loads (pallet of bricks or bundle of reinforcing bars) as these stacks of materials will be lifted by a crane and positioned before further distribution

- eccentric and out-of-balance loading – due to construction sequence and tolerances
- accumulation – wet concrete during placement, demolition rubble or spoil during excavation before removal
- construction machinery – on slabs or floors and often used for work at height access
- lifting – induced temporary stresses during lifting of items and a degree of impact loading when the object is landed on a structure
- accidental impact – from passing vehicles or a swinging suspended crane load (see section 2.10)
- lateral – horizontal loading due to wind, surge load from a concrete pump, vehicles braking, flowing water or reactions due to supports not being exactly vertical (see section 2.12).

Slabs and floors can be subject to significant localised point loads during construction, such as crane outriggers. The magnitude is likely to be significantly greater than any point load considered in the permanent condition (especially as there is a drive towards 'minimal materials' design and PWDs will focus on carbon in the final state rather than carbon overall). PWDs should provide guidance on the maximum magnitude of construction loads that can be permitted before additional temporary works measures are required (e.g. spreaders or back-propping).

BS 5975-2:2024 (BSI, 2024b) should be consulted for guidance, various loads should be considered and designers need to be familiar with how these loads are applied (i.e. from construction machinery operating). There have been some significant collapses contributed by a PWD's lack of appreciation / experience of buildability and the behaviour / loading of temporary works elements and insufficient control measures. Numerous media outlets and HSE widely reported on the collapse of a tower crane at Chandlers Wharf, Liverpool, on 6 July 2009. Collapse was attributed to inadequate foundations which were unable to cope with the forces generated by the crane during operation. Both designer and contractor were found culpable.

Confidential Reporting for Structural Safety (CROSS) is a useful source which highlights industry issues and concerns, some of which are related to temporary works. The Temporary Works Forum website also has a section on 'Causes for Concern'.

2.9. Vibration, noise and dust

Vibration, noise and dust (occupational health issues) can be significant considerations and designers need to become familiar with the project restrictions and understand / devise mitigation measures. Two scenarios should generally be considered.

- During the installation or removal process: projects may have restrictions based on location (all three will be an issue in city centres) or adjacent stakeholder (vibration will be a significant issue adjacent to railway lines) and these issues this could limit the installation methodology for sheet piles (see Pallett and Filip, 2019, section 11.3).
- While temporary works are in use, and this could include construction machinery: vibration is a particular concern (from machinery or passing vehicles). Designers should follow the principles of prevention by preventing items such as bolts working loose and specify an inspection regime to check.

2.10. Accidental impact

Progressive collapse must be avoided for permanent and temporary works (with structural redundancy by way of alternative load paths being created). A major factor contributing to collapse could be accidental impact. This can occur during the construction phase caused by items such as vehicles, falling wet concrete during placement, falling demolition rubble or a swinging load suspended from a crane. The assessment of accidental impact is essential. These can be low probability / high consequence events resulting in a catastrophic outcome and the general principles of prevention must be applied. The magnitude of an appropriate accidental impact load is subject to some debate and should be determined for each scenario.

Some sources of guidance on accidental impact:

- Eurocode 1 (BSI, 1998), called accidental actions
- Temporary Works Forum's rebar stability document (TWf, 2021c), section 5.3
- BS 5975-2:2024 (BSI, 2024b), clause 6.3.3.3, table 8, table 12, clause 7.4.4.1 and clause 11.18
- CIRIA C579 (CIRIA, 2003), section 8.12.

Typical horizontal forces that can be considered are between 1 kN and 10 kN in the most onerous location / direction, depending on likelihood and consequences of failure. Props supporting large excavations may be designed to be sufficiently robust to withstand a significant impact. The likelihood may be high as excavators are working near the props and the consequences could be catastrophic. For this reason, it is for the project team to agree whether redundancy is to be designed into a cofferdam frame, such that the accidental removal of any single prop will not result in complete failure. Many items of temporary work are slender and have low-impact resistance. It can be impracticable and uneconomic to design for any significant impact – for example, a timber hoarding or scaffolding cannot reasonably be designed to withstand a significant impact from a passing lorry. If the temporary works cannot be designed for an impact, then measures are necessary to prevent the impact – for example, traffic management, concrete protection barriers or preventing a suspended crane load from swinging by using 'tag lines'. Temporary props supporting road overbridge decks during, for example, pier replacement works cannot realistically accommodate highways accidental horizontal load requirements and consequently protection barriers become essential.

Impact should also be considered on any structure where a load is being landed by crane or forklift. It is customary to allow a localised additional 10 to 25% of the vertical load during positioning.

2.11. Deflection / movement / settlement

A critical consideration in temporary works design is deflection / movement / settlement, which often governs the design. It is common for a structural member to be satisfactory for applied stresses, but the properties may need to be enhanced to limit deflection. Anyone involved in the design of temporary works must be mindful of deflection / movement / settlement criteria often provided in performance specifications. If an item of temporary works deflects, moves or settles excessively, it could have an adverse effect on the permanent works or its surroundings. Below are some specific examples.

- Excessive deflection of formwork or settlement of falsework (plus elastic shortening) will adversely affect the appearance of the permanent concrete (it may appear rippled and not level).
- Excessive deflection of needling beams (see Pallett and Filip, 2019, chapter 33) will cause cracking of the brick wall being supported.
- Excessive deflection of a temporary retaining wall will result in ground movement behind the wall which could cause damage to any adjacent services or roads.
- Excessive differential settlement of a crane's foundations could destabilise the crane.
- Excessive deflection of a façade retention scheme could cause damage to the façade.
- Excessive deflection or movement or a propped slab or beam in the temporary state may induce locked-in stresses once connected back into the new permanent structure.

Designers should consider potential sources of movement which may be caused by:

- elongation of tie rods due to applied tensile load – tie rods are regularly used for formwork and if the ties are long, they will sag under their own self-weight and stretch when the concrete pressure is applied to the formwork
- elongation of large props due to thermal effects
- elastic shortening of props due to applied vertical load – many temporary works props are made from steel or aluminium (to save weight) components which can be fitted together to provide the required length. All joints in these props will compress and they will elastically shorten when the load is applied
- lateral movement (sway) due to applied horizontal loading such as wind, vehicle braking or horizontal components from a lack of verticality. BS 5975-2:2024 (BSI, 2024b) provides significant guidance on minimum horizontal loading to be considered. Designers must ensure their structures are stable by providing sufficient lateral stability (see section 2.12).

Clients, architects, designers and anyone responsible for performance specifications should realise that all temporary works items will deflect, move, settle and so on to some degree – and consider carefully the magnitude of deflection / movement / settlement that could be considered acceptable (not excessively detrimental) and provide realistic criteria which can be reasonably achieved. There is a balance to be made between specifying unrealistically low deflection criteria, and the high cost of the temporary works to achieve this, against the cost of any remedial work.

Those assessing calculations should realise that predicted deflections / movements / settlements are theoretical and there are inaccuracies / assumptions relating to:

- how the applied loads are determined
- support conditions
- structural member or soil stiffnesses
- dimensional inaccuracies caused by construction tolerances and material manufacturing tolerances
- inaccuracies in analysis methodology or software
- temperature effects.

A pragmatic approach is required with a degree of engineering judgement when assessing theoretical predictions.

There are several options if deflections, movements or settlement are determined to be excessive. Member stiffness can be increased, ground improvement techniques investigated, lateral stability improved, preloading (predeflecting) considered and large props painted white / provided with shade to limit thermal effects. Preloading can be an economic and practical solution and can be achieved for relatively small loads by using proprietary screw jacks which are manually adjusted. For larger loads, manually adjusted jacks are unlikely to be suitable due to limited capacity and problems with releasing the load (for a high applied load an operative is unlikely to be able to unwind the jack) and hydraulic jacking systems should be used. The amount of preload should be carefully considered as an excessive preload can be detrimental to the permanent works (cause damage). There is no specific guidance on preloading, but a rule of thumb is: preload not to exceed 2nd / 3rd of applied dead loading (live loading should be ignored).

2.12. Lateral stability

Designers must consider and ensure lateral stability of the permanent works and temporary works. This process needs to be carried out during construction, any subsequent alterations / repairs and eventual demolition. For stability of falsework, BS 5975-2:2024 (BSI, 2024b), clause 7.2.2.2 recommends a minimum horizontal disturbing force to be the greater of

- 2.5% of the applied vertical load
- all known horizontal load plus 1% to allow for workmanship tolerance.

These minimum rules are typically applied to many temporary works items. CIRIA C579 (CIRIA, 2003), section 8.12 also has guidance on lateral loads.

Some items of temporary works rely on permanents works elements to provide lateral stability. A common example is top-restrained falsework (see Pallett and Filip, 2019, chapter 22) which often relies on permanent works columns. Designers should be aware of the potential limitations. In the scenario of cast in situ slabs and columns, the PWD will consider the column to predominantly carry vertical loading, be restrained top and bottom by slabs and have achieved full design concrete strength. In the temporary condition the upper slab has not been cast so the column is a cantilever, it is often unlikely to have achieved its full concrete strength and is being asked to sustain a horizontal load applied at the top column. This scenario can be amplified if the concrete is curing slowly (cold weather or low-carbon concrete used), the columns are precast and not fully fixed at the base or the columns are very tall and slender. The above needs to be assessed if top-restrained falsework is to be used for slab construction.

2.13. Interfaces

Although the overall CDM 2015 (HMG, 2015) requirements described above are clear, they cannot cover every possibility and scenario. There can be many examples where interfaces need to be considered:

- an item of permanent works that is in a temporary condition
- items of temporary works that rely on stability or support from the permanent work
- temporary construction loadings that are imparted onto the permanent works.

Due to the fragmented nature of the construction industry, these interfaces can often result in 'grey areas of responsibility and liability', which can be dependent on the specific contract conditions and the parties involved. If not adequately identified and addressed, they could result in incidents and arguments between the parties involved. Ultimately it is the responsibility of the principal contractor, who has overall responsibility for health and safety on site.

There is significant industry debate regarding interpretation of responsibility and liability. The authors would suggest the following based on Figure 2.2, assuming a traditional contractual arrangement (not design and build) between client, PC and PWD.

- The completed bridge is designed by the PWD to span between the piers and support loads such as self-weight, traffic, snow and so on. In the temporary condition, the bridge span is in two cantilevers, with significant temporary works items at the end of each cantilever (temporary works rely on the bridge), plus temporary construction loads (machinery, live loading, storage etc.) on the bridge deck. During the initial design stage, the PWD should recognise this temporary condition but is not expected to know the exact nature of temporary construction loading. They should make a reasonable estimate and check for any adverse effects on the permanent works. They should also demonstrate how the bridge can be built safely (i.e. an assumed construction sequence).

Figure 2.2 Complex interfaces between temporary works and permanent works (editor's own)

- If the PC proposes an alternative methodology (which may be quicker, cheaper, safer or suit available machinery), they should develop their own methodology by employing specialist TWDs, but the PWD should be employed to confirm there are no adverse effects on the bridge during and after construction.
- During the construction phase, the PC determines the precise extent of the temporary loading on the bridge (machinery, live loading, storage etc.), as this can be dependent on availability. The PWD already has intimate knowledge of the bridge, so the PC should employ the PWD to confirm these finalised loads do not adversely affect the bridge. If there are issues, the PWD should discuss options with the PC.
- The TWD is responsible for designing the temporary works which are suspended from the bridge. However, they are unlikely to be competent (and would not accept the liability) to assess the effect of the temporary works on the bridge structure. The TWD should provide reactions to the PWD to confirm the adequacy of the bridge structure to support these loads.
- The PD should ensure there is adequate communication and cooperation between the various parties involved.
- The project should have a formally appointed temporary works coordinator (TWC) who should coordinate these interfaces and provide design briefs for the various parties involved. They should also ensure that designs are appropriately checked as per BS 5975-1:2024 (BSI, 2024a), table 2.

For 'design and build' projects, the PC has contractual control over the PWD (but not the PD unless the client appoints the PC as PD) and it is likely to be easier to take these issues into account.

Figure 2.3 is a simpler scenario where a temporary edge protection system is connected to permanent steel beams.

- The PWD should be aware that edge protection will be required and make allowance for this in their design, but they will not know the specific system the contractor will use.
- The edge protection system is designed, certified and hired to the PC or a subcontractor by an equipment supplier. The supplier provides reaction loads applied to the steel beams but does not generally take responsibility or liability for the connections to the supporting beams or the beams themselves.
- A relatively simple check on the interface could be carried out by the PWD or by a TWD (contractors in-house team or external specialist) but exact apportioning of responsibility / liability should be established.
- The PC (by way of the TWC) should ensure that the interface between the edge protection and permanent works is adequately managed.

2.14. Summary

PWDs should understand their duties and ensure that their overall design takes account of buildability and any temporary works and temporary conditions which might be expected

Figure 2.3 Simple interface between temporary works and permanent works (edge protection and steel beams) (editor's own)

during construction, regardless of who is to develop those works. They need to realise that their permanent works and temporary works can be mutually dependent. A design analysis on its own is meaningless without an overall construction methodology. The methodology must be determined first; the analysis then justifies specific details.

The PD, PC and other designers should ask: are we satisfied with the relevance and quality of information provided by the PWD? If the answer is 'no' then they should challenge the PWD.

Interfaces between temporary works and permanent works must be identified, with responsibility and liability clearly established.

If things go wrong, any subsequent investigation will try to identify whether suitably qualified and experienced people were involved and whether an approved safe sequence of works was being followed.

2.15. Temporary works design philosophy and principles

The philosophy and principles for designing an item of temporary works should not be too different from an item of permanent works. A simplified way for TWDs to understand the design process for an item of temporary works at project construction stage is highlighted in Figure 2.4.

Figure 2.4 Temporary works design flowchart (editor's own)

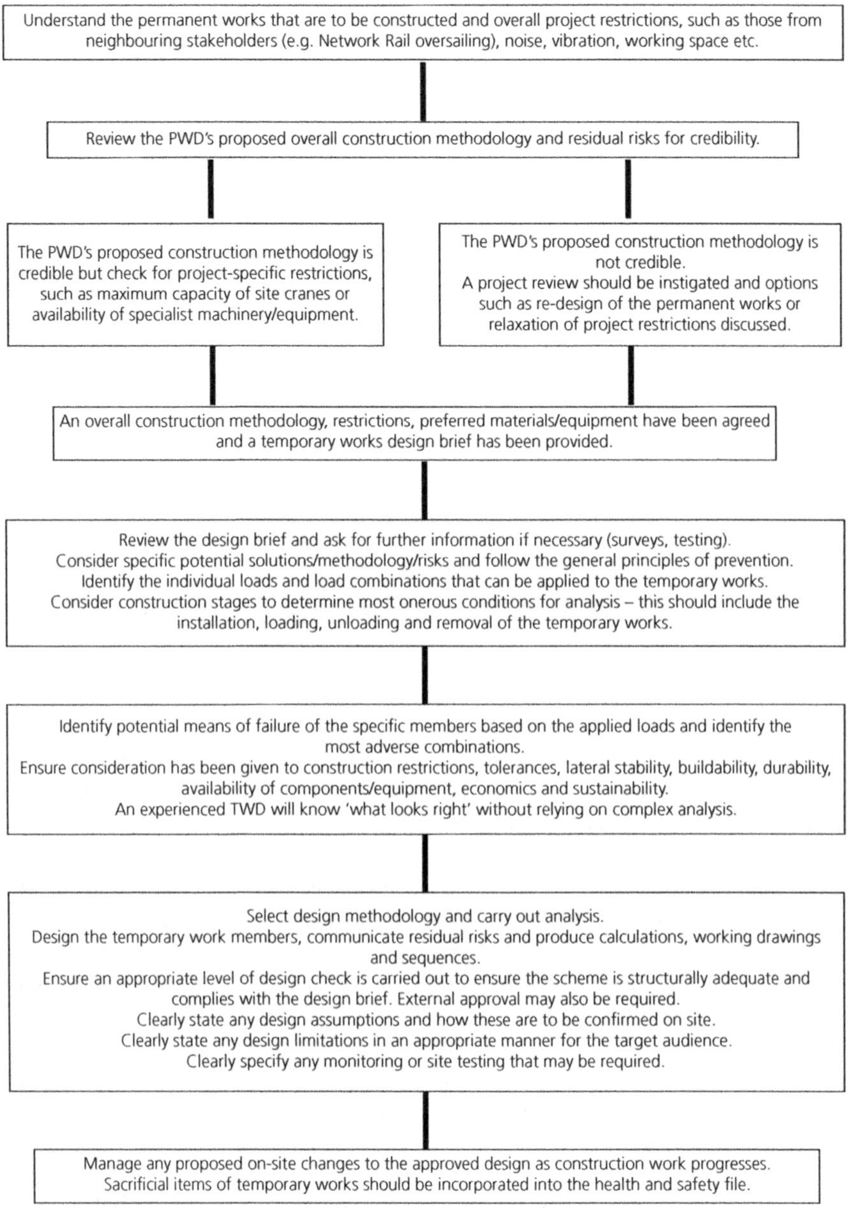

REFERENCES

Banks J (2016) *Temporary Works Toolkit Part 9: The Role of the Principal Designer with Reference to Temporary Works*. IStructE (Institution of Structural Engineers), London, UK.

BSI (1998) BS EN 1991-1-7: Eurocode 1: Actions on structures – Parts 1–7. General actions – Accidental actions. BSI, London, UK.

BSI (2024a) BS 5975-1:2024: Management procedures for the control of temporary works. Code of practice. BSI, London, UK.

BSI (2024b) BS 5975-2:2024: Temporary works, falsework: design and implementation. Code of practice. BSI, London, UK.

CIRIA (Construction Industry Research and Information Association) (2003) *C579: Retention of Masonry Facades*. CIRIA, London, UK.

Harris T and Angelino M (2017) *Temporary Works Toolkit Part 15: Designing Temporary Works to European Standards*. IStructE, London, UK.

HMG (2015) Construction (Design and Management) Regulations. The Stationery Office, London, UK. https://www.legislation.gov.uk/uksi/2015/51/contents/made (accessed 31 October 2024).

HSE (Health and Safety Executive) (2007) *L144: Managing Health and Safety in Construction, Construction Design and Management Regulations, Approved Code of Practice*. The Stationery Office, London, UK.

HSE (2015) *L153: Managing Health and Safety in Construction, Guidance on Construction Design and Management Regulations*. The Stationery Office, London, UK.

Markham P (2016) *Temporary Works Toolkit Part 3: Integration of Permanent and Temporary Works*. IStructE, London, UK

Moore-Bick M, Akbor A and Istephan T (2024) *Grenfell Tower Inquiry: Phase Two Report*. The Stationery Office, London, UK.

Network Rail (2023) Safe by Design Guidance Note: Early Focus on Constructability and Temporary Works, 4th revision. https://safety.networkrail.co.uk/wp-content/uploads/2023/11/HSbD-Guidance.-Early-Focus-on-Constructability-and-Temporary-Works.-Issue-4-16.11.23.pdf (accessed 28/07/2024).

Pallett PF and Filip RK (2019) *Temporary Works Principles of Design and Construction*, 2nd edn. ICE Publishing, London, UK.

TWf (Temporary Works Forum) (2014a) *TWf2014:01: The Use of European Standards for Temporary Works Design*. https://www.twforum.org.uk/viewdocument/the-use-of-european-standards-for-t .

TWf (2017) *Information Sheet 3: The Construction (Design and Management) Regulations 2015; Principal Designer: Guidance on Temporary Works*. https://www.twforum.org.uk/viewdocument/twf-information-sheet-no-3-the-c.

TWf (2020a) *TWf2020:02: Constructability: A Guide to Reducing Temporary Works*. https://www.twforum.org.uk/viewdocument/constructability-a-guide-to-reduci.

TWf (2020b) *TWf 2020:03: Temporary Condition of Reinforcement Cages Prior to Concreting: Part 1 – Management Guidance*. https://www.twforum.org.uk/viewdocument/rebar-guide-part-one-management .

TWf (2021a) *Information Sheet 13: PAS 91 Construction Prequalification Questionnaires*. https://www.twforum.org.uk/viewdocument/twf-information-sheet-no-13-pas-9 .

TWf (2021b) *Information Sheet 14: Considering Temporary Works Early in a Project: Reducing Cost, Improving Buildability and Reducing Risk.* https://www.twforum.org.uk/blogs/david-thomas/2021/10/05/considering-temporary-works-early-in-a-project.

TWf (2021c) *TWf 2021:01: Temporary Condition of Reinforcement Cages Prior to Concreting: Part 2 – Technical Guidance.* https://www.twforum.org.uk/viewdocument/rebar-guide-part-two-technical.

Webster M (2023) *Report RR1198: Implementation of the Principal Designer's Role within CDM 2015.* HSE, The Stationery Office, London, UK.

USEFUL WEBSITES

Andun Engineering Consultants – www.andun.co.uk

BSI (British Standards Institution) – www.bsigroup.com

CIRIA (Construction Industry Research and Information Association) – www.ciria.org

CROSS (Collaborative Reporting for Safer Structures) – www.cross-safety.org

HSE (Health and Safety Executive) – www.hse.gov.uk

IStructE (The Institution of Structural Engineers) – www.istructe.org

TWf (Temporary Works Forum) – www.twforum.org.uk

FURTHER READING

Andun Engineering Consultants – temporary works webinars (available for free at www.andun.co.uk)

APS (Association for Project Safety) (2015) *Principal Designers Handbook and Guide to CDM 2015.* Association of Project Safety.

Carpenter J (2010) Risk Management with ERIC. *The Structural Engineer Magazine* **88**: 7 April.

Carpenter J (2021) *Designing a Safer Built Environment: A Complete Guide to the Management of Design Risk.* ICE (Institution of Civil Engineers), London, UK.

Gilbertson A (2015) *C765: CDM 2015 Workplace In-Use Guidance for Designers*, 2nd edn. CIRIA, London, UK.

HSE (Health and Safety Executive) (2003) *Research Report 148: The case for CDM: Better Safer Design – a Pilot Study.* The Stationery Office, London, UK.

ICE (Institution of Civil Engineers) (2018) Design Risk Management Guidance. https://www.ice.org/ICEDevelopmentWebPortal/media/Documents/Disciplines%20and%20Resources/Best%20Practice/drm-guide-version-1-3pdf

IStructE (Institution of Structural Engineers) (2016) *Temporary Works Toolkit Part 1: An Introduction to Temporary Works for the Structural Engineer.* Lohman T, IStructE, London, UK.

IStructE (2016) *Temporary Works Toolkit Part 2: CDM and The Responsibilities of Permanent Works Designers.* Carpenter J, IStructE, London, UK.

IStructE (2017) *Temporary Works Toolkit Part 9: The Role of the Principal Designer with Reference to Temporary Works.* Banks J, IStructE, London, UK.

IStructE (2017) *Temporary Works Toolkit Part 13: Importance of Understanding Construction Methodology.* McBride D, IStructE, London, UK.

IStructE (2017) *Temporary Works Toolkit Part 16: Legal Requirements in the UK.* Underwood J, IStructE, London, UK.

O'Driscoll R (2019) *Construction Engineering a Guide to Temporary Works in Construction and BS 5975:2019.* www.constructionengineering.uk

Ove Arup & Partners and Gilbertson A (2015) *C755: CDM 2015 Construction Work Sector Guidance for Designers*, 4th edn. CIRIA (Construction Industry Research and Information Association), London, UK.

TWf (Temporary Works Forum) (2016) The Roles and Responsibilities of Permanent Works Designers with Regards to Temporary Works. https://www.twforum.org.uk/viewdocument/the-roles-and-responsibilities-of-p (accessed 07/01/2025).

TWf (2017) Information Sheet 3: The Construction (Design and Management) Regulations 2015; Principal Designer: Guidance on Temporary Works. https://www.twforum.org.uk/viewdocument/twf-information-sheet-no-3-the-c (accessed 07/01/2025).

TWF (2020) Information Sheet 10: Can a Client's Permanent Works Designer be the Contractor's Temporary Work Designer? https://www.twforum.org.uk/viewdocument/twf-information-sheet-no-10-the (accessed 07/01/2025).

Webster M (2019) CDM 2015 Designer Duties – Considering Temporary Works in Permanent Works Design. http://mpwrandr.co.uk/cdm-2015-designer-duties-temporary-works/.

emerald PUBLISHING ice Publishing

Ray Filip
ISBN 978-1-83608-525-6
https://doi.org/10.1108/978-1-83608-524-920251003

Chapter 3
Geotechnical site investigation

Ray Filip
RKF Consult Ltd, UK

3.1. Introduction

For any construction project that applies a load to the ground, involves an excavation or any potential changes to the groundwater regime, some consideration of the ground and groundwater conditions at the site must be undertaken, so that the project can be designed with an appropriate factor of safety.

Examples of such items of temporary works could include:

- pad and strip foundations for site cabins, scaffolding, propping and façade retention
- foundations for temporary access bridges
- access / haul roads to allow vehicles onto and off a site (could include ground improvement techniques)
- foundations for construction machinery, such as hoists, mobile elevated working platforms (MEWPs), excavators, mobile cranes, crawler cranes and piling rigs (could include piled foundations for tower cranes)
- embedment for king posts, hoarding posts and retaining walls
- battered or supported trenches and excavations (including cofferdams) for utilities or structures
- ensuring existing or new slopes on or around a project can remain stable during the works
- lowering or controlling groundwater for trenches and excavations so the work can be carried out in the dry
- headings, tunnels and shafts (these could also include some degree of lowering or controlling water)
- thrust blocks for pressure testing pipelines or providing restraint
- anchorage blocks for cable stays or hydraulic tirfors
- underpinning existing foundations (the underpins are likely to become permanent but the process of constructing them is temporary works)
- piled or other foundations which provide reactions or support for testing (e.g. foundation pile testing).

There are some common challenges (the same challenges could equally be applied to structural and dilapidation investigations; see chapter 4).

- There is often a misconception (especially among clients and cost controllers) that 'money spent on an investigation is not money well spent' and this leads to a reticence to invest in investigation work. The reality is that inadequate or inappropriate information increases risks, and the risks will result in conservative and not cost-effective design decisions being made. It could be argued that there is a direct correlation between the amount, quality and relevance of information provided and the cost-effectiveness of design decisions. (However, in the real world there are always budgetary constraints.) As the accuracy of the information improves then the designer can optimise the design and minimise the risk of unexpected issues during construction.
- The ground is not uniform and there is a balance to be made. No single process will provide all the necessary information and even an extensive investigation cannot provide a totally accurate picture of what is below the surface. Some level of risk will always remain, and these risks need to be identified and managed effectively.
- A geotechnical site investigation is generally carried out at preconstruction stage to allow the permanent works to be designed. The extent, location, depth and techniques used in the investigation will be determined by the requirements of the permanent works; often, little or no thought is given to temporary works requirements. Some mitigation could be argued as the precise nature of the temporary works is often not known at the preconstruction stage. However, as discussed in chapter 2, the permanent works designer (PWD) is required to justify a safe construction methodology for their design and consider any temporary works that might be required (BSI, 2024a, 2024b; HMG, 2015). The PWD should be expected to make reasonable assumptions on the types and locations of temporary works elements that might be required to construct their design.

Soils comprise particles, with water and gas (usually air or water vapour) filling the voids between the particles. This chapter will mainly focus on the investigation of soils, as these are most often the concern of temporary works designers (TWD). Rocks can pose different challenges.

3.2. Why is an investigation required?

To deliver a project successfully, an understanding is required of the zone of influence of the works; then an appropriate ground model should be developed that will enable an adequate, safe and economic design to be prepared. The scope and complexity of a ground investigation will depend on the complexity of the works and the potential level of risks that are present. Investigations (including testing) and their interpretation can be grouped together:

- an investigation carried out at preconstruction stage on behalf of the PWD; consideration is also given to temporary works requirements and a temporary works designer (TWD) will have sufficient information for the design of the temporary works
- an investigation carried out at preconstruction stage on behalf of the PWD; very little or no consideration is given to temporary works requirements and the TWD will not have sufficient information to design the temporary works
- an additional or supplementary investigation carried out during the construction stage on behalf of the TWD to provide relevant information to design the temporary works

- an additional or supplementary investigation carried out during the construction stage to check or confirm performance which could be specified in contract documentation (e.g. compaction testing for working platforms)
- an additional or supplementary investigation carried out during the construction stage to check or confirm design assumptions to reduce risks or to refine designs (e.g. hand shear vane could be used to check on clay formation beneath a tower crane base or pile testing could be used to refine pile design models)
- an additional or supplementary investigation carried out during the construction stage to classify materials for waste or hazard removal or to determine whether the material could be used on the project (e.g. contaminated soils for removal or using site-won materials as aggregates).

3.3. What are we trying to achieve?

An investigation should not be considered as a standard process that is followed just to meet an obligation. Time and money could be wasted collecting irrelevant and unnecessary information, while critical information that could optimise the design or reduce risks may not be obtained.

Before any investigation or testing work is undertaken, it is time- and cost-effective to understand why the work is required (see section 3.2) before specifying any intrusive techniques. For simple, low-risk projects where the elements are lightly loaded, the investigation could be relatively quick, simple and cheap. Often a simple assessment or description of the ground and groundwater conditions near to the proposed location and at shallow depths may be sufficient. However, designers should treat this information with caution as it might not be very accurate. These simple assessments could be further confirmed (to ensure conditions are as expected) as the construction work progresses. If the project is complex, the loadings are high, the perceived groundwater level is high and the ground is poor, then the risks will be high and a sophisticated investigation involving sampling and testing will be necessary.

For an existing PWD's preconstruction site investigation to be sufficient for temporary works purposes, the information must be suitable for temporary works purposes, taken from locations near to the proposed temporary works, to a relevant depth and with appropriate testing to enable an appropriate interpretation to be made.

3.3.1 Determine the purpose

For an investigation to be effective it should have clearly defined objectives based on the proposed construction for the site and understanding of how the construction may affect the ground and groundwater regime. Both the permanent and likely temporary works must be understood, even if based on preliminary proposals at the early stages of a project. Understanding the construction proposals and the potential engineering solutions that will be employed allows an assessment of how the project might influence the ground and a consideration of what information is likely to be required to design the works efficiently. If the construction proposals are unknown, it will not be possible to determine the scope of the investigation.

The purpose of an investigation for temporary works purposes should be established before specifying any investigation work and could include:

- understanding the history and previous usage of the site and the immediate surroundings
- topography of the site
- determination of the soil (and rock level if present) profile with engineering properties – especially areas that may pose a risk, such as 'soft spots', backfilled areas or weak soil layers beneath the surface (at relatively shallow depth)
- determination of groundwater conditions and any variations over time
- determination of the suitability of site materials (existing soils or demolition rubble) for use on the project
- classification of any possible contamination for hazards and waste removal
- identification of any below-ground services, hazards (could include voids or mine shafts) and potential below-surface obstructions (e.g. tree roots or buried structures)
- checking performance of subcontractors in achieving a pre-specified result
- determination of the extent of variations in profile and properties across the site.

3.3.2 Restrictions on investigation

There are likely to be several restrictions that need to be considered before specifying any investigation work:

- logistical restrictions (access, working space and headroom) which may limit the size of machinery that can be used or locations where investigation work can be carried out. Specific permissions from stakeholders may be necessary (e.g. when near Network Rail property)
- obstructions, such as existing foundations or services, which limit locations where investigation can be done
- restrictions on types of techniques (due to noise / vibration or potential for mess)
- availability of specialist machinery
- time and financial restrictions
- availability of trained personnel to carry out the investigation work and to correctly log the results
- will on-going monitoring be required (e.g. monitoring water levels over a period of time which may require on-going access permission)?

3.4. BS EN 1997

BS EN 1997-1 (BSI, 2013) classifies projects into three geotechnical categories:

- geotechnical category 1 structures – small and relatively simple structures that involve minimum ground risk in terms of overall stability and ground movement and may be in relatively straightforward ground conditions. If excavations are required, these are above the water table or where comparable local experience has demonstrated that excavations below the water table are straightforward
- geotechnical category 2 structures – conventional types of structures and foundations with no exceptional risks, difficult ground conditions or difficult loading situations
- geotechnical category 3 structures – involving high levels of risk or having complex ground or groundwater conditions.

Geotechnical category 1 structures can often be designed with qualitative geotechnical investigations. This approach is often used for the design and construction of strip foundations for housing. Building Regulations (HMG, 2010) provide a table of the minimum widths required for strip foundations and guidance on how to check that the actual ground conditions at the structure are as assumed in the design. Geotechnical category 2 and 3 structures must be designed with quantitative geotechnical data and analysis to demonstrate that the appropriate limit states are not exceeded. It is necessary to determine the characteristic strength of the foundation strata during a ground investigation and then demonstrate that the limit states are not exceeded by calculation. The geotechnical category should be reviewed at each stage of the design process and then amended where necessary.

The geotechnical triangle shown in Figure 3.1 provides a simple, holistic approach to geotechnical design. If the geotechnical triangle is in balance, the project is much more likely to be successful.

The first part of the process is the ground profile. The likely ground profile can be determined from the desk study and then validated with site investigation or, for certain cases, during construction. The ground profile should be as accurate as possible (based on the nature and risks involved with the element being designed), but the level of detail depends on the complexity of the project. For very small structures with limited risk, confirming that the formation is

Figure 3.1 Geotechnical triangle (based on Burland *et al.*, 2023)

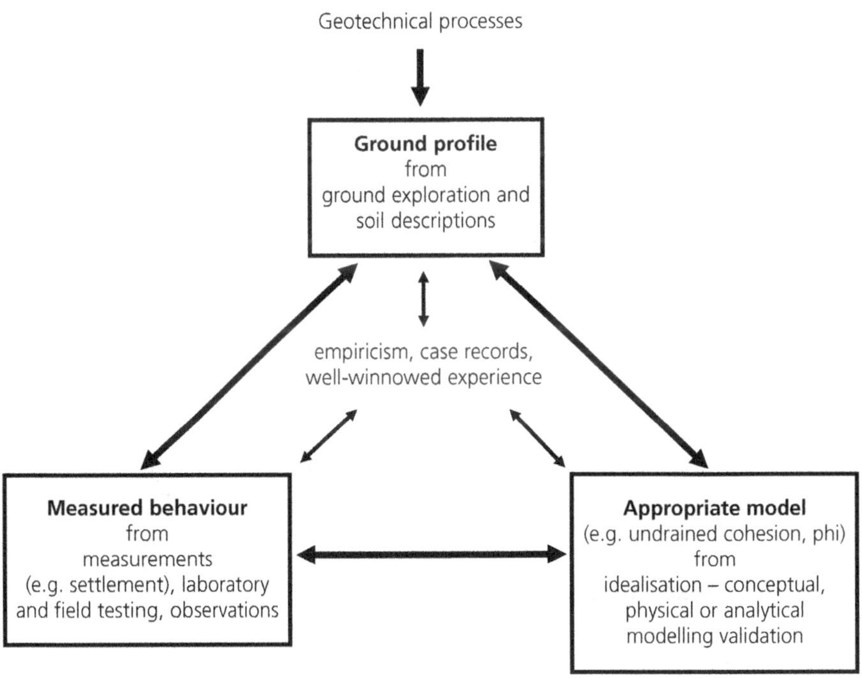

as assumed during the desk study may be sufficient. For larger structures, a more detailed investigation is often required so that the strata within the zone of influence of the structure can be confirmed. The next part of the triangle is the development of an appropriate model to determine how the ground will behave during the construction and service period. This could range from the use of simple models or tables to determine approximate foundation sizes for preliminary design to complex finite element models which predict how the ground will behave for a more complex structure. The third part of the triangle is the measured behaviour of both the ground and the structure. For all but the simplest structures, the geotechnical properties of the strata should be measured using in situ and laboratory test methods so that characteristic values of the essential ground properties can be determined for design.

3.5. Typical properties that may be required for geotechnical design

For design purposes it is essential to identify strata depths, soil properties and variations in these properties across a site. Below are some typical properties that should be established:

- soil description of each stratum and depth of each layer
- variations in stratum depth and description / properties; there may be some areas that have previously been excavated and backfilled
- density and strength of soil in each stratum
- angle of friction for granular soils and undrained shear strength for cohesive soils
- grading of soils
- permeability of soils
- stiffness of soils
- depth to water table and any variations in water table over time
- depth and extent of any existing foundations and below-ground structures such as basements
- potential adjacent surcharge loads from roads, railways, structures or spoil heaps
- below-ground hazards and contamination.

3.6. BS 5930 and BS 1377

BS 5930 (BSI, 2020) provides legal, environmental and technical guidance (it specifies tests and equipment) and is the recognised standard for conducting effective ground investigation on construction projects to ensure neighbouring land and properties are not adversely affected by the planned construction work and should be read with BS 1377 (BSI, 2020). BS 5930 (BSI, 2020) recommends best practice on gathering the right information and has developed to include newer and evolving investigative technologies, such as geophysical field testing, digital data capture and the increasing geo-environmental aspects, addressing potential contamination of brownfield sites. Familiarity with BS 5930 (BSI, 2020) will give those involved in investigation work an understanding of the appropriate approaches used to carry out ground investigations and a more in-depth comprehension of the different equipment and methods deployed to carry out the work.

3.7. Soil descriptions

There is a standard use of terms for the consistent classification and describing of soils which should be adopted; using the correct standard terminology allows designers to make reasonable assessments of some geotechnical properties for design purposes. It should be noted that these descriptions are subjective, so designers should treat them with some caution. Older investigation reports would have used soil descriptions from the superseded *Code of Practice 2001* (BSI, 1957). The current standard identification and classification of soils and rocks for engineering purposes can be found in BS EN ISO 14688 (BSI, 2017), BS 5930 (BSI, 2020) and is also replicated in BS 5975-2:2024 (BSI, 2024b), tables 40 to 44.

Soils are classified according to particle size. The main types that can be encountered are (starting with the smallest): clay; silt; sand; gravel – with strength and density being further sub-divisions.

- Clay and silt are generally classified as cohesive (fine-grained soils); sands and gravels (and larger-sized particles) are generally classified as granular / cohesionless (coarse-grained soils).
- Coarse-grained soils are generally visible to the naked eye whereas fine-grained soils are only visible with a magnifying glass or microscope.
- Figure 3.2. shows that coarse-grained soils do not clump together and will naturally want to form an angle. It can prove very challenging to excavate them to a vertical face as this face cannot be relied on to remain stable for any appreciable time.
- Fine-grained soil particles adhere (a property which is independent of friction and 'stick together') because of the presence of water (plus other effects such as bonding between particles) between the small particles, but over time this water will drain / dry out and cohesion diminishes. A cohesive soil can be excavated with vertical sides (see Figure 3.3), which can remain relatively stable for a time (note: it should not be assumed that any vertical face will remain stable where the consequences of failure could lead to fatalities, injuries or damage to buildings or services). However, the

Figure 3.2 Coarse-grained (editor's own)

Figure 3.3 Fine-grained (editor's own)

question to be asked of a vertical face in a cohesive material is: '*when* will a vertical fine-grained soil face fail?' not: 'will a vertical fine-grained soil face fail?'

- For temporary works purposes it is important to distinguish between soils which are likely to behave in a drained or undrained manner and to recognise this time-related difference for slopes, cuttings, trenches, excavations, retaining walls and foundations.
- Fine-grained soils can be prone to shrinkage and expansion. Coarse-grained soils can be prone to liquefaction.

Site engineers, site managers and temporary works coordinators / supervisors should become familiar with these standard descriptions as designers will often specify what soils they have assumed (often notes on drawings give assumed soil descriptions) in their designs and site teams will be required to verify these assumptions.

3.8. Investigation techniques

For many items of low risk, simple temporary works (foundations for single-stacked site cabins or lightly loaded scaffolding), a simple soil description with some basic testing carried out by hand at shallow depth would be adequate. However, for more complex projects, such as large and deep excavations below the water table, a far more comprehensive investigation with laboratory testing would be required (for both temporary and permanent works purposes).

An investigation will generally consist of two main activities:

- a desk study or preliminary investigation
- fieldwork and laboratory testing.

3.8.1 Desk study and preliminary work

A desk study can provide a great deal of useful information for little expense and its value should not be underestimated. It involves the gathering of information from existing historical documented sources (see Table 3.1) which should be used to develop an initial general understanding of the history and usage of the site, including potential hazards. An initial conceptual ground model with anticipated ground and groundwater conditions can be developed with some degree of accuracy. This information should be used to develop a preliminary design which should be refined as the desk study continues. An understanding of restrictions from adjacent stakeholders such as Network Rail and logistical constraints is also required. For example, a sheet pile retaining wall may not be feasible if the desk study identifies shallow rockhead at the site, or the use of piled foundations could be changed if the desk study identifies a high risk of unexploded ordnance (UXO). The initial ground model should then be used to determine the requirements for an intrusive ground investigation (see section 3.8.2) which will provide more detailed and specific information.

A desk study can identify potential hazards, such as:

- unstable slopes which have previously failed and leave visible evidence
- inadequate bearing capacity identified from settlement and damage to structures
- swelling or shrinking soils which could be caused by tree removal

Table 3.1 Some useful sources of information for a desk study

British Geological Survey	A wide range of geological information including: 1:50 000 and 1:10 000 scale geological maps for most of the UK and historical records of boreholes	www.bgs.ac.uk
National Library of Scotland	Some historical Ordnance Survey maps	www.maps.nls.uk
The Coal Authority	Details of coal mining areas, likely presence of shallow coal mine working and recorded position of many mine entries	www.groundstability.com or https://www.gov.uk/government/organisations/the-coal-authority
Historic England	Over 400 000 historical aerial photographs of England	www.historicengland.org.uk
Zetica	Unexploded ordnance risk maps for the UK	www.zeticauxo.com
Google Earth	Detailed images of UK with datums and contours	www.earth.google.com
Local planning authorities	Information about site or adjacent sites. Planning application submissions can include ground investigations undertaken as part of the planning process.	Website for each local authority

- high groundwater or potential for regular site flooding
- previous industrial usage could identify contamination – asbestos, hydrocarbons, methane, radon
- possible unexploded ordnance – Ministry of Defence properties or wartime bombing flightpaths
- buried structures, obstruction, cavities, sinkholes, mine-working and services
- aggressive ground
- possible presence of mineshafts or old mine workings, possible solution features or former landslips.

Some hazards, such as deep solution cavities or mine workings, would not present a major risk to many items of temporary works – for example, lightly loaded scaffold foundations – but could pose a major issue for deeper and heavier loaded items, such as sheet pile cofferdams or tower crane foundations.

A range of sources that could help provide a historical picture of the site and likely ground conditions is available and includes aerial photographs, previous investigations and historical maps. Table 3.1 provides examples, many of which are freely available.

Some caution should be used when consulting existing historical information and it should be treated as a guide only to the potential ground conditions at the site. There may be missing or inaccurate information and the standards for recording geotechnical information have changed over time. Some older information would have been logged before clear standards were available and could be misleading; this information should be used to develop a conceptual ground model that should be verified by an intrusive ground investigation.

Once the desk study has been completed and a conceptual ground model developed, a thorough visual examination of the site should be undertaken to supplement the findings of the desk study. During the examination, the proposals for the ground investigation should be reviewed to confirm that they appear appropriate, with modifications made where necessary. The field reconnaissance should note any natural or man-made exposures which reveal the soil or rock types and their stability characteristics. The presence of potential voids or cavities below the site, such as mine workings, basements or cellars, may also be identified. The presence and abundance of vegetation should be noted as this can provide an indication of the groundwater conditions, the presence of ground contamination or the presence of ground gas. The presence of any protected or invasive plant species should also be noted.

3.8.2 Intrusive fieldwork and laboratory testing

This involves the physical investigation of the site to validate (or otherwise) the conceptual ground model developed during the desk study and provide further detailed information about the physical characteristics of the ground and groundwater. The scope should be planned and clearly defined (to avoid collecting unnecessary and irrelevant information) and driven by the hazards identified in the desk study, together with a knowledge of how those hazards might affect the proposed construction work (permanent and temporary works). The spacing and depth of the investigation work is determined from having a knowledge of the likely information required for the design process (see section 3.5); BS EN 1997-2 (BSI, 2007) provides recommendations.

Before any intrusive investigation is carried out, the working area should be made safe (fencing / barriers) and a 'permit to dig' system should be in place to ensure it is safe to carry out the intended work. The location (coordinates) and a datum level (at surface level) should be provided for each investigation technique so the information can be related to the intended construction work. Those logging the information should be trained to do so.

Some assume that boreholes are the only investigation technique available and must be carried out on all projects. Although boreholes serve a purpose, they can be expensive, time-consuming to perform, require significant working space and messy. The main purpose of boreholes is to carry out investigations at a significant depth below surface level. For many temporary works purposes (such as foundations for cabins, scaffolding and small mobile cranes plus shallow trenches), investigating what is 20+ metres below the surface serves little purpose and there are many more cost- and time-effective methods that can provide relevant information at relatively shallow depth. A combination of investigation techniques is often advisable. The Association of Geotechnical and Geo-environmental Specialists (AGS) has produced a series of publications and guidance notes on a wide range of investigation, sampling and testing techniques (see Useful websites section at the end of this chapter).

Below are some examples of commonly used investigation and onsite testing methods.

- Trial pits – perhaps the most widely used technique for simple temporary works investigations. They can be excavated by hand but are commonly excavated by mechanical excavator, with the vertical or near vertical sides of the pits (or trenches) generally left unsupported. Access into the pits should be avoided unless deemed safe. They are useful for temporary works purposes as they are quick, cost-effective and easy to perform (excavators are readily available on many projects). They are used to record strata depths, potential for obstructions in the ground, groundwater level, extent of existing shallow foundations and basic descriptions of the disturbed excavated soils or the sides of the pit, at relatively shallow depth (typically 3 m to 5 m as maximum depth is limited by the excavator being used and stability of the pit). They provide little useful information below the water table. Several trial pits should be carried out at any location to indicate any significant variations. The stability of the pit sides can be observed over time, as can seepage, and they can indicate how easy or difficult the ground is to excavate at that location. Disturbed samples (from the excavated spoil) can be taken, along with jar samples for groundwater contamination and large bulk / box samples (less disturbed samples) for laboratory testing. Some on-site testing can be carried out using shear vanes or pocket penetrometers to verify and quantify soil properties, the specific resistance of the ground cannot be measured, and monitoring devices are generally not installed.
- Dynamic cone penetrometer (DCP) or dynamic probing – a metal cone (typically 50 mm diameter) attached to 1 m-long rods (typically 35 mm diameter) is driven into the ground using a known impact energy (from drop weight with known fall or small percussive impact hammer). It can be handheld or attached to small mechanical equipment. It can be quick, very cost-effective and easy to perform in a small working area or with limited headroom. The number of impact blows to drive the cone and rods into the ground is noted (typically for every 100 mm of penetration). The number of blows to penetrate the

ground can provide an indication of strata profile and strength / density; however, the specific soil type is not determined without correlation with other techniques. It is useful for profiling by 'filling in gaps' between other techniques, such as trial pits or boreholes. It is limited by depth of penetration into hard ground and buried obstructions.

- Windowless or dynamic sampling – 40 mm to 100 mm hollow steel sampling tubes are driven into the ground to a typical depth of 5 m to 6 m using relatively small, tracked machinery (handheld window sampling equipment is also available). Core samples of soil can be retrieved in a hollow plastic tube. It can be useful for temporary works as it can be quick, cost-effective, easy to perform, cause little ground disturbance and require limited working space (can be carried out within a building). It is not possible to obtain undisturbed samples and can be difficult to retrieve samples of granular soils below the water table. It can indicate groundwater levels, strata depths, strength of clay strata and density of sand if the 'standard penetration test' is carried out. It is limited by depth of penetration to around 10 m, depending on the strata type and buried obstructions.
- Cable percussion (shell and auger) boreholes – uses a tripod rig and winch, which requires significant working space and headroom (demountable versions are available for low headroom). This is perhaps the most common method for identifying soil strata, with descriptions and properties to a significant depth. For many simple temporary works purposes, it may prove too time-consuming and costly and will generally be carried out in conjunction with permanent works requirements. Typically, a 150 mm-diameter hole is formed using a variety of cutting tools, depending on soil type (clay cutter or shell for granular materials). When drilling, a casing may be required to line the borehole and this is an indication of collapsing soils. Water levels can be identified along with artesian pressures (pressurised groundwater) and water horizons at different depths can be sampled (casing will seal certain depths) with minimal possibility of cross-contamination. If hard layers or obstructions are encountered, then chiselling may be used. Disturbed samples can be recovered from the excavated material and undisturbed samples (for laboratory testing) by driving a 100 mm-diameter open sampler tube into the bottom of the borehole at regular depths as the borehole progresses. Standard penetration tests are regularly carried out as the borehole progresses.
- Rotary coring boreholes – specialist machinery with crew (which can be costly to mobilise and use) used to core into very hard strata and rocks, using compressed air, mud or a water flush, which can be messy. They require a significant working area and headroom but can penetrate to significant depth. The cores can be used to obtain depths and properties of rock strata, along with information regarding fractures and how intact the rock is. They are rarely used for temporary works purposes unless for specific investigations at depth for major projects, such as large excavations for nuclear work, highways or rail and tunnels in rock.
- Cone penetrometer testing (CPT) – the equipment is mounted on a vehicle (tyres or tracks for rough terrain). A cone (cross-sectional area of 10 cm^2 or 15 cm^2 with rods) is pushed into the ground at a controlled rate of typically between 1.5 cm to 2.5 cm per second, with the vehicle acting as kentledge. As the cone is pushed into the ground, the resistance of the ground is measured on the cone and shaft behind the cone. The correlation of shaft and cone resistance indicates the type and strength of strata to a significant depth (in favourable ground, 20 m+ is common). CPT has enhanced accuracy over some other methods, provides a continuous readout (data can be recorded

every 1 cm if required) and can identify thinner strata, which could be important for overall permeability of a site. It is much quicker (hence more cost-effective) to reach a significant depth than boreholes and commonly used for deep excavations (for temporary and permanent works) – for example, where sheet piling may be anticipated. It is limited in dense or hard soils as it may struggle to penetrate through the layer. The equipment has developed to allow a variety of additional subsurface information to be obtained, especially to determine soil behaviour under strains and vibration. It can be used to determine density, friction angle of granular soils, undrained shear strength, skin friction for pile design and bearing capacities.

- Standard penetration test (SPT) – a simple, cost-effective, dynamic test carried out at predetermined depths at the bottom of an advancing borehole (typically every 1.5 m depth) or during sampling. A 50 mm-diameter, 650 mm-long, 35 mm-thick-walled tube is driven 450 mm into the ground by a known impact. The blow count for every 75 mm is recorded, with the initial 150 mm being ignored (known as seating blow) as the base of the borehole will be disturbed soil. The final 300 mm gives a value known as SPT 'N' value unless there is effective refusal if 50 blows have not achieved the required penetration. Correlations for the 'N' value indicate the relative density, strength, effective angle of friction in granular soils and potential bearing capacity. It is useful for determining driveability of king posts or sheet piles (Pallett and Filip, 2019, section 11.5.14). Disturbed samples can also be collected for soil identification or moisture content. In gravels and gravelly sand, the driving shoe can be replaced by a solid cone. Thin-walled sampling tubes can be used and these result in less soil disturbance.
- Pocket penetrometer – a portable, quick, cheap and simple site tool, which is pushed into the ground to confirm approximate soil type and to determine consistency, strength and compressibility. It should be used with some caution as the tip only tests a very small area. Erroneous results can be common (tests can be carried out on a stone or soft spot). A single test is meaningless: many tests should be carried out with erroneous results being discounted.
- Shear vane – Figure 3.4 shows a hand shear vane (borehole versions are also available) which is commonly used for measuring the undrained shear strength of relatively uniform cohesive soils. It is a portable, quick and simple site tool which comprises a cruciform paddle on a rod (extension rods are available) which is pushed into the ground (beyond any zone of disturbance) and rotated at approximately 1 revolution per minute. The torque resistance is measured on a dial until the soil around the paddle fails (paddle will spin); the dial reading is converted into shear strength. It is useful for shallow temporary foundations, testing subgrades beneath cranes or working platforms and confirming soil strength for trenches and excavations. By remoulding the failed cohesive soil and repeating the test, an indication of the residual strength can be determined. As with the pocket penetrometer, a single test would be meaningless: many tests should be carried out, with erroneous results being discounted.
- California bearing ratio (CBR) – a relatively quick, cost-effective and simple test which is carried out in a laboratory or on site. It measures the pressure required to penetrate a 50 mm-diameter plunger at a rate of 1.25 mm / minute into a material. On site the reaction force to the penetration is provided by a vehicle such as a 4x4. The CBR is the ratio of the bearing load that penetrates a material to a specific depth compared with the load giving the same penetration into a high-quality crushed stone in a laboratory and

Figure 3.4 Hand shear vane (editor's own)

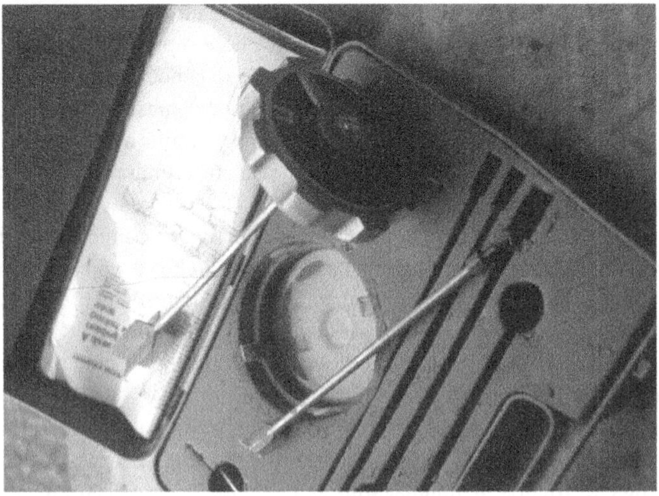

is expressed as a percentage. The test can be used to give an indication of stiffness and shear strength of subgrades at the surface. It is useful for estimating pavement thickness for flexible road construction and how thoroughly specific layers have been compacted when constructing a site haul road or working platform. The harder the surface, the higher the CBR value. CBR is used for fine-grained materials, with the maximum size of the particles being tested not exceeding 25 mm. The purpose of the test can often be misunderstood on site, as the scale of the test is small and does not replicate a pressure bulb created by a mobile crane outrigger pad or tracks beneath a large piling rig. CBR values can be obtained with depth by using DCP.

- Plate bearing test – an in situ field test carried out at surface level or in a shallow trench to examine the performance of a near surface formation under vertical loading. It can be relatively quick, cost-effective and easy to perform. It is used to determine the bearing capacity and anticipated settlement beneath the imposed load for shallow foundations. A steel plate is jacked into the ground using a load cell, generally using an excavator for reaction, and displacement measured with a gauge. The plate sizes can vary and are significantly larger than for CBR; hence, it is more suitable for larger particle sizes and better replicates (though still limited by size) larger pressure bulbs from heavy site machinery. The larger the plate, the more representative the result, but the size may be limited by the size of site machinery available for reaction. Similar to CBR, the scale of the test does not exactly replicate the pressure bulb created by a crane or piling rig and may not identify potentially weaker strata at a depth below the influence of the test (CPA, 2014, section 11.2.2). These tests are performed for a relatively short time so do not provide information on longer-term settlement and consolidation.
- Standpipe piezometers – plain or slotted plastic pipes can be installed into boreholes and used to determine groundwater levels (by dipping the levels) and permeability. They are cost-effective and simple but need careful installation to prevent clogging.

The above list only gives brief descriptions of some common investigation and testing techniques; many other techniques are available, and they are evolving. Site engineers and designers should familiarise themselves with the full range, along with specific uses and limitations and the relevant standards that apply to ensure the work is carried out correctly.

3.9. Interpretation of investigation and determination of ground parameters for design

The results of the investigation can be presented as a factual report with or without some interpretation, depending on the scale and complexity of the investigation. One of the principal objectives of any investigation is to enable a reliable estimate of the properties of each stratum (plus variations across a site, which may be significant on larger sites) to be undertaken, which are generally determined from a combination of the results of field and laboratory tests, complemented by well-established experience. A principal requirement of BS EN 1997-1 BSI (2013) is that a ground investigation should provide adequate information about the properties of the ground to enable a reliable assessment of soil parameters to be undertaken. No amount of complex analysis and design calculations can substitute for a lack of basic information. For some simpler, low-risk temporary works schemes, it may be appropriate to use the suggested values from a wide variety of geotechnical engineering textbooks and standards for quantitative geotechnical analysis. However, BS EN 1997-1 BSI (2013) recommends that the established values of geotechnical parameters are compared with relevant published data. The following properties are commonly determined: internal angle of friction for granular soils, undrained shear strength for clays, unit weight of soils, soil stiffness and variations in water table.

A simple tool is to plot the strata and water levels plus other investigation information on a series of sections, which form a simple picture of what is below the surface. Hazards such as services, tunnels or existing foundations can be added. Specific geotechnical values can also be plotted so the variations with depth and across the project can be determined. Site engineers and designers should be encouraged to carry out these simple tasks, which can also be used as part of a risk register, design brief or toolbox talk for operatives. On larger projects where sufficient time is available, statistical methods can be used to determine values. However, for many geotechnical category 2 structures, there is usually insufficient data for statistical methods to be used.

Designers should be cautious when estimating specific soil properties as the ground is naturally variable and any investigation cannot provide all the required information or be totally precise. Using unrealistically optimistic values based on one sample or test can lead to unsafe assumptions, whereas using unrealistically pessimistic values will lead to solutions which are not cost-effective. In general, the more relevant information that can be provided, the more realistic the ground model will be. Once designers have decided on design parameters, a check should be carried out to determine the sensitivity of their design to variations in these parameters – for example, strength (bending moment) and stability (embedded depth) of a sheet pile retaining wall to variations in the coefficients of active and passive earth pressures. The design can be further refined or verified by onsite observations or monitoring, such as deflection of a sheet pile retaining wall as excavation progresses or settlement of foundations beneath temporary props for a bridge refurbishment project. Monitoring is often specified in contract documentation with monitoring trigger values being established, typically for a large excavation as follows:

- green – excavation can progress as deflections are within agreed limits
- orange – deflections are approaching, or slightly exceeding, agreed limits and the frequency of monitoring can be increased or checks carried out to eliminate erroneous readings
- red – deflections exceed agreed limits and excavation work should be stopped, safety checks carried out and remedial measures such as additional propping considered.

Monitoring of noise, vibration, dust and variations in groundwater levels are also common.

3.10. Managing geotechnical risks

The purpose of undertaking a site investigation is to identify and manage ground risks. If only limited geotechnical information is available to design the foundations of a structure that imparts a significant load on the ground, this is a ground risk that needs to be managed. Ground engineering and geotechnical design can sometimes appear to be complex to those not familiar with the processes that need to be followed to ensure that all the geotechnical risks are managed. But, if a methodical approach is followed, the potential ground risks can be identified and managed.

A risk assessment should be used to document the identified hazards; the principles of prevention should then be followed. BS 5930 (BSI, 2020) recommends that residual risks are identified on project drawings. CD 622 (HE, 2020) provides a good example of the risk management process. The process is designed so that ground hazards are identified as early as possible in the project lifecycle. The hazards are documented in a table and the likelihood and severity of the hazard are assessed. As the project moves from initial concept to site works, the significant hazards can be systematically managed or eliminated.

3.11. BS 5975

BS 5975-2:2024 (BSI, 2024b), section 10 and annex M provide some guidance on the process to be followed for the consideration of foundations and ground conditions for falsework and can be applied to other simple items of temporary works, but the guidance has some limitations for more complex schemes. Designers should be aware that the approach may not be consistent with other standards (such as BS EN 1997, BSI, 2007, 2013) and some of the terminology may differ. However, BS 5975-2:2024 (BSI, 2024b) is still widely referred to for simple items by many designers who specialise in temporary works.

3.12. Geotechnical reporting and longer-term monitoring

BS EN 1997 (BSI, 2007, 2013) discusses the importance of local experience in comparable situations. It is important for designers to keep records of their projects so that this experience can be developed and shared. To fulfil the third part of the geotechnical triangle (see section 3.4), it is important to make some observations or measurements of the performance of a structure. In this way, experience can be developed and the methodology used can be refined. The simplest method for providing data about the performance of a structure is to undertake deflection / settlement monitoring during construction and within the early part of the service life of a structure. This can provide valuable feedback about the design and the chosen parameters.

REFERENCES

Burland J, Chapman T, Skinner H and Brown M (2023) *ICE Manual of Geotechnical Engineering*, 2nd edn. Emerald, Leeds, UK, vol. 1.

BSI (1957) CP 2001: Code of practice for site investigation. BSI, London, UK.

BSI (2007) BS EN 1997-2: Eurocode 7: Geotechnical design, ground investigation and testing. BSI, London, UK.

BSI (2013) BS EN 1997-1:2004+A1:2013: Eurocode 7: Geotechnical design. General rules. BSI, London, UK.

BSI (2017) ISO 14688: Geotechnical investigation and testing. BSI, London, UK.

BSI (2020) BS 5930: 2015+A1:2020: Code of practice for ground investigations. BSI, London, UK.

BSI (2022) BS 1377: Methods of test for soils for civil engineering purposes (9 parts). BSI, London, UK.

BSI (2024a) BS 5975-1:2024: Management procedures for the control of temporary works. Code of practice. BSI, London, UK.

BSI (2024b) BS 5975-2:2024: Temporary works – Falsework: design and implementation. Code of practice. BSI, London, UK.

HMG (Her Majesty's Government) (2010) Building Regulations. The Stationery Office, London, UK.

HMG (2015) Construction (Design and Management) Regulations. The Stationery Office, London, UK.

HE (Highways England) (2020) *Design Manual for Roads and Bridges: CD 622 Managing Geotechnical Risk, Revision 1.* Highways England, Transport Scotland, Welsh Government and Department for Infrastructure. https://www.standardsforhighways.co.uk/tses/attachments/ff5ed991-71ed-4ff2-9800-094e18cd1c4c?inline=true.

Pallett PF and Filip RK (2019) *Temporary Works: Principles of Design and Construction*, 2nd edn. ICE Publishing, London, UK.

Strategic Forum for Construction Plant Safety Group (2014) *CPA 1402: Good Practice Guide Ground Conditions for Construction Plant.* CPA, London UK.

USEFUL WEBSITES

AGS (Association of Geotechnical and Geo-environmental Specialists) – www.ags.org.uk
Andun Engineering Consultants – www.andun.co.uk
ASUC (Association of Specialist Underpinning Contractors) – www.asuc.org.uk
BSI (British Standards Institution) – www.bsigroup.com
CIRIA (Construction Industry Research and Information Association) – www.ciria.org
CPA (Construction Plant Association) – www.cpa.uk.net
DFI (Deep Foundations Institute) – www.dfi.org
GSL (Geological Society of London) – www.geolsoc.org.uk
HSE (Health and Safety Executive) – www.hse.gov.uk

FURTHER READING

AGS (Association of Geotechnical and Geo-environmental Specialists) (2016) *Guidelines for Good Practice in Geotechnical Ground Investigation.* https://www.ags.org.uk/item/ags-guidelines-for-good-practice-in-geotechnical-ground-investigation/ (accessed 10/01/2025).

AGS (2017) *Guide to Ground Investigation Reports.* https://www.ags.org.uk/item/ags-guide-to-ground-investigation-reports/ (accessed 10/01/2025).

AGS (2019) *Guidance on the Safe Excavation of Trial Pits*. https://www.ags.org.uk/2019/07/trial-pitting-still-fit-for-purpose/ (accessed 10/01/2025).

Andun Engineering Consultants – Temporary works webinars (available for free at www.andun.co.uk)

ASUC (Association of Specialist Underpinning Contractors) (2011) *Recommended Minimum Site Investigation for Typical Domestic 1, 2 and 3 Storey Buildings*. ASUC, Borden, UK.

CIRIA (2010) *X525: Geotechnical Issues in Construction*. CIRIA, London, UK.

CIRIA (2010) *C812: Good Practice Guide to Using Geophysics in Ground Investigation: A Client's Guide*. CIRIA, London, UK.

CPA (Construction Plant-hire Association) (2014) *Ground Investigation for Construction Plant*. CPA, London, UK.

Liff JF and Clayton CRI (1986) *Special Publication 45: Recommendations for the Procurement for Site Investigation*. CIRIA, London, UK.

Liff JF and Clayton CRI (1991) *Special Publication 73: Roles and Responsibilities in Site Investigation*. CIRIA, London, UK.

Ray Filip
ISBN 978-1-83608-525-6
https://doi.org/10.1108/978-1-83608-524-920251004

Chapter 4
Structural surveys and structural alterations

Ray Filip
RKF Consult Ltd, UK

Simon Smith
Contractors Design (CDS) Ltd, UK

4.1. Introduction

This chapter will focus on commercial / industrial and residential buildings, some of which may be of historical or architectural importance. It will cover two aspects:

- the structural survey information required to undertake a temporary works engineering design
- some of the potential risks and techniques involved in carrying out structural alterations (refurbishment, partial demolition and structural repairs).

Many of the principles can also apply to civil engineering projects, such as replacing bridge bearings or repairing retaining walls. In the UK, a significant amount of existing infrastructure (canals and subsequently railways) date back to the earliest days of the industrial revolution.

Understanding how a building or structure works is a fundamental requirement when seeking to undertake structural alterations, which could include:

- creating or enlarging openings in structural walls or creating openings in floors (e.g. for retrofitting disability access lifts or fire escapes)
- removal of walls, beams, columns or floors or sections of floors, including changing the level of a floor to create more headroom
- change of usage, such as converting commercial / industrial buildings into flats; converting or extending part of a building, such as loft conversions, due to growing demands on the building or structure
- increasing structural loading by adding additional floors, extending floors, changing the use of areas (e.g. changing an office area into an archive area for heavy storage) or adding heavy machinery / equipment
- adding a basement level, lowering an existing ground floor or underpinning foundations
- alterations to the façade, including façade retention schemes

- undertaking structural repairs or strengthening works after damage, such as water ingress, overloading (excessive deflection), fatigue (e.g. due to vibration), fire or vehicle impact; this may include the addition of temporary weather protection
- undertaking exploratory works
- allowing for imposed temporary works loading from items such as back-propping, access scaffolding (including ties), crane or hoist ties, propping or mechanical access / lifting machinery.

It is important to recognise that historical buildings may not have been 'designed' (in the sense of the word as used today) but built using traditional rules of thumb (sometimes questionable rules). In the UK, the history of some buildings may be traced back many centuries, when construction rules were based on experience, often using locally available materials and labour, and they are unlikely to conform to modern building regulations. Traditional buildings may need visionary and determined people to take a complex project to completion. Dealing with such buildings can often be risky and challenging, as they may have achieved a state of equilibrium over many years; unintended instability, leading to partial or total collapse, which could also affect adjacent properties, must be avoided. 'On paper' some structural items may have theoretically failed but the building remains standing. The work should be carried out in a sympathetic and careful manner (especially when modern materials and techniques are incorporated) to preserve historical elements; however, this can be demanding as many of the traditional construction skills are now in very short supply.

The key to providing a practical, cost-effective engineering solution is accurate information gathering, which should therefore form an integral part of the design process. Sufficient time and resource should be allocated to this activity, which applies equally to both permanent works and temporary works engineering designs. It should be noted that the permanent and temporary works design solutions should not be completely rigid / fixed and a certain degree of flexibility is required. Old buildings and structures have a habit of throwing up unusual and unexpected issues as the work progresses and design solutions often evolve over time. Dealing with this uncertainty often precludes 'fixed price' contracts and 'schedules of rates' are often preferred. Clients should be aware that they generally need to allow for significant contingency costs and these projects inevitably take longer to complete than initially anticipated.

4.2. Initial information gathering

The starting point for any engineering design is the client's brief. This should clearly state the task objectives but, in many temporary works design commissions, this is rarely the case, which can lead to frustration and wasted effort. As mentioned in chapter 3, much of the surveying effort is often focused on permanent works requirements and temporary works requirements can be overlooked. Clients and local authorities may hold useful information.

An initial desk study should be carried out to determine the following.

- Locate any 'as-built' drawings, other historical drawings or previous surveys. It is unlikely that all such information is available; often only historical fragments are available and some may not match the current situation.
- Determine the history of the structure, such as age (often one of the most useful pieces of information), which can indicate likely architectural style, original construction techniques

and materials (CIRIA, 1994, section 1.2 has an informative table). Internet searches of previous planning applications, historical maps satellite images and street view images often provide some information, but occasionally should be treated with some scepticism.

- Determine whether any alterations / extensions have been carried out over time. However, be aware that some historical work may not have been recorded (e.g. removal of chimney breasts or replacement of lintels). It should not be assumed that alterations / extensions have been carried out to an adequate standard with appropriate materials.
- Previous usage of the building / site can also indicate possible original construction methods, materials and anticipated loadings. If the building was used for an industrial purpose (e.g. mill), it may indicate the possibility of heavy machinery which may also have led to vibration damage within the building fabric.
- Identify any legal constraints, such as site boundaries and party wall arrangements (consultation with party wall surveyors is required; see section 4.4). Basement construction may require an impact assessment.
- Identify any listed building or heritage status, archaeology, planning constraints or conservation area constraints, including tree protection orders. There are often rigid requirements relating to preservation, use of materials and construction techniques.
- Identify any imposed project restrictions, such as access across a different owner's property, out-of-bounds areas which may affect where temporary works can be positioned, working hours, deliveries (time, weight and size restrictions). There may be specific restrictions from neighbours, especially Network Rail.
- Are there any adjacent works occurring or anticipated that could impact the project? These can cause sequencing issues, additional constraints or additional temporary loads. The client should give some indication of expectations of preferred project completion date.
- The anticipated permanent works details, indicative sequence of works, residual risk information, deflection / movement limits and any major design loads acting on the temporary works should ordinarily be provided by the permanent works designer (PWD). However, it is always prudent to validate any loading information by undertaking independent load assessments as the temporary state loadings will almost certainly differ from that of the permanent condition (which is generally what the PWD will be concerned with). Furthermore, the PWD may have overlooked something fundamental or made an incorrect assumption.
- Is the site subject to regular flooding?
- Identify what services may be present from utility companies' information.

The London Building Acts (HMG, 1930) (originating from HMG, 1774) are a useful source of early information. From the early to mid-19th century, building design rules started to become more formalised and formal codes of practice were later introduced; these can give an indication of techniques and material properties (e.g. minimum percentages of reinforcement in concrete). They also shed light on potential design loads, including permissible stresses and typical section sizes.

An initial site visit is necessary to get an overview of the structure and surroundings, which cannot be wholly determined by merely using 'Google Earth'. Local knowledge should be sought where possible. During a site visit after a de-watering system had been installed, one of the authors of this chapter was chatting to an elderly local resident who expressed surprise at the attempt to build at this location as it had been the site of a pond many years ago.

All relevant information is not always readily available and is less likely to be available the older the structure is. Figure 4.1 shows a temporary works scheme in Rome where the historical structure could date back many centuries; it is questionable how much existing information is available. A desk study and initial site visit are the starting point; these cannot gather all the necessary information but will give a steer on where further investigation should be focused to understand how the structure 'actually works' at 'this point in time' – hence the need for a detailed survey.

There is a need for good forward planning to answer four key questions.

- Why is a survey needed? – to fill in missing information from the desktop study so a safe and efficient design can be produced (permanent and temporary works design).
- What are we trying to find? – how the building has achieved equilibrium over time and its current state of repair.
- How will the survey be carried out? – combination of methods, using remote technology but also requiring visual inspection and measurement.
- When is it best to undertake the survey? – before work commences, when safe access is available and necessary permissions have been obtained. However, some survey work may have to be ongoing as work progresses, as new areas are 'opened up' and the structure becomes progressively exposed.

Once the desk study is completed, a survey brief should be compiled to determine the missing information required, which should contain all pertinent information that could influence the final design.

Figure 4.1 Temporary works scheme to a historical building in Rome (editor's own)

4.3. Surveys

The value of undertaking a detailed survey cannot be over-emphasised: the more information is known, the less the risk of unforeseen problems arising during the works. It is important to look beyond surface finishes, such as ceilings, floorboards and plaster, to examine the structural fabric / carcass of the building. However, not all finishes can be removed (some may be listed), so any survey can only provide a glimpse of the overall structure and often have to be undertaken simultaneously with the main contract works.

Before an intrusive survey can be carried out, all necessary permissions should be obtained (this may take some time and money to organise), and a plan of action determined including an outline brief. Safe access into the building is an essential requirement and the site area should be secured, especially if the structure is damaged (after fire, impact or explosion etc.), derelict or dilapidated. Aerial surveys are a useful starting point to provide an overview and some details. Entry into the building will eventually be required for closer and detailed investigation / sampling / testing. However, there are occasions where the structure is in such poor condition that entry to carry out a survey can be hazardous. An example of this is shown in Figure 4.2.

The intention of a survey should be to build up an insight of the building at the current point in time. Often the initial part of the survey involves visual inspection and measurements. A brief should be provided and should ordinarily include most of the following.

Figure 4.2 An abandoned and dilapidated historical building that required a detailed structural survey (editor's own)

Note: Some parts of this building dated back to the 16th century; access into the structure to carry out detailed survey work and dating was precarious. A method of progressively propping the structure (akin to a tunnelling shield) was developed. A subsequent line of propping was installed from beneath a previously installed line (a safe line), until the structure was sufficiently propped and supported to allow a survey team to gain safe entry.

- Dimensional surveys – floor plans (datums, coordinates, line and level), sections, heights, position of openings, balconies, corbels and any other overhangs or protrusions. How vertical and horizontal are structural elements (any wall out of plumb by more than 10% of its thickness should be a cause for concern)? Some of these can often be carried out with great speed and accuracy by drones and 3D laser scanning; however, a tape measure will be necessary for many specific internal items.
- Structural framing / composition – wall types (solid, cavity or rubble fill, cement or lime mortar) and thicknesses; floor types and thicknesses; iron or steel section sizes (many are likely to be discontinued and imperial sizes). Are walls and floors solid or partially hollow? Are perpendicular walls tied together or built with straight joints? Type of roof structure. Position of flues or voids and venting arrangements.
- Type of lintels and bearing of lintels (sometimes visible) – are there any relieving arches visible?
- Structural arrangements – direction of primary spans, load-paths and supports. Sometimes these arrangements may have changed from the original concept as alterations or repairs have been carried out. Details of connections, bearing of structural members and restraint conditions.
- Position of ties, basements / vaults (especially if they extend beyond the building line under pavement – they may have previously been used for coal deliveries) or arches and how vault / arch thrust is buttressed.
- How is overall stability achieved – location of shear walls, moment frames, bracing or cores? Is the structure part of a terrace that mutually support each other or does the structure provide support to a neighbouring structure? An overview of the structure and surroundings is necessary.
- Position, type and extent of foundations (plus any evidence of excessive settlement or undermining of foundations) and ground / groundwater conditions. Extensions are likely to have been built by different contractors and it cannot be assumed they used the same foundation details as the original. Often shallow trial pits can be adequate for this exploration purpose.
- Proximity and extent of tree roots should be identified.
- Exact position / type of services that are present and are they safe (any evidence of damage)?
- Condition and potential strength of structural elements (will affect structural performance) – any signs of distress, such as excessive deflection of floors, cracks, bowing of walls or sagging of lintels. Any obvious signs of previous repairs or alterations as the quality of the work carried out and the materials used may be questionable. Any areas of visible structural damage due to fire, impact or vandalism.
- Neighbouring buildings, structures, other assets (roads, tunnels, pipelines, railways, basements etc.) – these may be under different ownership.
- Possibility of hazardous materials such as asbestos.
- Are there any existing temporary works that have been previously erected by a third party – for example, because of emergency measures after an impact / fire or as part of a previous contract? Details of these should be sought, including design and any ongoing inspections. These temporary works will form part of the new project, so the principal contractor should be satisfied they are adequate for their intended purpose before adopting the site. Where this verification is not possible, the existing items of temporary works may need to be modified or replaced. An example is shown in Figure 4.3.
- Identify possible locations (or 'no go' areas) for temporary works items such as cranes, propping and scaffolding.

Figure 4.3 Historical existing temporary works (editor's own)

Note: This timber shoring was installed in central London during the Second World War because the adjacent building was demolished after being bombed. The area was used as a public car park for the subsequent years until being sold for development in the 1990s. No records of design or maintenance for the timber shoring could be found; hence, it had to be replaced as part of the new contract.

Note – the survey should pay particular attention to areas that are obviously distressed or damaged (could include areas of partial collapse), or where alterations / repairs have been carried out and where the structure is not weathertight.

R111: Structural Renovation of Traditional Buildings (CIRIA, 1994) has extensive information on the following:

- section 3 – masonry construction, potential defects and chimneys
- section 4 – timber construction, decay, traditional connections and repairs
- section 5 – cast iron, wrought iron and early mild steel

4.3.1 Follow-up investigation

An initial report should be produced based on the outline scope of works from the client, desk study, and visual and dimension surveys. This will then determine whether any more intrusive works are required, which may involve testing.

Follow-up exploratory works and testing / monitoring may comprise:

- core drilling to determine precise element thickness, depth (where not previously identified) and for material strength sampling. These are commonly carried out for permanent works purposes, but the information should be made available to temporary works designers (TWD) as it may be useful

- ultrasound surveys to detect voids in walls or below ground
- trial pits to ascertain foundation depths, profiles, ground conditions, presence of groundwater
- identifying precise connection details – steelwork connections, concrete to steel, timber to timber joints and so on
- robustness provision – tie details between elements (floor to walls, wall junctions, especially masonry party walls to façades, timber joists to walls / beams, especially flat roofs to prevent wind uplift, restraint to exposed masonry gable walls and façades etc.)
- identifying reinforcement content in structural concrete members
- identifying levels of fire protection provided and testing existing services for safety
- borescopes and endoscopes for intrusive investigations of cavities or where coring large diameter holes is not permitted
- it should not be assumed that all parts of the building share the same construction methods – alterations, extensions, repairs and so on may have used different materials and construction methodology. The survey should be representative of the whole structure and investigate different areas.

Where stability cannot be established precisely, the engineer must be able to make a judgement with a reasonable level of confidence. This is especially important during works involving significant structural alterations as any reliance on secondary load-path elements may have been lost due to their removal or diminished function. As mentioned previously, possessing a good level of historical building knowledge is crucial when making such judgement calls.

4.3.2 Monitoring

Various types of continuous monitoring can be utilised, starting before a project commences, and can continue into the construction phase and occasionally beyond the end of the construction work. Monitoring can be used to check on performance and as evidence to counter any spurious claims lodged from any disgruntled neighbours. Daily visual monitoring is recommended as a minimum requirement. In some cases, it will be necessary to measure and record structural movements more accurately (the degree of accuracy required should be established and agreed) using modern surveying equipment or laser technology. Time-lapse photography is also popular and may be useful in certain instances. With all monitoring, an existing / baseline level (prior to works being carried out) needs to be established and further measurements checked against this starting point. Some typical examples can include the following (also see chapter 17).

- Existing cracks in structural members or paved surfaces can be accurately monitored using 'Demec' studs, 'tell-tales' and photographs.
- Movement monitoring is generally carried out during the construction phase to check on performance. It can include checking verticality of retained façades, movement of retaining walls during basement excavation, level of floors during loading (storage of materials or general construction operation loading).
- Settlement of foundations beneath scaffolding, propping, façade retention schemes and neighbouring buildings can be monitored.
- Noise, vibration, light pollution and dust monitoring is generally carried out during the construction phase. Limits are often agreed and set out in contract documentation.

Realistic trigger levels should be agreed prior to commencement, with a traffic light system for action. Setting realistic levels is subjective and can be the source of many protracted negotiations, especially where party walls are concerned. The current trend is to impose far too stringent movement criteria, which take no account of natural building movements throughout the year. Adopting such practices is ill-advised as it will increase the cost and complexity of the temporary works and create an unnecessary adversarial environment. Understanding the systemic tolerances of the selected monitoring method and the likely range of thermal movements of the retained elements and the associated temporary works are key factors to be considered when setting the trigger levels and reviewing movement data.

The following ranges are a reasonable starting point for negotiations:

- green = movements well within tolerable limits (< 50%): works may continue with no further restrictions
- amber = movements > 50% < 75% tolerable limits: review required, but works may continue with certain restrictions e.g. amended methodology, more frequent monitoring, additional temporary works and so on
- red = > 75% tolerable limits: works that have a direct impact on the performance of the temporary works to stop pending outcome of a detailed review.

Tiltmeters, levels and strain gauges which can be linked to remote monitoring and recording equipment can be set up to record continuously, alerting users by email or text message when trigger levels have been reached.

4.3.3 Testing

Even if a full and precise investigation is carried out there may still be some degree of uncertainty and testing may be useful. Often this testing may be beyond the experience of a structural engineer and specialist advice / services should be commissioned. Examples include:

- material sampling and testing – for example, concrete compressive strength (there are several in situ methods available, such as using ultrasonic pulse echo), carbonation levels, iron or steel composition and strength / grade, corrosion levels (may be obscured by paint), presence of asbestos, masonry unit and mortar types and compressive strengths
- moisture content of timber and presence of dry / wet rot
- testing soils to determine properties to design retaining walls and foundations – for example, shear strength of clays (see chapter 3)
- weld testing – non-destructive testing is discussed in chapter 15
- ultrasound / radar surveys of party walls in old, terraced properties are considered essential where the works entail knock-throughs or where such walls will be required to support new structural elements or temporary works (e.g. beam bearings, walings or flying shores). Many of these walls may comprise a continuous series of flues, some of which may still be in use by the adjoining owners and their location needs to be identified.

Physical in situ load tests may also be undertaken to determine certain properties and capacities. A common test is to load test unusual floor constructions, or where the connections or degree of dilapidation are uncertain. Kentledge can be added to the floor, but the preferred safer solution is to use a controlled jacking system where the structure is fully exposed beyond the finishes (see chapter 17.5). Care needs to be taken when using the data obtained to determine the floor load capacity. This is especially pertinent to demolition works, where removal of neighbouring elements may strongly influence the load-carrying characteristics of the tested floor section. Furthermore, this form of testing may overlook the lack of shear provision around columns. Sudden floor collapses due to punching shear failure around column heads have been reported during demolition works which have then caused the knock-on collapse of lower floors. In addition, such testing may not be fully representative of the structure. There is also a danger of unrecorded localised alterations (e.g. infilling former stairwells) or poorly constructed repairs, which can affect capacity. When dealing with old structures it is often prudent to assume the worst case scenario.

Tension and shear testing of installed fixings is commonly undertaken on site to demonstrate that a satisfactory margin of safety exists (see chapter 17). Again, caution needs to be exercised as the test needs to replicate the end use as closely as possible. Regardless of the results achieved, the designer will still need to make a global assessment as to whether the structure itself can accommodate the loads applied.

4.4. Neighbouring properties (party wall awards)

Works associated with shared structural elements or where neighbouring buildings are in proximity need special consideration and will be subject to the Party Wall etc. Act (HMG, 1996). The Act gives legal requirements relating to:

- informing the adjacent property owner, even if the work does not extend into their property
- making good any damage
- dispute resolution
- access to neighbouring properties.

Before commencing any work, including opening-up investigations, the status of the award must be established through the party wall surveyor who should ordinarily be appointed by the client / building owner. The award process can be lengthy, particularly where several different neighbours are involved. Reasonable time and resource need to be allocated.

Detailed information on the adjoining buildings, such as construction methodology, floor levels, load paths, foundations and basements, overall stability provision, schedule of condition and presence of flues will be required; this should be included in the list of information. Obtaining this information can prove extremely difficult, especially where adjoining buildings are occupied under several separate tenancy agreements. A good relationship with adjacent property owners / tenants should be maintained and they should be consulted and kept informed.

4.5. Works adjacent to highways and other key assets

Works that may have an impact on a public asset, such as a local authority highway or footpath, railway land, pipeline, services, tunnels and so on, will be subject to additional scrutiny. Each will entail a lengthy approval process with associated costs and these need to be taken into consideration at project concept stage. In most instances, separate approvals in principle

(AIPs) will be required for both the permanent works and the temporary works. The term 'AIP' is a little misleading as the information required to complete the submission and obtain the requisite approvals is virtually that needed for a full construction status. Quite often, achieving such consent falls on the critical path and therefore becomes a priority for the project. Consequently, this puts additional pressure on the design engineer as it is unlikely that all relevant information will be available at this time.

4.6. What to do with the information

Under the Construction (Design and Management) Regulations (CDM 2015) (HMG, 2015), the client and the principal designer (PD) have an obligation to freely and early enough provide all relevant information they have regarding the project. When it comes to temporary works information, this is often not the case, and the TWD is often given very little time to develop a fully compliant design, often with minimal information. Where crucial information is lacking, the TWD must state clearly the assumptions and residual risks to the client and a strategy for dealing with these must then be put in place.

On the basis that all the pieces of the puzzle are readily available, the TWD will be able to provide the most appropriate solution with a high degree of confidence and, most importantly, minimal residual risk. Inevitably, some unknowns will remain, especially works involving alterations to existing occupied buildings. In these circumstances, the TWD will rely on 'well-winnowed' experience, taking cognisance of the potential risks and how these are best addressed.

There will be occasions when it is not possible to determine all information precisely and assumptions need to be made. These should be based on engineering principles, experience and knowledge (gained from previous similar projects) and assumptions based on engineering judgement, such as considering standards and techniques that were used at the time the building was constructed or altered – for example, the age of the building may indicate the minimum percentage of reinforcement used in concrete, based on the code of practice in use at that time.

4.7. Structural alteration philosophy

There are many reasons to alter rather than demolish the whole and rebuild. Structural alteration is a broad term which may include other terms, such as partial demolition, refurbishment (improvements, change of usage, structural additions / extensions or removals etc.), renovation / restoration, cut and carve (where an element of existing structure is retained); this chapter is concerned with buildings, although some of the principles can equally apply to civil engineering structures. This type of work requires compliance with the law and regulations, requires appropriate insurance and warranties, is often challenging, and clients (commercial and domestic) should seek professional advice. Further information is available in *C740: Structural Stability of Buildings During Refurbishment* (CIRIA, 2017). The complexity and uncertainty involved in this type of work makes planning and pricing challenging.

Temporary works for these projects should be managed by competent persons in accordance with the recommendations of BS 5975-1:2024 (BSI, 2024a). Roles and responsibilities should be clearly identified and performance should be continually monitored. Temporary works should be inspected with a permit system used for safely loading and removing temporary works.

Before significant structural alterations are carried out, the following philosophy should be adopted.

a. Ensure all organisations / individuals involved are competent to carry out the work they are being asked to do.
b. The health and safety plan should include fire safety and evacuation procedures.
c. Secure the site, protect the public and provide suitable access – provide fencing / hoarding / warning signage and scaffolding / access machinery and so on. Obtain all relevant consents / permissions (Note – carrying out unauthorised work on a listed building is a criminal offence). It is likely that preliminary method statements and drawings will need to be submitted to obtain listed building consent.
d. Understand the site constraints, how the structure works and its condition – carry out a desk study, site investigation, testing, monitoring (if safe to do so) and so on and assess the results to formulate a picture of the building in its current state of equilibrium and repair.
e. Understand the risks involved (including potential temporary works risks) and try to eliminate them where possible. If risks cannot be eliminated, identify them as residual risks.
f. Protect any listed items and make safe / weathertight the existing building in preparation for the main works – identify critical areas such as fire-damaged areas, bowing walls, weathered chimney stacks, areas of major water ingress, dangerous services and so on and carry out repairs (some temporary works may be necessary).
g. Once the existing building has been made safe, formulate a construction methodology (including temporary works requirements) and carry out the structural alterations / partial demolition / renovations and so on in an agreed safe sequence. The methodology and sequence may need to be adapted (designers should be consulted) as the work progresses due to unforeseen circumstances. It may be necessary to salvage / restore certain items as the work progresses.
h. Remove any temporary works in an agreed sequence once the permanent works can support themselves. Consideration should be given to how loading is transferred from temporary works items onto permanent works.

Desk studies and physical surveys are the starting point to understanding a building or structure, but if it has stood for many decades (or even centuries), it is likely that stability might not be compromised providing that:

- the original support conditions are in good repair, remain unchanged and the materials are in good condition; the aim should be to retain the original supports or provide temporary supports which replicate the original supports as closely as possible
- the loading and stress / strain conditions (including those due to temporary loading and wind) remain unchanged
- vibration (due to demolition, pile driving etc.) or weathering (structure may require new roof) during the construction process is carefully controlled.

4.8. Planning and coordination issues

CDM 2015 (HMG, 2015) requires that alteration, demolition and dismantling are planned and carried out by competent people to avoid unplanned structural collapse. Commercial clients must provide contractors with relevant information about a building's structure, including stability, structural form, significant design assumptions, suggested work methods and sequences. The principal contractor (PC) must use the information to plan and carry out the work safely.

The following should be addressed prior to commencing a project that involves structural alterations.

- Ensure there is coordination between permanent and temporary works requirements. This should include establishing a clear division of responsibility between the various parties involved. Have overall and partial stability in the temporary condition(s) been established?
- Ensure site constraints have been understood and addressed. Is there provision for public protection, offloading deliveries, removing waste, storing and moving materials around site?
- What provisions have been made for inclement weather (temporary weatherproofing measures), protection of any listed finishes, impact, fire protection and evacuation procedures?
- What procedures are in place if unexpected items or conditions are found during the demolition or re-building?
- What is the preferred temporary works methodology / equipment? How is equipment to be erected, maintained and dismantled? Are road closures necessary to aid the work?
- If the temporary works scheme relies on the permanent works for stability / support, has the permanent works been checked for adequacy? Temporary works should seek to reduce disturbance to the existing building.
- Has a clear temporary works management procedure been established (preferably in accordance with BS 5975-1:2024 (BSI, 2024a))? Clearly identify who has authority to issue permits for demolition, loading and dismantling of temporary works.
- Identify who will be responsible for any testing or monitoring and who will interpret the results. Have realistic limits on movement and settlement in the temporary and permanent condition been agreed? Have trigger levels and remedial action procedures (if levels are exceeded) been established? The procedure should also clearly set out how any alterations to the agreed temporary works scheme are to be managed.

On site, the PC is responsible for ensuring that risks are managed and that the work is carried out by competent persons. Under CDM 2015 (HMG, 2015), temporary works designers (TWD) have the same designer's duties as permanent works designers (PWD) – also see chapter 2.

Often design responsibility may be split between different organisations – for example, a structural engineer calculates the magnitude of the load to be supported and an acceptable deflection limit, an equipment supplier may then be responsible for ensuring a suitable structural item is provided. Whichever organisation carries out the design, the principal contractor should ensure that any 'grey areas' of responsibility are identified and appropriately managed.

Even though thorough surveys may have been carried out, not all details may have been established, and once site work commences some items may be found to be different from those expected. The design and site management process should anticipate dealing with unexpected items. To prevent partial or full collapse of the structure (or adjacent structures) during construction, careful thought must be given to correctly sequencing the demolition and rebuilding, ensuring that critical supports are not removed without temporary support being provided. Installing and removing temporary works can be challenging, as generally they are installed prior to main demolition and remain in place until support is provided by the new permanent works. The temporary works needs to be positioned to provide the required amount of support, while allowing the work to be carried out with as little obstruction as possible. Often it may be of economic benefit to delay certain parts of the demolition until part of the new structure is built, which could allow saving on temporary support requirements but may make the sequencing more challenging. Limiting deflection and movement are essential to prevent cracking and other significant damage.

4.9. Loads to be considered and structural form

CDM 2015 (HMG, 2015) requires that all foreseeable loads are accurately assessed. These will generally comprise:

- self-weight of roof, floors, walls, columns, insulation, finishes and services
- self-weight of people and objects within the building – service class 1 (access) is 0.75 kN/m^2, service class 2 (general construction) is a minimum of 1.5 kN/m^2
- construction material storage, construction machinery or existing plant (heating, water storage tanks, air conditioning units etc.) and accumulation of demolition rubble within the building
- loads from temporary propping / back-propping
- the elements: wind loads, rain, ice and snow loads
- earth, surcharge and water pressures acting on retaining walls and basements
- dynamic and cyclic loads – such as nearby passing trains or working machinery
- accidental impact loads – during construction works caused by swinging loads on a crane or falling debris during demolition. If the item of temporary works is positioned on a pavement or road, then accidental vehicle impact should be considered to prevent progressive / catastrophic collapse. Scaffolding and adjustable props (e.g. Acrow props) cannot realistically be designed for any significant impact so impact prevention measures should be provided (concrete barriers, traffic lane closures etc.)
- an allowance for a percentage of the vertical load acting horizontally to ensure sideways stability; this may be in addition to other horizontal loads, such as wind
- temperature effects, especially on long structural steel members where the increase in compression force can be considerable (see CIRIA C760, 2017b), (e.g. flying shores and props in large basement excavations).

As discussed in Table 2.1, for temporary works, the effects of thermal expansion / contraction can often be ignored (except large structural props) and fatigue is also often ignored, as all temporary works items / components should be inspected for signs of fatigue or other damage prior to use. With most temporary works schemes, limiting deflection / settlement will be primary considerations to avoid damage and expensive repairs to the building.

Any temporary works scheme should:

- be positioned to provide the required level of support / restraint and avoid hazardous areas / services

- consider the weight, to allow manual handling if necessary, and dimensions of members / components to ensure they fit in the available space when carried out in areas of restricted or difficult access (e.g. lofts or basements)
- be stiff enough to prevent excessive deflection or settlement, which may cause cracking
- allow sufficient working space to complete the new works
- keep the bearing stresses on the remaining structure, within acceptable limits
- carry the loads either directly or indirectly to the ground
- maintain the lateral stability of the structure (i.e. when removing shear walls)
- be installed and removed in an agreed sequence
- allow the making good afterwards to be as easy as possible.

Collaboration with the PWD is essential to ensure the overall scheme is safe, efficient and the agreed sequence of works can be followed.

Figure 4.4 shows a simplified schematic diagram of a building. In this diagram, the roof spans horizontally between the outer walls or intermediate columns. The top (sloping) chord is in compression and the bottom chord acts as a tie (intermediate members may be used to reduce the spans). The floors span horizontally between load-bearing walls, columns and beams. Floors support self-weight and any applied live loads and provide restraint to walls and columns (reduce the effective length) and act as a 'plate' to transfer horizontal loads to bracing or shear walls (the external walls will generally be shear walls). Beams support floors and span between columns or walls. Walls and columns support floors and carry vertical loads (shear walls may also carry horizontal loads) down to the foundations. Shear walls or bracing provide the lateral

Figure 4.4 Simplified schematic diagram of structural building form (editor's own)

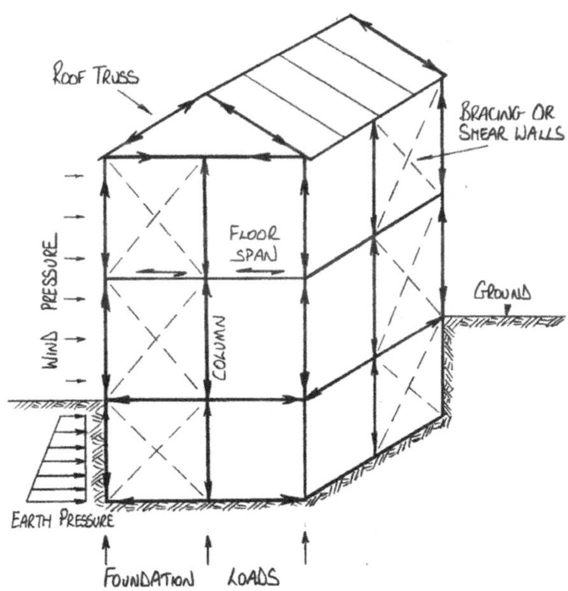

stability to the building. Basement walls are retaining walls which span between the basement and ground floors. Strip footings spread the loads into the ground and must limit settlement. Where the loads are high or the near surface ground conditions poor, then piled foundations may be necessary. The load path to the ground must be established for the overall building.

The remainder of this chapter will discuss some common temporary works techniques used during refurbishment projects.

4.10. Vertical propping

Vertical propping is likely to be required:

- when existing supports are removed or are replaced / strengthened because the supports are no longer capable of supporting the existing or new loads
- to provide temporary support until any new structure becomes self-supporting (e.g. concrete curing)
- to limit deflection which may cause damage
- to gain access for work into hazardous or weak areas
- to provide temporary support beneath a slab or floor, to allow for construction plant, operatives, materials or demolition rubble to prevent overloading; back-propping subsequent levels below may then also be necessary to carry these loads through the building to foundation level or to a more solid part.

These schemes can utilise timber, scaffolding, adjustable proprietary props, proprietary soldiers or structural steelwork. Temporary vertical supports are installed beside the member to be removed to support the structure above and to carry the loads to the ground by way of back-propping if required. The loads to be supported can be substantial and punching shear, settlement and elastic shortening of the support system should be considered. It is often necessary to shore the full height of the building, particularly older structures, to deal with problems of punching shear or overloading existing connections. Pre-loading of the temporary propping is often used to limit settlement.

The magnitude of the load to be supported and effective length of the prop should be determined so a suitable structural member can be designed. The prop could be bespoke (timber, steel, aluminium etc.), proprietary equipment (e.g. Acrow prop) or scaffolding. In addition to supporting the vertical load, the designer / site team should also consider the following.

- Eccentricity of load (for temporary works 25 mm is typically used) – this will cause a moment in the prop unless concentric loading can be guaranteed.
- Is the prop being installed 'out of plumb' (for temporary works 1.5° is typically used)?
- Sequencing of the works – including a means of adjustment so props can be installed and removed. The removal sequence should allow loads to be transferred through the structure as per the load paths in the permanent condition (i.e. generally from the middle of a span working back to the supports). For complex systems, a load transfer system (from temporary props to permanent supports) should be provided. A safe means of releasing the load from temporary props also needs to be determined, especially when the applied vertical loads are large (jacking systems should be provided).

- Foundation options – existing foundations may be utilised, or new temporary foundations provided. When the prop is positioned above a suspended slab or floor then back-propping may be required.
- Stability of the item being supported (e.g. deep beams may become unstable when permanent supports are removed) and overall stability of the structure if shear walls are removed – propping systems should be braced to provide the required lateral stability. Large window openings should also be braced with timber if the window frames are removed.
- Elastic shortening of the prop and foundation settlement – leading to movement which may be significant enough to cause damage. The degree of acceptable movement should be agreed by the. parties involved. Deflection can be minimised by using stiffer props or pre-loading (using the mechanical advantage of threads, pairs of folding wedges or jacking systems). In general, the authors recommend a maximum pre-load of 2/3rd of the applied vertical dead load. The likely deflection limits and façade / building movements are discussed in more detail in CIRIA C579 (2003).
- Excessive stress (crushing) between the top of the prop and the member being supported. Punching shear should also be considered; suitable spreaders will resolve this issue.
- Capacity of any existing connections should be checked.
- Exclusion zones for safe working and protection measures to prevent significant accidental loading – will crash decks be required for any falling items during demolition and how will rubble be removed?
- Handling the props (if lifting equipment cannot be used then props need to be 'manhandled'), provision of sufficient working space and safe access for any work at height.

The designer should ensure any temporary works are robust and designed so progressive collapse is prevented.

4.11. Forming openings in walls

Very minor openings may be required for services (e.g. extractor fans) and the brickwork will arch over the new hole so a lintel or temporary supports would not be required.

Larger openings in single or double skin brickwork (but still relatively minor e.g. for new doors / windows or minor house extensions) where the load to be supported is relatively small require permanent bridging by using a lintel (steel beam, precast concrete, timber beam or in situ reinforced concrete beam) or by forming an arch. There are many proprietary systems for providing temporary support to such openings (e.g. 'Strong-boys', 'No More Props', 'Brick-brace'). Users should become familiar with the manufacturer's requirements which highlight the correct use of these systems and, more importantly, the limitations. These systems can be misused or used in an inappropriate manner / situation, leading to structural problems. They are generally not suitable for 'thick walls', where a large opening is to be formed, where the load to be supported is large or where lateral stability could be compromised. If in doubt, a professional temporary works engineer should be consulted.

For large openings in thicker walls (e.g. to open an area for increased space or to allow more light into an area), a temporary works scheme (needling) will be required to support the wall above the opening until the permanent support is in place.

Further information for a typical needling scheme (plus alternative options to 'standard' needling) can be found in:

- Building Research Establishment (BRE, 1992a, 1992b, 1992c)
- Institution of Structural Engineers – *Temporary Works Toolkit 10* (IStructE, 2017)
- CIRIA C740 (CIRIA, 2017a)
- Pallett and Filip (2019), chapter 33.

Thick walls which appear to be solid may in fact be partially filled with rubble. These pose an issue when forming openings as the rubble fill could 'fall out' during the work. Small holes can be drilled through the mortar joints with resin / grout / expanding foam pumped into the wall at relatively low pressure, above the proposed new opening prior to needling. Alternatively, use of tensioned through ties to compress the inner fill materials may be viable.

4.12. Removing floors in buildings

Floors in buildings perform several roles. The most obvious is that they span between walls or columns carrying live loads. Less obviously, they:

- provide a plate action for the transfer of horizontal loading, such as wind loading, to shear walls or bracing
- reduce the effective length of walls and columns by providing restraint (especially external walls and below-ground retaining walls). The connections must be capable of carrying 2.5% of the vertical load at that point to provide adequate restraint.

There are many different types of floors which can include:

- regularly spaced timber joists with floorboards above and plasterboard ceilings below; occasionally there may be infill material, such as ash between the joists to limit noise transmission
- reinforced concrete slabs with drop beams, beams in slab or two-way spanning flat slabs; these may also be post-tensioned
- filler joist floors – 'RSJ' (rolled steel joist) beams with mass concrete (known as clinker concrete) infill between the beams; a similar method uses 'clay pots' between the RSJs and a concrete topping
- precast concrete beams with block infill between the beams
- arched or vaulted floors
- steelwork beams supporting steel decking with a lightly reinforced concrete topping.

Floors may need to be removed because of damage (e.g. fire or flood), replaced because of increased loading (e.g. due to change of usage) or to change headroom. It may be possible to remove and replace small sections of floor progressively without affecting the overall performance / stability of the building, but this methodology is slow, can be difficult to carry out and is not possible for areas of extensive existing damage. Where large sections of floor have collapsed due to damage, the adjacent walls may also show signs of distress (bowing / leaning), which may need to be addressed by tying opposite walls together. The position (level) of the supports requires careful consideration to minimise obstructions during demolition and rebuilding. The authors are aware of several projects where new floors were not at the same level as existing floors and the temporary supports were positioned at the same level as the new floors.

Often it is necessary or preferable to remove entire floor(s) and replace them; hence, temporary support is necessary.

- Flying shores – (see Figures 4.5 and 4.6) installed before floors are removed and removed after new floors are installed so positioning needs careful consideration. They are used to replace the plate action of floors by providing horizontal support, as well as when part of a row of terraces is removed, where they will strut walls apart. They can be made of timber, scaffolding, proprietary soldiers and ties or fabricated steelwork. When the length of the flying shores becomes significant, the self-weight (sagging) pulls the supported walls inwards and additional support should be provided. Lacing and bracing with tube and fitting scaffolding can reduce the effective length of members but the braced scaffolds can be obstructive when carrying out the demolition and rebuilding. Flying shores are also used when return walls are available and the façade and return wall can be tied together to mutually provide support. Care must be taken to ensure loads are safely transmitted to the ground and the system is not trying to 'hold itself up by its own boot laces'.
- Raking shores – (similar to those shown in Figures 4.1 and 4.3) transfer horizontal loading from a building or structure to a lower level or the ground. They are commonly used for retained façades, shoring unstable external walls or basement walls. They can be timber, scaffolding, proprietary soldiers and ties or fabricated steelwork, which are used for supporting an external party wall when an adjacent

Figure 4.5 Flying shores (editor's own)

Note: Figure 4.5 shows flying shores which comprise proprietary soldiers laced and braced with scaffolding due to their length. They are positioned to allow the new works to progress as easily as possible.

Figure 4.6 Detail of a flying shore (editor's own)

Note: The long length and significant self-weight of the flying shores in Figure 4.5 would lead to excessive sagging in the flying shores, if not for the support detail shown in Figure 4.6.

structure is removed. Substantial thrust blocks may be required to resist the applied forces (can be surface-mounted relying on friction or partially / wholly buried to mobilise passive resistance). Particular attention should be paid to wall plates and shear stops arrangements.

- Horizontal frames – (see Figure 4.7) generally proprietary soldiers and ties or fabricated steelwork. Horizontal frames are used to maintain the 'plate action' and 'support' provided by the floor and to reduce the effective length of walls. The frames are installed prior to removal of the floor and remain in place until the new floor is installed.

Figure 4.7 Horizontal frame replacing each removed floor (editor's own)

4.13. Vaults and arches

Arching is one of the oldest methods of bridging an opening. An arch is in compression as a series of wedges and these wedges must be supported by vertical reaction and restrained from spreading horizontally (buttressing). The line of thrust must remain within the body of the arch. Tension can develop if the line of thrust is below the bottom of the arch and the arch potentially becomes unstable. A temporary former is required when constructing an arch and a similar arrangement required when demolishing an arch to prevent an uncontrolled collapse.

Figures 4.8 and 4.9 show two possible options for providing temporary support to carry any out-of-balance thrust when vaults or arches are removed. Support would also be necessary at the end of a row of arches if the buttressing is removed.

4.14. Façade retention

Building façades are retained because they are of architectural or historical importance, or to maintain the visual aspect of the area. A shoring scheme (façade retention) is required to support the façade while the work is carried out. These schemes can be complex and high risk to design and construct.

This subject is extensively covered in:

- CIRIA R111 (1994)
- CIRIA C740 (2017a)
- Pallett and Filip, (2019), chapter 8.

4.15. Crash decks

Crash decks are provided during demolition to prevent falling rubble causing damage. The crash decks should be positioned as close to demolition level as possible, ideally as a

Figure 4.8 Option 1 for arch removal (editor's own)

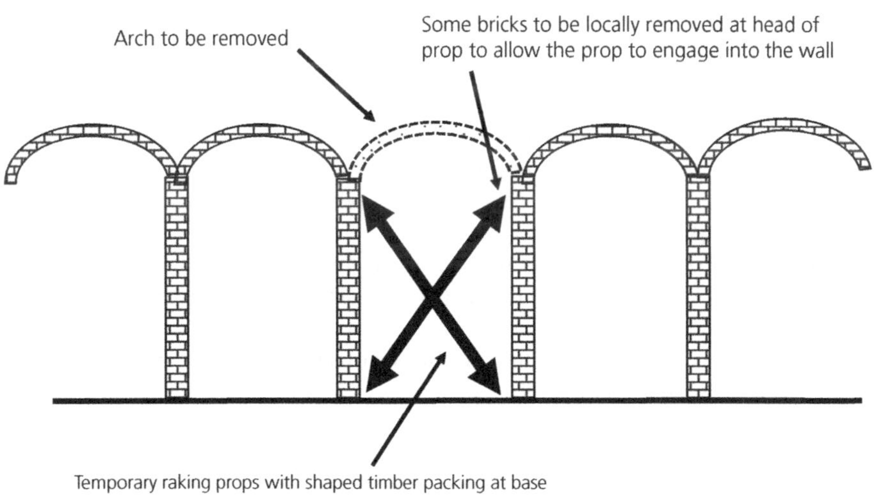

Figure 4.9 Option 2 for arch removal (editor's own)

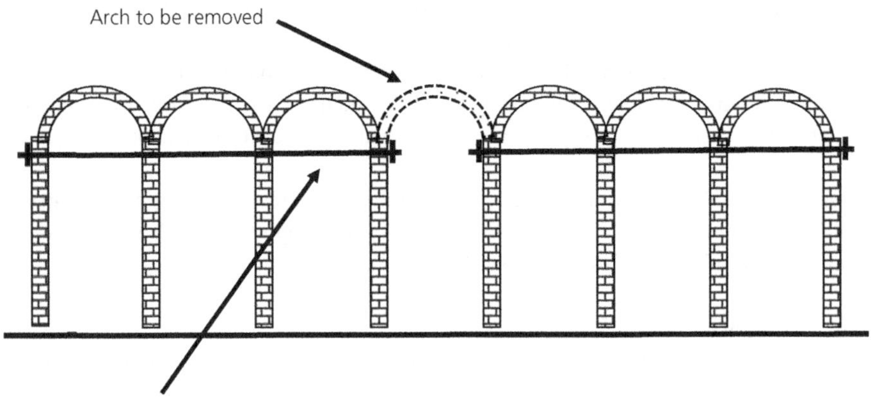

falsework scheme to prevent rubble falling any significant distance to reduce impact loads. Rubble should be progressively cleared to prevent overloading the crash deck. The floor beneath must be able to support the self-weight of the crash deck, the demolition rubble and any live loading. The following items influence the impact force on a crash deck (from basic laws of motions):

- mass of falling object
- distance object falls vertically under gravity or velocity it travels at some other angle
- how quickly it comes to rest on the crash deck (i.e. deceleration) – this is related to how the crash deck material can absorb the energy of the falling object.

Polystyrene is often used to absorb impact loading as it can be easily shaped for vaults and arches.

Filler joist floors comprise mild steel joists positioned at relatively close centres with concrete arching between the joists. A crash deck positioned directly beneath the floor can allow for easy demolition of the concrete and the joists can subsequently be cut out.

4.16. Underpinning

Underpinning an existing foundation involves excavating and replacing the removed soil with concrete to make the foundation deeper (and occasionally wider) because:

- the foundation is undermined by the construction of a new basement, lift pit, services or other adjacent excavation
- excessive settlement has occurred
- the load on the foundation has increased – for example, additional floors are added or load-bearing walls are replaced by beams and piers / columns and the load becomes concentrated.

Figure 4.10 Undermining an existing foundation (editor's own)

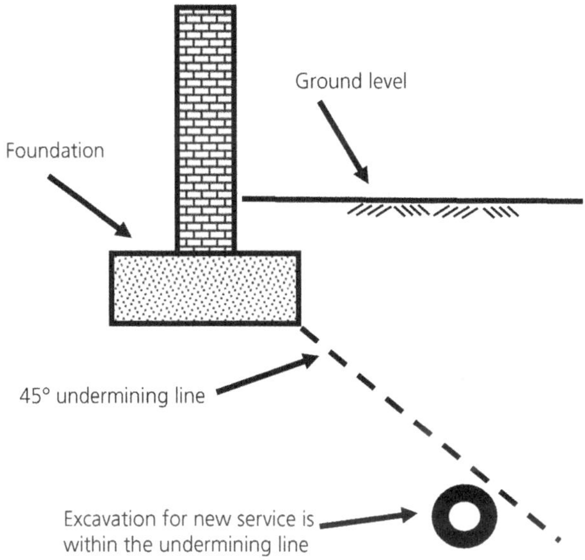

The underpinning should be taken to a level below the undermining zone (see Figure 4.10) or to a more competent stratum (when the load is increased or where settlement has occurred). An existing foundation (strip footing or pad) is deemed to be undermined if a new excavation encroaches within a 45° line drawn from the foundation, as the excavation process will cause the surrounding ground to relax and the existing foundation may settle. To avoid undermining an existing foundation:

- move the excavation outside of the 45° line
- underpin or pile the existing foundation
- carry out the work in segments to allow arching of the foundation (not possible for small pad foundations)
- design an excavation support scheme to support the full surcharge from the foundation
- move the structure (a complex and costly process).

Underpins are used to extend the existing foundation loads to a deeper level. This process is very distinct from merely increasing the size of a foundation at the same bearing level (usually by doweling new sections into the existing foundation). Shallow underpins (up to 1 m deep) usually may only require nominal excavation support for safety reasons. Deeper underpins (can be 3 m or more) require more careful consideration. It is likely that these deep underpins will require a designed excavation support scheme and the stability of the underpins must be checked in the temporary condition. Horizontal earth pressure acts on the back of the underpins and is resisted by a nominal 'toe in'. If deemed unstable, the 'toe in' may have to be increased or temporary propping (or anchors) provided.

Figure 4.11 Underpinning an existing strip footing (editor's own)

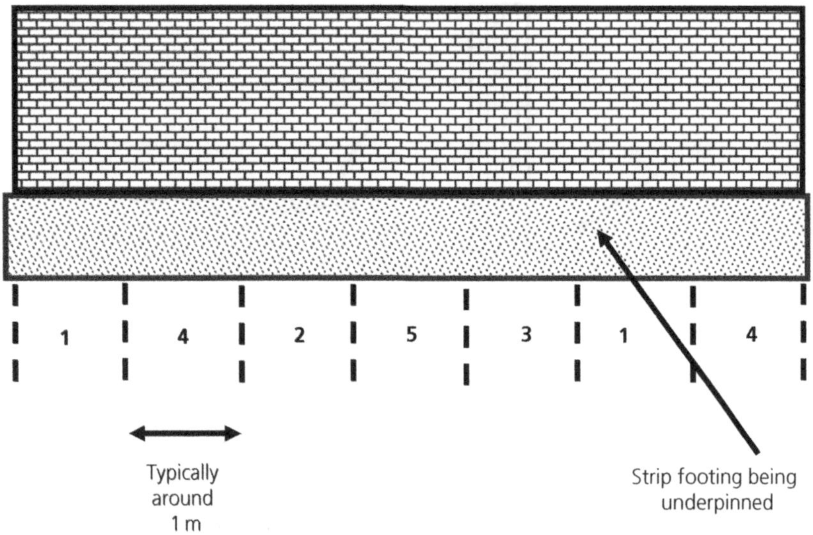

Underpinning is a slow process which must be carried out in a planned sequence on a 'hit and miss' basis, with each pin typically 1 m to 1.5 m wide (this is to allow the wall and ground to arch over the excavation). Typically, it would be hit one, miss three (or four), as shown in Figure 4.11.

Once excavated, the concrete is poured to within about 75 mm of the underside of the existing foundation (using a one-sided shutter). The gap is then filled using non-shrink grout or dry pack to ensure an even bearing on the underside of the existing foundation. Underpinning can be further complicated if continuity is required (by the permanent works engineer) between individual pins. This may be required to allow for the distribution of horizontal loading on the back of the underpins. Shear keys are sometimes specified between individual mass concrete underpins. If the underpins are reinforced, then reinforcing couplers are perhaps the easiest way to provide the required continuity. Openings near to the foundation to be underpinned may need to be temporarily bricked up to allow a more even load spread. Unsupported excavations for underpinning should not be left for any significant amount of time as collapse may occur.

A slightly simplified methodology relies on 'flood pouring' concrete to the underside of the existing foundation, thus eliminating the need for packing any gap. However, there can be issues with trapped air pockets at the interface of pin and foundation and it can be difficult to verify the contact interface.

Alternatives to traditional underpinning could include the following.

- Mini piling (see Figure 4.12) can be installed utilising piling rigs that can work in very constricted areas and headroom. The existing foundation is drilled through and steel

Figure 4.12 Mini piling instead of underpinning (editor's own)

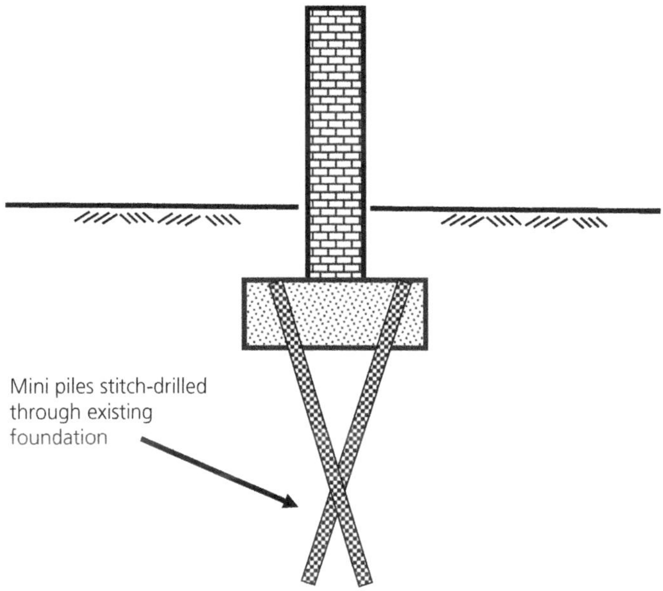

casings (or self-drilling mini piles known as Pali Radice) are driven into the ground at relatively close centres (depending on the load to be carried) and filled with grout or concrete (to form a pile) or concrete mini piles formed. The piles formed are small (typically from 150 mm to 200 mm diameter) and have relatively small capacities (typically 100 kN to 150 kN). A drawback is a reliance on the bond stress developed between the existing foundation and the pile to transfer the load from the foundation into the pile. Small foundations may not be suitable as the thickness of foundation available to develop the bond may not be sufficient.

- Horizontal beams and piles at relatively close spacing (depending on the load to be carried) can be installed to carry heavy loads to much deeper levels (see Figure 4.13). The beams are installed like needles and the bearing stress above the beams should be checked.

Underpinning piers and columns can be extremely challenging. Unless the column foundation is very large (in plan) and the underpin is very shallow (perhaps less than 300 mm), the column will be undermined by the excavation process. For deeper underpins it is claimed that the excavation and underpinning can be carried out beneath one quarter of the foundation at a time (often called 'in quarters'). However, the excavation process for these deeper underpins is likely to undermine the whole pier / column footing and settlement is likely as the vertical load in the pier / column will be large. Realistically, the only way to carry out such work is to support the load in the pier / column independently. This may involve connecting large steelwork sections (to limit deflection) into the column and spanning onto piles (the connection detail will be challenging) or installing raking props at the head of the pier column (vertical and horizontal reactions must be considered).

Figure 4.13 Horizontal beams and piles (editor's own)

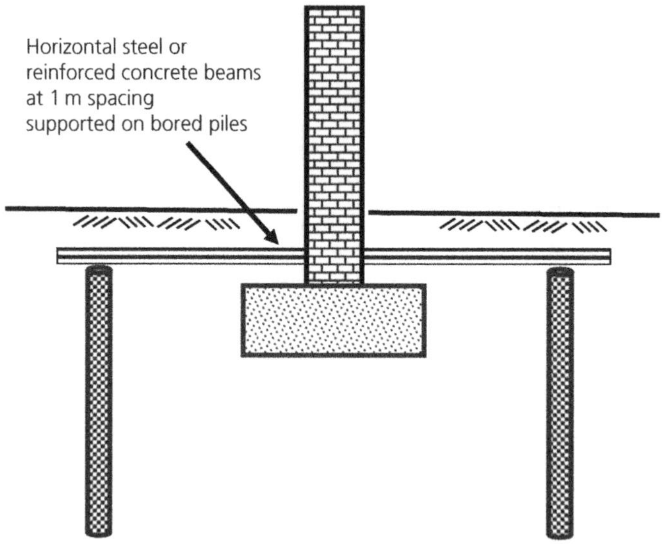

Further information can be found in:

- Temporary Works Forum (2023)
- CIRIA R111 (1994)
- CIRIA C740 (2017a)
- The Association of Specialist Underpinning Contractors (ASUC) have produced a wide range of guidance documents (see Useful websites section at the end of this chapter).

4.17. Scaffolding

Projects involving partial demolition and structural alterations / repairs often require scaffolding. A sheeted general-purpose scaffold around the building / structure is common to allow demolition, repairs, replacements, protection from weather and to control noise / dust. In the UK the dominant scaffolding is 'tube and fitting' and standard (pre-designed) details are available in National Access and Scaffolding Confederation's TG20:21 (NASC, 2021).

The scaffold should allow the work to be carried out safely and with minimal obstruction. A significant consideration can be how the scaffold is tied to the building. The preferred option is to use drilled ties into the outside brickwork (these require testing; see chapter 17.4). This solution is not always possible, as the brickwork may be in poor condition or listed (especially the façade). Alternatives can include buttressing (if there is sufficient space around the scaffold), through ties which utilise door or window openings (these will be an obstruction when replacing windows or doors) or clamping / drilling into steel / iron / timber beams. The number and location of tie positions requires careful consideration to avoid sensitive areas or clashes. The scaffold and ties may need to be altered or repositioned as the work progresses and this process should be carefully controlled and can only be carried out by competent scaffolders. The structure to which the scaffold is tied also requires an assessment to justify overall stability and not just locally at the tie points.

The scaffold may also have a sheeted temporary roof (if the roof of the building / structure needs extensive repair or replacement) to protect the inside and particularly sensitive or listed finishes. Such roofs are subject to uplift and require restraint which may require anchorage to the ground and / or kentledge positioned on the scaffold.

Scaffolding components may be used for a structural purpose, such as propping, flying shores or façade retention, and these scaffolds should always be designed and the design checked (see BS 5975-1:2024, BSI, 2024a, table 2) and installed in accordance with the design.

The National Access and Scaffolding Confederation (NASC) produces a wide range of scaffolding guidance (see Useful websites section at the end of this chapter) and the National Federation of Demolition Contractors (NFDC) has produced a guide to demolition scaffolding (NFDC, 2020).

Brickwork chimneys and towers are exposed to the elements and can often be in a poor state of repair. Repointing and flashing repair are common activities and scaffolding is provided for access and as a working platform. If the chimney or tower is in very poor condition, it may need to be demolished and rebuilt. The brickwork should be demolished carefully and cleaned so it can be reused. This work will often be carried out by operatives using hammer and bolster to progressively remove bricks. The scaffolding will be erected around the chimney or tower and should be designed for operatives, tools and some storage of bricks. The loading limits should be clearly communicated. If there are concerns that the progressive demolition could destabilise the chimney or tower, the scaffolding could also be used to provide support. Figure 4.14 shows a typical detail.

Figure 4.14 Scaffolding supporting brick tower which is being partially demolished and then rebuilt (editor's own)

Note: polystyrene blocks can be used to help spread any point loads and offer as much support as possible to the brickwork by deforming to take up the shape of the brickwork.

REFERENCES

BRE (Building Research Establishment) (1992a) *Good Building Guide 1: Repairing or Replacing Lintels*. BRE, Watford, UK.

BRE (1992b) *Good Building Guide 10: Temporary Support for Openings in External Walls; Assessing Loads*. BRE, Watford, UK.

BRE (1992c) *Good Building Guide 15: Providing Temporary Support During Work on Opening in External Walls*. BRE, Watford, UK.

BSI (2024a) BS 5975-1:2024: Management procedures for the control of temporary works. Code of practice. BSI, London, UK.

BSI (2024b) BS 5975-2:2024: Temporary works – Falsework: design and implementation. Code of practice. BSI, London, UK.

CIRIA (Construction Industry Research and Information Association) (1994) *R111: Structural Renovation of Traditional Buildings*. CIRIA, London, UK.

CIRIA (2003) *C579: Retention of Masonry Facades*. CIRIA, London, UK.

CIRIA (2017a) *C740: Structural Stability of Buildings During Refurbishment*. CIRIA, London, UK.

CIRIA (2017b) *C760: Guidance on Embedded Retaining Wall Design*. CIRIA, London, UK.

Filip RK (2017) *Temporary Works Toolkit Part 10: Propping and Needling*. IStructE (Institution of Structural Engineers), London, UK.

HMG (His Majesty's Government) (1774) George III. Fire Prevention (Metropolis) Act (better known as The Building Act). The Stationery Office, London, UK.

HMG (1930) The London Building Act. George V. The Stationery Office, London, UK.

HMG (Her Majesty's Government) (1996) Party Wall etc. Act. The Stationery Office, London, UK.

HMG (2015) Construction (Design and Management) Regulations. The Stationery Office, London, UK.

NASC (National Access and Scaffolding Confederation) (2021) *TG20:21: Good Practice Guidance for Tube and Fitting Scaffolding*. NASC, London, UK.

NFDC (National Federation of Demolition Contractors) (2020) Safety video on scaffolding in the demolition industry. https://www.youtube.com/watch?v=gr8KIjI (accessed 28/07/2024).

Pallett PF and Filip RK (2019) *Temporary Works: Principles of Design and Construction*, 2nd edn. ICE Publishing, London, UK.

TWf (Temporary Works Forum) (2023) *TWf2023:01: Underpinning: Good Practice Guide*. https://www.twforum.org.uk/viewdocument/underpinning-good-practice-guide (accessed 07/01/2025).

USEFUL WEBSITES

Andun Engineering Consultants – www.andun.co.uk
ASUC (Association of Specialist Underpinning Contractors) – www.asuc.org.uk
BRE (Building Research Establishment) – www.bregroup.com
BSI (British Standards Institution) – www.bsigroup.com
CIOB (Chartered Institute of Building) – www.ciob.org
CIRIA (Construction Industry Research and Information Association) – www.ciria.org
CROSS (Collaborative Reporting for Safer Structures) – www.cross-safety.org
HSE (Health and Safety Executive) – www.hse.gov.uk
IStructE (Institution of Structural Engineers) – www.istructe.org

National Access and Scaffolding Confederation (NASC) – www.nasc.org.uk
National Federation of Demolition Contractors (NFDC) – www.demolition-nfdc.com
TWf (Temporary Works Forum) – www.twforum.org.uk

FURTHER READING

Andun Engineering Consultants – temporary works webinar: Cut and Carve Demolition on 27/01/2024 (available for free at www.andun.co.uk).

Andun Engineering Consultants – temporary works webinar: Propping and Needling Design Considerations on 29/04/2024 (available for free at www.andun.co.uk).

Bates W (1984) *Historical Structural Steelwork Handbook*, Publication No.11/84. BCSA, London, UK.

Beckmann P and Bowles R (2012) *Structural Aspects of Building Conservation*. Routledge, London, UK.

BRE (Building Research Establishment) (1996) *Good Building Guide 20: Removing Internal Loadbearing Walls in Older Buildings*. BRE, Watford, UK.

BRE (2015) *Sustainable Refurbishment: How to Better Understand, Measure and Reduce the Embodied Impacts*. BRE, Watford, UK.

Curtin WG, Shaw G, Beck JK and Bray WA (1982) *Structural Masonry Design Manual*. Granada, London, UK.

European Commission (2015) *New European Technical Rules for the Assessment and Retrofitting of Existing Structures*. Joint Research Centre Publications Repository, European Commission, Brussels, Belgium.

Giaccio C (2024) *Provisions for Assessment of Existing Structures in the Second-Generation Concrete Eurocode*. IStructE, London, UK.

Harper RH (1985) *Victorian Building Regulations Summary*. Mansell Information, London, UK.

Historic England (2008) *Conservation Principles, Policies and Guidance*. English Heritage, London, UK.

Historic England (2012 to 2014) *Practical Building Conservation, Technical Handbooks, Series of 11 Titles*. Routledge, Aldershot, UK.

HMG (Her Majesty's Government) (1972) Defective Premises Act. The Stationery Office, London, UK.

HMG (1979) Ancient Monuments and Archaeological Areas Act. The Stationery Office, London, UK.

HMG (1983) National Heritage Act. The Stationery Office, London, UK.

HMG (1984) Building Act. The Stationery Office, London, UK.

HMG (1990) Planning (Listed Buildings and Conservation Areas) Act. The Stationery Office, London, UK.

HSE (Health and Safety Executive) (2006) *Research Report 463: Avoiding Structural Collapses in Refurbishment*, prepared by Loughborough University. The Stationery Office, London, UK.

IStructE (Institution of Structural Engineers) (2008) *Surveys and Inspections of Buildings and Associated Structures*. IStructE, London, UK.

IStructE (2010) *Appraisal of Existing Structures*, 3rd edn. IStructE, London, UK.

IStructE (2023) *Circular Economy and Reuse: Guidance for Designers*. IStructE, London, UK.

IStructE (2023) *Subsidence*. IStructE, London, UK.

IStructE (2024) Assessment of Existing Structures Safety Tool, free video available on IStructE website https://www.istructe.org/resources/training/assessment-of-existing-structures-safety-tool/ (accessed 05/08/2024).

IStructE (2024) *Structural Robustness and Disproportionate Collapse in Buildings*. IStructE, London, UK.

Lamden BS (1988) *Technical Information Sheet 89: Remedying Defects in Older Buildings*. CIOB (Chartered Institute of Building), Bracknell, UK.

Melville I, Gordon IA and Boswood A (1992) *Structural Surveys of Dwelling Houses*, 3rd edn. Estates Gazette, London, UK.

Richardson C (1985) *A Guide to Structural Surveys*. Architectural Press, London, UK.

Seeley IH (1995) *Building Technology*, 5th edn. Houndmills, Basingstoke, UK.

The Pyramus and Thistle Club (1996) *The Party Wall Act Explained*. P&T, Newry, UK.

Thorburn S and Littlejohn GS (1993) *Underpinning and Retention*. Blackie Academic and Professional, London, UK.

Ray Filip
ISBN 978-1-83608-525-6
https://doi.org/10.1108/978-1-83608-524-920251005

Chapter 5
Hoarding, fencing, security and public protection

Ray Filip
RKF Consult Ltd, UK

David Thomas
Retired Former Secretary, Temporary Works Forum; Formerly HM Specialist Inspector Health and Safety Executive UK

5.1. Introduction

The principal contractor (PC) has a critical role in managing health and safety risks during the construction phase of a project. Securing the site and protecting members of the public are critical elements that need to be considered. These issues should be identified at planning / tender stage and the principal designer (PD) should ensure adequate consideration is given to the potential risks. Hoarding, fencing and public protection measures are often the first items to be built on site and may well be the last to be removed.

The Health and Safety at Work etc. Act 1974 (HMG, 1974) requires that employers and the self-employed take all reasonable steps to ensure the health and safety of their workers and members of the public:

> Section 3. General duties of employers and self-employed to persons other than their employees
>
> It shall be the duty of every employer to conduct his undertaking in such a way as to ensure, so far as is reasonably practicable, that persons not in his employment who may be affected thereby are not thereby exposed to risks to their health or safety.

The Construction (Design and Management) Regulations (HMG, 2015) require the client to provide health and safety information at the preconstruction stage, including arrangements for site security and hoarding requirements:

> Regulation 12. (1) During the pre-construction phase, and before setting up a construction site, the principal contractor must draw up a construction phase plan, or make arrangements for a construction phase plan to be drawn up.

Regulation 13 requires the principal contractor to ensure that the necessary steps are taken to prevent access by unauthorised persons to the construction site. They should liaise with the contractors on site to physically define the site boundaries by using suitable barriers which take account of the nature of the site and its surrounding environment. The principal contractor should also take steps to ensure that only those authorised to access the site do so. Special consideration will be required for sites that have:

- rights of way through them
- other work areas next to them – for example, a shop refurbishment in a shopping centre
- occupied houses next to them, especially on new build housing estates
- children or vulnerable people nearby – for example, schools or care homes located near the site.

Regulation 18 has further guidance on site security:

> Regulation 18. (2) Where necessary in the interests of health and safety, a construction site must, so far as is reasonably practicable, and in accordance with the level of risk posed, comply with either or both of the following—
>
> (a) have its perimeter identified by suitable signs and be arranged so that its extent is readily identifiable; or
> (b) be fenced off. (HMG, 2015)

Site security and hoarding arrangements should be reviewed and altered (if necessary) as the project progresses.

Health and Safety Executive HSG151 (HSE, 2009) describes how to

> plan, provide and maintain suitable perimeters and barriers at locations where it is necessary to separate the public and others from the work.

It states that a risk assessment should be carried out to determine how a site perimeter is defined and the type of hoarding required to protect the public.

Security is an important consideration for sites because theft, vandalism and spurious accident claims are common. Temporary lighting towers and closed-circuit television (CCTV) cameras have been available for many years and require temporary foundations (or temporary connection to structures) and temporary power supplies. More recently, smart alerting technology utilising artificial intelligence analytics technology has been developed to identify intruders. Security turnstiles are used to control entry and exit from sites, along with secure security stores for valuable items, such as tools, and secure areas for machinery. Security personnel may be employed at night and holiday periods.

There are legal, moral and financial reasons for securing and protecting a site, those who work on that site or visit it and members of the public who may pass by it. Temporary works items are often the only part of a construction project where there is a direct interface with the public and the financial reasons may include the possibility of both genuine and spurious accident claims against contractors.

This chapter discusses some of the most common items of temporary works which are used on virtually all construction projects and these items should be controlled, as other items of temporary works, in accordance with BS 5975-1:2024 (BSI 2024a). The HSE have produced a variety of useful guidance on protecting the public, which are available on their website (www.hse.gov.uk/construction/safetytopics/publicprotection.htm).

5.2. Site hoarding

The client should provide information on the precise boundary of the project, adjacent land usage, rights of way and potential access routes before construction work commences. Construction sites should have safe and controlled entry / exit points with security gates (plus security personnel and turnstiles on larger projects) to ensure only authorised persons and vehicles can access / exit and to prevent unauthorised entry by members of the public. Signs must be displayed to warn and inform the public about the potential dangers on the site, with safe vehicle routes and safe routes for operatives established. Securing the site is usually the responsibility of the PC, who should consider groups which may be more at risk / more vulnerable, such as children, elderly, disabled or visitors to the site. There may be occasions where additional measures may be required (e.g. military, nuclear or 'controversial projects'), where the client may take responsibility for securing the site.

The PC is usually responsible for providing a solid site hoarding (Figure 5.1), which is commonly used around projects in populated areas and should be at least 2 m high. Most are generally around 2.5 m high (standard sheet size plus skirting), unless specified differently by the client, local authority or for specific security reasons. The hoarding could also be used for acoustic purposes to reduce noise emanating from the site, in which case it may be higher. The hoarding should be sufficiently robust to prevent unauthorised access and withstand foreseeable forces. A bespoke design can be carried out, whereby it may comprise plywood sheets

Figure 5.1 Hoarding and advertising signboard in a city centre (editor's own)

connected to horizontal timber rails which are in turn connected to timber posts (or occasionally steel sections); alternatively, proprietary systems using steel sheeting or recycled plastic can be used. With any proprietary system, the users should follow the manufacturer's instructions and fully understand any limitations for the use of these systems.

There are several useful documents which provide guidance:

- Temporary Works Forum (TWf) – *TWf2012:01: Hoarding: A Guide to Good Practice* (TWF, 2012)
- National Federation of Demolition Contractors (NFDC) – *Temporary Works Hoardings* (NFDC, 2019)
- Institution of Structural Engineers (IStructE) – *Temporary Works Toolkit Part 18* (Khan and Dixon, 2018).

A hoarding should be designed for a combination of loads which can be determined by risk assessment.

- Wind – wind pressure can vary significantly (depending on factors such as postcode, topography, altitude etc.) and consideration should also be given to probability based on the length of time of exposure (this has been the subject on much recent discussion). TWf2012:01 (TWf, 2012) and BS 5975-2:2024 (BSI, 2024b) use a simplified methodology rather than the more in-depth and complex method in BS EN 1991-1-4 (BSI, 2005) and the UK National Annex. Designers should be aware of localised funnelling wind effects, such as those that can be experienced adjacent to tall buildings or where building density is high. Wind load will also increase at the corners of the hoarding and at openings (see TWf2012:01, TWf, 2012, Figure B1). Designers can consider shielding effects but for hoardings it can be difficult to justify and it is usually taken as 1.0. Hoardings can be in use for a prolonged time (several years on major projects) and designers should carefully consider the wind probability factor in their calculations.
- Crowd loading – can vary but TWf2012:01 (TWf, 2012), clause 4.2.3 recommends a minimum value of 1.5 kN/m applied at 1.1 m above ground level. It is reasonable to assume that crowd loading need not be considered for a project in a field in the countryside, but a higher crowd loading (crush load) value would be appropriate adjacent to a football stadium (Figure 5.2) or busy railway station.
- Notional horizontal load – if crowd loading is not considered than TWf2012:01 (TWf, 2012), clause 4.2.1 recommends a value of 0.74 kN/m at 1.2 m above ground level.
- Robustness – TWf2012:01 (TWf, 2012), clause 4.2.4 recommends that the face material for the bottom 1.1 m should be designed for a minimum of 1.5 kN/m^2.
- Impact – from swinging load on a crane or vehicles. A risk assessment should be carried out to determine the likelihood of impact. It is generally not practicable to design hoardings for any substantial impact loading (e.g. impact from a lorry travelling at 50 mph) and such impacts should be prevented. This may involve concrete (or similar) traffic barriers to protect the hoarding, traffic management measures, such as lane closures, or controlling loads hanging from a crane to prevent them swinging into the hoarding.
- Security gates – openings will need to be formed in the hoarding to allow access onto and off the site. Designers need to allow for the self-weight (some may be heavy) of the

Figure 5.2 Hoarding designed for football stadium crowd loading (editor's own)

gates and consider how they will be connected to the posts. Often steel posts (square hollow sections) may be used either side of the opening to allow hinges to be welded to the posts for the gates. BS 1722-14 (BSI, 2017) provides some guidance on the design of mesh site gates and appropriate net pressure coefficients should be used (see TWf2012:01, TWf, 2012).

- Other loading – if a hoarding is adjacent to a railway line, then additional loading from passing trains should be considered. Further guidance is available in TWf2012:01 (TWf, 2012), clause 4.2.6.

The site temporary works coordinator (TWC) should ensure that an adequate design brief is provided. Designers should be competent but are free to choose their own preferred design methodology – for example, limit state or permissible stress with appropriate factors, although the methodologies should not be mixed. TWf2012:01 (TWf, 2012), table 1 and table 2 provide information for different load combinations. Designers should ensure the hoarding is structurally adequate and robust. Stability can be ensured by using appropriate factors and 'planting' the posts into augured or excavated holes in the ground. Where this is not possible (due to proximity of buried services or to prevent damage to a finished surface) then a frame with kentledge can be used. Kentledge solutions should be checked for overturning and sliding. Sliding resistance is provided by friction (see BS 5975-2, BSI, 2024b, table 15) for recommended values of coefficient of static friction. The amount of kentledge required should be determined by the most severe of these two modes of failure. Often, sliding resistance may govern and the coefficient of friction can be increased by providing rubber mats between the hoarding frame and surface beneath. Consideration should be given to the choice of kentledge. Water can leak / evaporate and may become a biohazard over time and items such as bags of sand or cement could be stolen. Reusable concrete blocks are often preferred;

their positioning should optimise resistance to overturning, they should have designed lifting eyes (or slots if being lifted by forks) and their weight should be clearly marked.

Many hoarding designs can be checked as category 1 (BS 5975-1:2024, BSI, 2024a, table 2). However, at some locations, such as adjacent to a railway, it is more likely to be category 3, with client approval / assurance also required due to the specific risks associated with railways. Failures of hoardings are generally due to a lack of design, or a poor design that does not adequately consider the potential loads, possibly combined with poor construction. TWf2012:01 (TWf, 2012) provides full design examples. A hoarding should be inspected after construction and a permit issued. There may be occasions where a hoarding would have to be altered; again, robust temporary works procedures should be followed.

Project teams and designers should consider the longevity (design life) and environmental impact of their hoarding. A hoarding should not be a single use item where the materials used inevitably end up as waste. Plywood and timber should be sustainably sourced and reused / recycled where possible. Materials such as recycled plastic may be more appropriate. Some companies advertise 'net zero carbon hoardings' and these should be considered. Once erected, the site hoarding should be regularly inspected for damage and deterioration. A maintenance schedule should be provided to ensure it remains in good condition for its design life; this would include ensuring the aesthetics of the hoardings (i.e. occasional painting).

A question that temporary works designers are often asked is: why is there not a standard hoarding detail for all UK locations? The variations in loading (wind, crowd, impact etc.) can be very large. For example, there could be at least a fourfold difference in peak velocity wind pressure if a simple comparison is made between a hoarding in a city centre site in the southeast of England and a hoarding at an exposed coastal site on the northwest coast of Scotland. Even before considering other variables, it would be extremely difficult to determine a suitable single value that could combine a safe and economic solution.

If a client has not specified their preferred choice of materials, it is often desirable for designers to provide a 'shopping list' by specifying different options for posts (including post spacing and depth of any foundations beneath), different options for rails, fixings (screws, nails) and face material and then allowing the client to 'price' their preferred option.

5.2.1. Case study

A major civil engineering project in central London has a 4.8 m-high hoarding which is used for security and as a noise barrier for nearby residential properties. The hoarding is required for the 6- to 7-year duration of this project. The hoarding will remain after this project is completed, as the site is to be redeveloped as a residential block of flats, which is likely to take a further 3 years. The design life of this hoarding is 10 years.

5.2.2. Signboards

Signboards (as shown in Figure 5.1) can be used for providing information regarding the project or to advertise the project completion. Signboards will generally be significantly higher than a hoarding, but similar planning and design principles apply. Tall signboards will pose a work at height risk during construction or removal and consideration could be given to prefabrication (offsite or onsite if sufficient space is available) and lifting the completed signboard into its final position and connecting to foundations. Designers should provide lifting points, ensure the signboard can be lifted as a whole and consider necessary tolerances for connections to foundations. Some of the principles of BS 559 (BSI, 2023) may apply.

5.3. Mesh fencing

Mesh fencing systems can comprise

- lightweight systems (commonly known by the brand name Heras fencing) – light-gauge tubular frame with mesh infill, supported by lightweight blocks (or similar) at the base
- heavier duty and more rigid engineered systems – comprise heavier mesh infill (can be double wire mesh) with posts which may be 'planted' or use kentledge. Panels may be marketed as 'security / anti-cut' and can be temporary or used for permanent fencing at schools, airports, sports facilities and so on. Some systems utilise recycled materials to reduce carbon footprint.

BS 1722-14 (BSI, 2017) covers the design of open mesh fencing. Temporary Works Forum has produced guidance on lightweight temporary open mesh fencing (TWf2024:01, TWf, 2024a).

Lightweight open mesh systems are commonly used on construction sites (on almost every project) for a variety of different purposes (see Figures 5.3 and 5.4). There are numerous similar open mesh systems (including Heras), but the most robust have a single tube forming the frame, which is bent into right angles at the corners (rather than two posts and a crossbar with corner brackets which are welded together).

They should be selected for suitability (for the specific purpose) by risk assessment and are generally considered suitable for shorter-term usage as demarcation, not edge protection (see chapter 6), and are intended to allow the wind to blow through them. They can be engineered to meet some additional requirements, but these options are limited and may require additional space, so alternative solutions should be considered.

A very common issue with lightweight fencing is when items such as netting, large signage, banners and so on are added to sites without engineering consideration of the increased wind loading on stability (overturning and sliding). There is also a lack of precise design information – for example, choice of solidity ratios and wind coefficients for various types of netting. Counterweights or kentledge are required for stability, but the structural capacities of fencing components are often not known and often some degree of engineering judgement must be used. If the structural capacity of the fencing can be verified, additional kentledge blocks can be specified for increased stability but all the additional blocks must be mobilised, which again can prove challenging and sufficient space must be available. TWf2024:01 (TWf, 2024) recommends further research, testing and the need for manufacturers / suppliers to provide detailed engineering information. It also recommends a risk-based approach to factors of safety (TWF, 2024, table 1) and provides design guidance, potential stability solutions and warnings regarding common misconceptions.

Sites should ensure that these systems are used appropriately in accordance with manufacturer's recommendations and any departures should be checked by competent designers.

Figure 5.3 Lightweight open mesh fencing positioned around construction site (editor's own)

Note: the fencing is used in lieu of a solid site hoarding. The use of this system at this location (adjacent to a road with no footpath and away from residential areas) and for the specific purpose must be justified by a suitable risk assessment.

Figure 5.4 Ineffective temporary fencing in a town centre (editor's own)

Regular inspections and maintenance are essential. The base blocks may become trip hazards if used incorrectly in areas adjacent to the public and screw anchors or ground pegs could be viable alternatives.

5.4. Scaffolding and falling objects

Occasionally, working above areas used by the public may be unavoidable and there will be inherent risks. Consultations must take place with local authorities, emergency services, highways, utility companies, neighbours and so on to get an understanding of the potential restrictions and obtain the necessary permissions. If an external site scaffold is founded on a public footpath, adjacent to a road or station platform, then it will have a major interface with the public. The erection, use, alteration and dismantling should be carefully planned to minimise disruption and ensure public safety. Specific risks can be encountered when offloading scaffolding components, as well as falling objects during assembly / alteration / use / dismantling, unauthorised access onto the scaffold and pedestrians walking into standards or tripping over sole boards. Pavement scaffolds in urban areas often have chutes for the disposal of debris into skips. Their use should be controlled and the area around the base should be protected.

If there is a risk of falling objects, ideally the principles of prevention should be followed, with exclusion zones created beneath such work – that is, fencing off / closing pavements, railway platforms, roads and so on. Where this is not practicable / not possible, then protection measures such as sheeting, netting, brick guards, scaffolding fans (Figure 5.5) or covered pedestrian walkways should be provided and tools tethered to prevent them being dropped.

Scaffolding components can be protected with foam or fencing or 'boxed in' with plywood with anti-climb measures, such as fencing, alarms and ladder guards, used to prevent

Figure 5.5 Substantial scaffold fan in a city centre (editor's own)

unauthorised access. Temporary traffic management measures (may include pavement and road closures) should be provided during offloading, assembly, alteration and dismantling.

The following documents provide further useful information:

- Health and Safety Executive: *HSG151* (HSE, 2009); *Safety Bulletin CON1* (HSE, 2018)
- National Access and Scaffolding Confederation (NASC): *SG4:22: Safety Guidance* (NASC, 2022); *TG20:21* (NASC, 2021).
- The Health and Safety Laboratory published a report on scaffold fan testing (HSL, 2003); however, this is no longer on the HSL website.

5.4.1. Pedestrian walkways

Some examples of covered pedestrian walkways are shown in Figures 5.6 and 5.7. Risk assessments should identify the possibility of falling objects. Pedestrian walkways should be designed for all applied loads (cabins, working platforms, wind etc.); however, it is unlikely / impracticable that they could be designed for the impact from a substantial object falling a significant distance – for example, a precast stair unit falling from a crane during lifting operations. They should, however, be designed to be sufficiently robust for impact loads from smaller falling objects, such as bolts, scaffold fittings or hand tools. Careful attention should be given to wheelchair access, avoiding potential trip / slip hazards, avoiding protruding items (fittings, nails, screws etc.) and adequate signage and lighting (including electrical safety measures). In accordance with the procedures in BS 5975-1:2024 (BSI, 2024a), these items should be regularly inspected and maintained (a maintenance schedule should be drawn up), which should include replacing light bulbs or any damaged or worn items. Fire safety, aesthetics and company branding / image are often considered along with longevity, as these items may be in use for several years.

In Figure 5.7 the design of steel gantry would consider:

- imposed restrictions on height and width of gantry
- vertical loading, such as self-weight, live load and snow
- horizontal loading, such as wind and accidental impact
- lateral stability
- provision of access / egress and services to the cabins
- safety of pedestrians
- concrete foundations – these would spread the vertical load into the ground and accommodate accidental impact loading
- scaffolding walkway and staircase – these would become a fire escape in an emergency and an alternative means of escape would need to be provided
- fire protection of steelwork
- fabrication / assembly tolerances and connection of steelwork to foundations
- speed and ease of assembly and removal – the steelwork is prefabricated off site, delivered and then assembled.

Note: The risk of fire needs to be considered; see section 5.6.

Figure 5.6 Covered pedestrian walkway in a city centre with working scaffold above (editor's own)

Figure 5.7 Temporary steel gantry supporting cabins over footpath and cycle path in a city centre (editor's own)

5.5. Delivery vehicle and construction machinery management

Vehicular deliveries and loading / offloading can create risks to the public outside the perimeter of the site, particularly in an urban environment. As mentioned in section 5.4, various consultations must be had to determine the appropriate type and extent of traffic management measures. These may include measures which would not be considered as temporary works, such as painted lines, signs and cones which have standard configurations. These should be deployed correctly by competent persons and rely on appropriate driver behaviour. These systems are subject to regular inspection and maintenance to ensure their continued performance. Other measures may include lane closures, creating temporary lay-bys, avoiding the need for vehicles to reverse (set up a one-way traffic system), planning deliveries to avoid busy times and traffic marshals. Particular attention should be given to projects where delivery vehicles are offloaded on the kerbside outside the site by a site-based crane (e.g. tower crane). Exclusion zones should be enforced while the offloading occurs. Some projects may have a bespoke designed temporary scaffolding loading / offloading bay, some with built-in lifting beams (see Figure 5.8). Proprietary systems (decks which protrude / cantilever from the building) are also available; these rely on the existing structure for support and restraint. The existing structure should be assessed, with additional temporary propping provided as necessary to prevent overloading the existing structure.

Figure 5.8 Temporary scaffold loading / offloading bay in a city centre (lifting beams can be incorporated) (editor's own)

A risk assessment may determine that physical traffic barriers are a more appropriate solution. Temporary vehicle restraint systems (TVRS) are specified in the *Design Manual for Roads and Bridges* (DRMB) (National Highways, 2020) which can be accessed from the Highways website (see Useful websites section at the end of this chapter). They provide a segregation between live traffic and the works area.

Construction machinery such as cranes and excavators have traditionally been controlled by traffic marshals, reversing cameras, exclusion zones, proximity detectors and slew restrictors. Anti-collision technology and digital safety screens / safety envelopes can now also be created to provide fully encompassing coordination and control measures.

5.6. Fire safety plan

The construction team should develop a fire safety plan which considers the site, surrounding buildings and members of the public who may be affected. Preventing fire should be paramount by ensuring electrical safety, enforcing hot works permits procedures, correct storage of gas cylinders and chemicals, avoiding the use of flammable materials (where possible) and considering safe storage of any flammable materials / waste and recycling. Fire suppression measures with trained operatives should be provided.

The fire safety plan should include emergency evacuation routes which may require temporary scaffold walkways, temporary staircases (including secondary staircases for alternative escape routes), ramps and assembly points. These routes should be clearly marked and maintained and be appropriately designed for emergency escape loading and fire rating as necessary.

5.7. Services

Services should be identified as there are significant risks if services are inadvertently hit and damaged, which could endanger operatives and the public (as well as being costly and time-consuming to repair). Temporary works may be required to support, divert or protect services. These could include the following.

- Any excavations for buried services should have a barrier / fence / edge protection or should have a temporary cover to prevent persons falling into the excavation. If an excavation is in a public area (e.g. street works) there is a significant risk to the public out of working hours. A risk assessment should be carried out to determine whether there is a risk of ground collapse and support provided as necessary.
- Scaffolding, timber or steel beams can be used as support if excavating beneath a buried service.
- Safety clearances and zones must be identified and maintained for overhead services to prevent contact and damage. Temporary ground barriers and overhead gates (made from scaffolding, timber or steel, with concrete blocks as foundations to prevent overturning) can create clear passageways for vehicles and machinery and cranes may require slew restriction. Tied back (anchored into the ground) scaffolding with protection nets (see Figure 5.9) can be used for protection from falling objects and their use / design needs to be agreed with the relevant utility company and other stakeholders.

Figure 5.9 Temporary scaffold with netting beneath power lines over a road (reproduced with kind permission of RKF Consult Ltd)

REFERENCES

BSI (1998) BS EN 1317: Parts 1 to 3: Road restraint systems. BSI, London, UK.

BSI (2005) BS EN 1995-1-4:2005+A1:2010: Wind actions. BSI, London, UK.

BSI (2017) BS 1722-14:2017: Specification for open mesh steel panel. BSI, London, UK.

BSI (2023) BS 559-1:2023: Interior and exterior signs. Design, manufacture and installation of signs. Specification. BSI, London, UK.

BSI (2024a) BS 5975-1:2024: Management procedures for the control of temporary works. Code of practice. BSI, London, UK.

BSI (2024b) BS 5975-2:2024: Temporary works – Falsework: design and implementation. Code of practice. BSI, London, UK.

HMG (Her Majesty's Government) (1974) Health and Safety at Work etc. Act 1974. The Stationery Office, London, UK. https://www.legislation.gov.uk/ukpga/1974/37/contents (accessed 31/10/2024).

HMG (2015) Construction (Design and Management) Regulations. The Stationery Office, London, UK.

HSE (Health and Safety Executive) (2009) *HSG 151: Protecting the Public: Your Next Move*, 2nd edn. HSE Books, London, UK. https://www.hse.gov.uk/pubns/books/hsg151.htm (accessed 31/10/2024).

HSE (2018) *Preventing Unauthorised Access Onto Scaffolding and Other Work Platforms*. HSE, London, UK.

HSL (Health and Safety Laboratory) (2003) *Report Number HSL/2003/17: Scaffold Fan Testing*. HSL, Buxton, UK.

Khan A and Dixon R (2018) *Temporary Works Toolkit Part 18: Introduction to Site Hoarding*. Institution of Structural Engineers, London, UK.

NASC (National Access and Scaffolding Confederation) (2021) *TG20:21 Good Practice Guidance for Tube and Fitting Scaffolding*. NASC, London, UK.

NASC (2022) *SG4:22: Preventing Falls in Scaffolding Operations*. NASC, London, UK.

NFDC (National Federation of Demolition Contractors) (2019) *DRG115:2019: Temporary Works Hoardings, Guidance Notes*. NFDC, Hemel Hemspstead, UK.

NH (National Highways) (2020) *Design Manual for Roads and Bridges*. National Highways, Birmingham, UK.

TWf (Temporary Works Forum) (2012) *TWf2012:01 (revised 2020): Hoardings – A Guide to Good Practice*. https://www.twforum.org.uk/viewdocument/hoardings-a-guide-to-good-practic-1 (accessed 07/01/2025).

TWf (2024) *TWf2024:01: Lightweight Temporary Open Mesh Fencing – Design Considerations*. https://www.twforum.org.uk/viewdocument/lightweight-temporary-open-mesh-fen-1 (accessed 07/01/2025).

USEFUL WEBSITES

BCGA (British Compressed Gas Association) – www.bcga.co.uk
BSI (British Standards Institution) – www.bsigroup.com
CIRIA (Construction Industry Research and Information Association) – www.ciria.org
CLOCS (Construction Logistics and Community Safety) – www.clocs.org.uk
Fence-Lok – www.fence-lok.com
Fire Safety Advice Centre – www.firesafe.org.uk
HERMEQ – www.hermeq.com
HSE (Health and Safety Executive) – www.hse.gov.uk
IStructE (Institution of Structural Engineers) – www.istructe.org
NFDC (National Federation of Demolition Contractors) – www.demolition-nfdc.com
NH (National Highways) – www.nationalhighways.co.uk
Panthera Group (EnviroHoarding) – www.pantheragroup.com
TWf (Temporary Works Forum) – www.twforum.org.uk

FURTHER READING

Aviva Insurance and Fire Protection Association (2022) *Fire Prevention on Construction Site*, 10th edn (incorporating amendment 1 January 2023). www.static.aviva.io/content/dam/document-library/broker/aviva_jcop_2022_v10.pdf (accessed 03/08/2024).

BCGA (British Compressed Gas Association), Liquid Gas UK (2022) *Code of Practice 7: Storage of Full and Empty LPG Cylinders and Cartridges*. BCGA, Derby, UK.

BSI (2010) BS 7375: Distribution of electricity on construction and demolition sites. BSI, London, UK.

BSI (2006) BS 1722-5:2006+A1:2018: Fences. Specification for close boarded fences and wooden palisade fences. BSI, London, UK.

BSI (2014) PAS128: Specification for underground utility detection, verification and location. BSI, London, UK.

BSI (2020) BS EN ISO 7010:2020+A6:2023: Graphical symbols. Safety colours and safety signs. BSI, London, UK.

CIRIA (Construction Industry Research and Information Association) (1995) *SP121: Temporary Access to the Workplace*. CIRIA, London, UK.

HSE (Health and Safety Executive) (1997) *Information Sheet CIS51: Construction Fire Safety*. HSE, London, UK.

HSE (2009) *Guidance HSG144: The Safe Use of Vehicles on Construction Sites*. HSE, London, UK.

HSE (2013) *Guidance GS6: Avoiding Danger from Overhead Power Lines*, 4th edn. HSE, London, UK.
HSE (2014) *Guidance HSG136: A Guide to Workplace Transport Safety*, 3rd edn. HSE, London, UK.
HSE (2022) *Guidance HSG168: Fire Safety in Construction*, 3rd edn. HSE, London, UK.
HSE (2023) *Guidance HSG141: Electrical Safety on Construction Sites*. HSE, London, UK.
HMG (Her Majesty's Government) (1980) Highways Act. The Stationery Office, London, UK.
HMG (1991) New Roads and Streetworks Act. The Stationery Office, London, UK.
HMG (2004) Traffic Management Act. The Stationery Office, London, UK.
HMG (2009) *Traffic Signs Manual Part 8: Traffic Safety and Signs for Roadworks and Temporary Situations*. The Stationery Office, London, UK.

Ray Filip
ISBN 978-1-83608-525-6
https://doi.org/10.1108/978-1-83608-524-920251006

Chapter 6
Work at height, preventing falls, demarcation and edge protection

Ray Filip
RKF Consult Ltd, UK

David Thomas
Retired Former Secretary, Temporary Works Forum; Formerly HM Specialist Inspector Health and Safety Executive UK

6.1. Introduction

Falls from a height are one of the largest causes of death and injury in the workplace and account for over 50% in construction. Designers and specifiers have a duty under Construction (Design and Management) Regulations (HMG, 2015 and see chapter 2) to avoid, through good planning and design, the creation of potentially dangerous situations, such as work from height, work over water, forming holes and voids in structures or, if not possible to avoid these risks, to adopt the 'principles of prevention' and then highlight any significant residual risks. The principal designer (PD) should take a 'hands-on' approach to controlling the preconstruction phase and ensure any work at height risks are adequately considered and highlighted in the construction phase plan.

An example is the design of a multi-storey steel frame building that considers how temporary edge protection systems can be connected to the steel beams to allow for safe construction. Several issues may make work at height more dangerous, including difficult access, poor weather (e.g. high winds, snow / ice or poor visibility) and how materials / tools are moved to and from a height. Work at height issues should be identified as early as possible so that effective solutions or equipment can be selected (e.g. prefabrication at ground level and lifting into final position with a crane which can utilise a remote release shackle to release the load).

Where work at height is unavoidable, it is essential that measures are taken to protect workers. These can include measures taken at design stage to facilitate a safe system of work. It is equally essential that any fall protection measures adopted are appropriate to the situation, are correctly maintained and that users have appropriate training (BS 8437, BSI, 2022).

Temporary works may be required to facilitate work at height and prevent falls. There are many options and proprietary systems which have their own standards and guidance. The purpose of this chapter is to provide a summary for readers, with signposts to further information.

6.1.1 Definitions

- Work at height – work in any place where, if no precautions are in place, a person could fall a distance (regardless of how far and even if it is below ground level) which would be liable to cause personal injury – for example, on a ladder or formwork or where a person could fall through a fragile surface or could fall into an opening or into an excavation. Note – this does not include access or egress by a staircase in a permanent workplace (Work at Height Regulations (WaHR), HMG, 2005).
- Demarcation – the establishment of boundaries or limits (lines) separating two areas, groups or things to prevent access to an area.
- Edge protection – measures used to prevent workers falling from a roof edge (HSE, 2021).
- Edge protection system – set of components intended to protect people from falling to a lower level and to retain materials; see BS EN 13374 (BSI, 2013a).
- Temporary demarcation system – a simple, continuous barrier to discourage access to an open edge and to identify safe areas and access routes and not to provide fall protection or debris containment. Note – a temporary demarcation system is not the same as a temporary edge protection system. Its purpose is to provide a visual barrier that should never be crossed unless additional fall protection measures are in place. A temporary demarcation system must never be relied on to stop a person from falling. Bunting, tape or markings at foot level, such as a painted line, are not sufficient other than for walkways (EPF/FASET, 2022). In the context of roof work, HSG33 (HSE, 2012, paras. 128 to 130) considers demarcation of access routes and work areas.

6.1.2 Work at height regulations

WaHR (HMG, 2005) are specific legal requirements which apply when any work takes place in a situation where a fall could lead to injury. They apply to all work at height where there is a risk of a fall which could cause personal injury. Regulation 6 requires employers to do 'all that is reasonably practicable to prevent anyone falling' (HMG, 2005). They include a hierarchy for duty holders for planning, managing and selecting equipment for work at height.

- Avoid work at height where it is reasonably practicable to do so (do not expose operatives to work at height).
- Where work at height cannot be easily avoided, prevent falls using either an existing place of work that is already safe or provide work equipment or other measures to prevent falls – for example, edge protection or fall arrest systems. It states: 'collective protective measures to be given priority over personal protective measures, and measures that prevent a fall given priority over those that minimise the height and consequences of a fall' (HMG, 2005).
- Where risk of a fall cannot be eliminated, use work equipment or other measures to minimise the distance and consequences of a fall – for example, netting or airbags.

Correctly selecting work equipment for any activity at height means being able to justify why safer alternatives required by the hierarchy have been ruled out (see Table 6.1).

Table 6.1 Hierarchy of fall protection measures given in BS 8437 (BSI, 2022), based on Work at Height Regulations 2005 (as amended) (HMG, 2005)

Priority		Highest	Lowest
	Category of work equipment (see Note 1)	Examples of protective measures	
		Collective	Personal
Highest	Work equipment which prevents a fall	Working platforms with barriers Advanced guardrail systems Barriers Mobile elevating work platforms (MEWPs)	Personal fall prevention systems (e.g. valley frames) Personal fall protection equipment (work restraint systems)
	Work equipment which minimises the distance and consequences of a fall	Safety nets at high level (rigged close to the work) Soft landing systems installed close to the work	Other personal fall protection equipment (rope access, work positioning and fall arrest systems) (see Note 2)
	Work equipment which minimises the consequences of a fall	Soft landing systems Safety nets rigged at low level (≤ 6 m below the work)	Inflatable injury prevention systems e.g. air jackets Other personal protective equipment not normally associated with fall protection e.g. lifejackets for work over water
Lowest	Work equipment which does neither (ladders, step-ladders, hop-ups, trestles etc.)	Instruction, supervision and training of users to minimise the risk of them suffering a fall	

Note 1:

Within each category:

a) collective protective measures take priority over personal protective measures

b) appropriate work equipment (and its order of priority) needs to be determined by taking into account the work to be undertaken and by considering the risk to those installing, using and removing the equipment and the implications for rescue associated with the work equipment being used.

Note 2:

The 'fall factor' has to be kept as low as possible, with systems with a fall factor of zero being given priority.

Several principles underpin safe work at height.

- There are two essential elements in the protection of persons against falls from a height: primary support and fall protection systems.
- The selection of a personal fall protection system should follow a hierarchy of protective measures, in which the risk of a fall from height should be eliminated (where possible).
- Collective protective measures should take precedence over personal protective measures.
- Wherever possible, a personal fall protection system that prevents a fall should be used in preference to a fall arrest system.
- Those planning and managing work should consult with employees or their representatives.
- The installation, use and / or dismantling of protective measures should not expose personnel to a greater risk than not using the system.

Note: a collective protective measure is equipment that protects a number of people simultaneously – for example, guardrails and toe boards. It is usually referred to as 'passive' – that is, the user does not have to do anything.

Note: a personal protective measure is equipment that protects only an individual – for example, full body harness and fall arrest lanyard. It is usually referred to as 'active' – that is, the user has to do something – for example, attach to an anchor point.

Duty holders have responsibilities to ensure that:

- all work at height is properly planned and organised
- all work at height takes account of weather conditions that could endanger health and safety
- those involved in work at height are trained and competent
- the place where work at height is done is safe
- equipment for work at height is appropriately inspected
- the risks from fragile surfaces are properly controlled
- the risks from falling objects are properly controlled.

This advice outlined below should be followed.

- Do as much work as possible from the ground.
- Ensure workers can get safely to and from where they work at height.
- Ensure equipment is suitable, stable and strong enough for the job, maintained and checked regularly.
- Do not overload or overreach when working at height.
- Take precautions when working on or near fragile surfaces.
- Provide protection from falling objects.
- Consider emergency evacuation and rescue procedures.

Regulation 5 (HMG, 2005) requires anyone working at height to be 'competent' or, if being trained, to be supervised by a competent person.

6.1.3 Guidance

There is no Health and Safety Executive approved code of practice or detailed guidance to accompany the WaHR. Basic advice is given at https://www.hse.gov.uk/work-at-height/the-law.htm and in INDG401 (HSE, 2014). The provision of detailed advice and guidance was left to industry. An alternative presentation of the hierarchy is given in INDG401 (HSE, 2014) (see Table 6.2).

Table 6.2 Hierarchy of all protection: step-by step diagram (based on INDG401, HSE, 2014))

Avoid	Prevent	Minimise *(sometimes referred to as 'MITIGATE')*	
Can you AVOID working at height in the first place? If NO, go to PREVENT	Can you PREVENT a fall from occurring? If NO, go to MINIMISE	Can you MINIMISE the distance and / or consequences of a fall?	
Do as much work as possible from the ground	You can do this by: ■ using an existing place of work that is already safe e.g. a nonfragile roof with a permanent perimeter guard rail or, if not ■ using work equipment to prevent people from falling	If the risk of a person falling remains, you must take sufficient measures to minimise the distance and / or consequences of a fall	For tasks of low risk and short duration, ladders and stepladders can be a sensible and practical option
	NOTE: Example(s) are provided of: ■ collective protection when using an existing place of work ■ collective protection using work equipment to prevent a fall ■ personal protection using work equipment to prevent a fall	NOTE: Example(s) are provided of: ■ collective protection using work equipment to minimise the distance and consequences of a fall ■ personal protection used to minimise the distance and consequences of a fall	NOTE: Follow HSE guidance on safe use of ladders and stepladders at www.hse.gov.uk/work-at-height/index.htm

6.1.4 Industry guidance

There are numerous industry associations dedicated to promoting safe work at height. Each produces guidance and information relevant to its own sector; these are listed in Useful websites and Further reading sections at the end of this chapter.

6.2. Demarcation

Section 6.1.1 provides a definition of demarcation. Demarcation does not require a design as it represents a point / line beyond which a person could be exposed to a higher risk area (such as a fall from height, falling objects, moving machinery, trip hazard or fragile / slippery surface). It could also be used to define fire escape routes, parking bays and storage areas where supplies are to be separated to prevent cross-contamination. Demarcation should be clearly visible and is commonly seen as painted / taped access walkways in workshops; however, on construction sites this would generally be considered inadequate, so fencing / proprietary posts with chains and warning signs (or road pins with tape) may be used. Typically, a minimum distance of 2 m away from the hazard is used to position the demarcation line. However, this may vary in certain industry sectors such as motorways, railways, airports, demolition and nuclear. Specific guidance should be sought for each of these sectors.

6.3. Scaffolding

Scaffolding is perhaps the most common method of providing access and a working platform where work at height is required. Scaffolding can comprise 'tube and fitting' or a proprietary system. For tube and fitting, the National Access and Scaffolding Confederation (NASC) has published TG20:21 (NASC, 2021) plus other extensive guidance documents (see NASC website). For proprietary systems, each manufacturer has produced their own specific guidance documents (see each manufacturer's website). Both options are covered in *ICE Temporary Works: Principles of Design and Construction* (Pallett and Filip, 2019, chapter 21) and the Temporary Works Forum has produced a management guide (TWf, 2020). A plan should be in place to ensure there is sufficient working space to erect, inspect, use, alter (if necessary) and dismantle scaffolding safely by competent persons. NASC (2022a) has produced guidance on preventing falls in scaffolding operations.

Scaffolding is also commonly used for items such as edge protection (see section 6.6), loading bays for offloading or storage of materials at height, protective fans above pavements or roads (for small falling objects), access stairs, access towers (see section 6.4) and for purpose-made gantries (in plan often a 'C' shape) to allow for safe loading and offloading of lorries.

6.4. Towers

Temporary access towers can be mobile proprietary systems or made from tube and fitting (that can also be mobile) and are often used for inspection and simple work operations at relatively low height (often inside a building) where a full access scaffold would not be feasible (due to space constraints, time or cost). They are relatively light, easy to assemble and suitable on floors with weight restrictions. However, their lightness can cause stability problems.

Tube and fitting versions should be designed to accommodate the anticipated loadings; TG20:21 (NASC, 2021) provides guidance.

Many types of proprietary tower are available, with the Prefabricated Access Suppliers and Manufacturers Association (PASMA) providing recognised training for their safe assembly, inspection and use. The manufacturer or supplier has a duty to provide an instruction manual which explains the erection sequence, safe heights, loading limitations and any bracing or tying requirements. BS EN 1004 (BSI, 2020) specifies the requirements for prefabricated mobile and access working towers to service class 2 and 3 only and requires a horizontal load to be allowed for in the design. It is the responsibility of the manufacturer to establish how the horizontal load is to be restrained; the often-used height-to-base ratios are no longer appropriate. If sheeting needs to be added to a tower, then a specific design is necessary.

6.5. Powered access machinery

Various types of machinery are available to enable safe work at height (e.g. scissor lift and mobile elevating working platform). An assessment should be made to ensure the most appropriate type for the work is selected. They have the advantage of manoeuvrability and adjustable working position. The selection process should include consideration of the type of work, number of operatives, any materials or tools, competence of operatives, access / terrain, space restrictions, ground conditions beneath the machine, overhead or buried hazards, wind exposure and refuelling provisions. They are commonly used for activities such as erecting steel frames, bridge inspections and maintenance, working on transmission or telecoms towers and installing high-level services in warehouses.

Machines themselves are not items of temporary works (they are classed as plant) but there are items of temporary works associated with their use. An assessment (and design if necessary) is required for foundations of all machines. Some machines have tracks, some tyres and some also have outriggers / stabilisers, which should be considered similarly to mobile crane outriggers. There may be examples where these machines cannot reach the required position; an alternative is to use a basket suspended from a crane (crawler or mobile). This method is common for inspections and repairs to very tall structures such as cooling towers.

The site temporary works coordinator (TWC) should ensure a foundation design / check is carried out where necessary for all machine foundations (see Figure 6.1).

The International Powered Access Federation (IPAF) promotes the safe and effective use of equipment, providing technical advice, safety guidance and training (see Further reading); the Health and Safety Executive (HSE) has also issued guidance (see Further reading).

Hoists and mast-climbers (for goods only or good / passengers) can be utilised to move equipment, materials and personnel up to a height safely and efficiently. They are commonly used for the installation of cladding to multi-storey office / residential buildings or replacing non-compliant cladding (fire rating) on existing buildings in city centres. They can also be combined with scaffolding if necessary. Similar to other machinery, they require a temporary works design for their foundations and mast ties to the structure. The manufacturer / supplier should provide information on foundation / tie loading and tie positioning. Careful consideration is required when positioning the ties to avoid clashes. Design / check responsibility / liability for the fixings into the structure (could be clamped to structural steel or drilled fixings into edges of concrete slabs) and the reaction on the structure (generally permanent works)

Figure 6.1 Lack of consideration for machinery outrigger foundations (outriggers positioned on channel drain) (editor's own)

needs to be apportioned. A methodology for removing the ties and completing the cladding as the ties are removed should be established.

Many modern buildings have roof-mounted cradles (e.g. for cleaning windows) which can be utilised for inspection and light duty repairs. Their load capacity should be established as many will not be suitable for moving heavy items of machinery or materials. All machinery should comply with Lifting Operations Lifting Equipment Regulations (LOLER) (HMG, 1998a) and Provision and Use of Work Equipment Regulations (PUWER) (HMG, 1998b).

6.6. Edge protection

Temporary edge protection can comprise tube and fitting scaffold, a proprietary system or utilising the permanent edge protection. Some examples where edge protection for temporary works purposes may need to be provided include:

- edges of slabs, floors and bridge decks during construction of structures or buildings
- around voids (for services or a lift shaft) in slabs and floors
- formwork working platform for pouring concrete and around falsework deck
- constructing or repairing roofs
- assembling, inspecting or maintaining large construction machinery
- scaffolding, access stairs and materials loading bays
- around trenches / excavations or at the top of a slope / batter
- placing precast units, profiled metal decking and steel frame construction
- loading and offloading lorries.

BS EN 13374 (BSI, 2013a) sets out the requirements for temporary edge protection systems on construction sites, including guardrails, barriers and other systems designed to prevent falls from height. The standard does not provide requirements for edge protection systems intended for:

- protection against impact from vehicles or from other mobile equipment
- protection from sliding down of bulk loose materials, snow and so on
- protection of areas accessible to the public.

The standard does not apply to side protection on scaffolds according to BS EN 12811-1 (BSI, 2003b) and BS EN 1004 (BSI, 2020), although this exclusion does not prevent these systems being used on temporary structures. The standard specifies three classes of temporary edge protection, as shown in Figure 6.2. Some systems are fixed to the structure and others rely on gravity and friction on flat surfaces.

- Class A systems provide resistance to static loads only, based on the requirements to support a person leaning on the protection or provide a handhold when walking beside it and collectively stop a person who is walking or falling towards the protection. They may be used if the angle is less than 10° and should resist a horizontal force of 0.3 kN anywhere on the top of the guardrail arrangement, while maintaining elastic deflection

Figure 6.2 Edge protection classes from BS EN 13374 (BSI, 2013a) (reproduced with kind permission of Combisafe Ltd)

Note: If the angle is more than 60°, or more than 45° and the falling height is more than 5 m, edge protection systems are not appropriate as protection. At greater falling heights, the system can be placed higher on the sloping area – for example, every 2 m and 5 m of falling height for system class B and C, respectively.

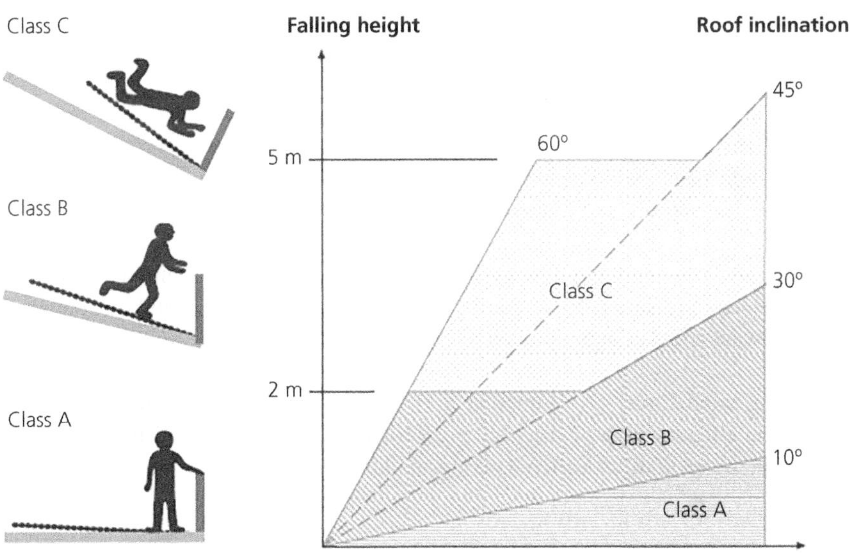

within 55 mm. They must also accept a vertical force of 1.25 kN applied as an accidental loading. These systems typically take their support from clamping, drilled anchors, counterweighted methods or compression (friction) posts.

- Class B systems provide resistance to static loads and low dynamic actions only, based on the requirements to support a person leaning on the protection, or provide a handhold when walking beside it and collectively stop a person who is walking or falling towards the protection and collectively stop a person sliding / falling down a sloping surface. They may be used if the angle is less than 30° without limitation of the falling height, or 60° and the falling height is less than 2 m. They should resist the same static loading requirements as class A and, additionally, must pass a low dynamic test using a swing bag. This test applies horizontal impact energy of 0.5 kJ to the top rail and 1.1 kJ to the lower areas.
- Class C systems provide resistance to high dynamic forces, based on the safety requirements to prevent the fall of a person sliding down a steep sloping surface, and collectively stop a person sliding / falling down a steep sloping surface. They may be used if the angle is between 30° and 45° without limitation of the falling height, or between 45° and 60° and the falling height is less than 5 m. Class C systems should resist a test impactor, comprising a 75.5 kg cylindrical mass (1 m long and with a diameter of 0.3 m), rolling 5 m down an inclined test ramp (60°). This test applies impact energy of 3.185 kJ to the system. The prime consideration is to prevent a person falling, irrespective of the initial damage sustained by the barrier system.

The standard (BSI, 2013a) sets out:

- the general requirements for an edge protection system – for example, principal guardrail height (at least 1000 mm), toe-boards (at least 150 mm), gaps (no more than 20 mm)
- the additional requirements for individual classes – for example, inclination
- the material requirements – for example, steel, timber, material for counterweights
- the static and dynamic design requirements for each class.

6.6.1 Wind loading

At ultimate limit state, the maximum wind load should act perpendicular to the edge protection system in the outward and inward directions, separately, and is 600 N/m^2.

Note: 600 N/m^2 covers most wind conditions in Europe. More onerous conditions may occur. The wind velocity pressure is based on 40 m height and an exposure period of 6 months and represents a wind speed of approximately 32 m/s.

The resulting wind force, Q_W (Equation 6.1), assumes that a wind velocity pressure is applied on an effective area of the edge protection system (which, in general, is the projected area in the wind direction), not taking shielding into account, as shall be determined as follows:

$$Q_W = \Sigma (C_{FI} \times C_{CO} \times Q_I \times A_I) \tag{6.1}$$

where:

C_{FI} is the aerodynamic force coefficient for the edge protection components I (C_{CO} may be used uncorrected). Note: the aerodynamic force coefficient C_{FI} appropriate to the cross-section of the edge protection components in question is given in BS EN 1991-1-4 (BSI, 2005). For any cross-sections not included in EN 1991-1-4, the aerodynamic force coefficient may not be assumed to be less than 2.0, unless it has been established by testing.

C_{CO} is the force coefficient of a component with infinite slenderness ratio.

Q_I is the wind velocity pressure acting on the edge protection components I and shall be taken as 600 N/m^2.

A_I is the reference area of the edge protection components.

The National Foreword to BS EN 13374 (BSI, 2013a) states that the wind loading in parts of the UK may be in excess of the requirements in BS EN 13374; therefore, site-specific calculations should be undertaken in accordance with the current version of BS 5975-2: 2024 (BSI, 2024), BS EN 1991-1-4 (BSI, 2005) and the appropriate national annex, in order to demonstrate product suitability.

6.6.2 Connection details

Temporary edge protection systems require a secure connection which could be:

- bolts or clamps to structural steelwork, aluminium, timber or concrete
- drilled fixings into brickwork
- drilled or cast fixings into concrete
- preformed pockets into concrete for the edge protection posts to slot into.

It is generally preferable for edge protection to be attached on the ground to items being lifted into position by crane.

The temporary edge protection imparts a load onto whatever it is connected to and often this detail can become an issue if it is not identified at an early stage. Careful positioning is required to avoid clashes as this connection can often be to an item of permanent works. Often there is an issue for this detail with responsibility and liability between the supplier of the edge protection and the permanent works designer. The site TWC should ensure that, where necessary:

- responsibility for specifying / checking the connection is apportioned
- a design brief is provided
- a simple check or bespoke design (and where necessary a design check) is carried out
- if the design requires testing, ensure this is carried out as specified
- any limitations or residual risks are communicated, with control measures being identified.

6.6.3 Industry guidance

Detailed industry guidance is available in the Edge Protection Federation (EPF)'s *Code of Practice* (EPF, 2014) (currently being revised) as well as the NASC's 'standard designs' in TG1:22 (NASC, 2022b) and general guidance in SG27:22 (NASC, 2022c).

Note: TG1:22 (NASC, 2022b) only covers the requirements for class A edge protection (up to 10°) constructed in 'tube and fitting' in compliance with BS EN 13374 (BSI, 2013a), together with the basic elements of construction. This document includes 15 generic design drawings, which may be used without the need for a 'bespoke design', provided that environmental conditions and floor / roof structures are suitable.

Where fall arrest safety netting is to be attached to edge protection (whether proprietary or not), the system should be designed to take the loads indicated in BS EN 1263-2 (BSI, 2014b).

6.6.4 Summary of issues

Below are some questions / issues for the site TWC to address.

- Has the correct class of edge protection (A, B or C) been provided at the correct location?
- Is the connection of edge protection to permanent works (or other) adequate? If not, then a design / check should be carried out and, where necessary, accompanied by testing.
- Plan ahead for differences in height and level as work progresses – for example, after a concrete slab is poured or when insulation / roofing sheets have been installed, when the height of the edge protection and gaps might not be appropriate.

6.7. Counterbalanced free-standing edge protection

BS 13700 (BSI, 2021) provides guidelines for the design and installation of counterbalanced free-standing (relying on mass and friction and not fixed to the structure) edge protection systems installed as a permanent solution to provide safe access to areas where operatives need to carry out activities at height – for example, future maintenance activities required at roof level.

The scope excludes:

- temporary counterweighted guardrails (see BS EN 13374, BSI, 2013a)
- warning chains / posts used for demarcation
- netting or infill panels used as barriers
- guardrails that penetrate the structure and are secured to the substrate
- guardrails designed to support material-handling equipment.

The maximum roof pitch at which a permanent free-standing guardrail system should be installed is 5°.

The design of a system needs to consider the wind conditions for the actual location and roof height where the system is to be installed. Wind loading calculations should consider the worst-case scenario based on the maximum velocity pressure for that location. Compared to temporary systems, this may result in reduced stanchion spacing, or additional weights on the stanchions, to ensure stability.

The height of the principal guardrail should be a minimum of 1100 mm above the walking surface. Intermediate guardrails shall be located so that the gap between any two rails does not exceed 470 mm. The gap between the lowest intermediate rail and the walking surface shall not exceed 530 mm.

6.8. Safety netting

BS EN 1263-1 (BSI, 2014a) is the standard on temporary works equipment that considers safety nets to ensure protection to workers against workplace hazards. It specifies safety requirements and test methods and is based on the performance characteristics of polypropene and polyamide fibres. Materials used in nets should have no significant reduction in mechanical properties between −10°C and +40°C. Four types and classes of net are specified (see Table 6.3).

BS EN 1263-1 (BSI, 2014a) is supplemented by BS EN 1263-2 (BSI, 2014b) that specifies safety requirements for the positioning of safety nets in accordance with the manufacturer's instruction manual and product specifications and for the testing of systems in accordance with EN 1263-1 (BSI, 2014a).

Note: small safety nets of system S according to EN 1263-1 (BSI, 2014a) (less than 35 m^2 and 5.0 m on the shortest side) are not dealt with in BS EN 1263-2 (BSI, 2014b).

Taking each system:

- System S – at each anchorage point, the characteristic load (P) shall be at least 6 kN with an applied angle of 45°. For a supporting framework – for example, a steel beam – three characteristic loads (4 kN, 6 kN and 4 kN) should be considered in the most unfavourable way. The distance between the anchorage points should be less than 2.5 m.
- System T – should be erected in accordance with the instruction manual supplied. The support frame should be positioned so that the person falling into the safety net does not impact with the frame.
- System U – for positioning limits, reference is made to BS EN 13374 (BSI, 2013a).
- System V – upper edge of the safety net should be located at least 1.0 m above the working area. Gaps should not exceed 100 mm and there should be no overlapping.

Table 6.3 The four types of safety netting and classes

Class	Energy absorption: kJ	Mesh size: mm
A1	2.3	60
A2	2.3	100
B1	4.4	60
B2	4.4	100
System	**Safety net**	
S	with border rope	
T	attached to brackets for horizontal use	
U	attached to supporting framework for vertical use	
V	with border rope attached to a gallows type support	

Fall Arrest Safety Equipment Training (FASET, 2016) has a good practice guide that gives general guidance on the various access methods available for rigging safety net fall arrest systems, plus advice and guidance on the principal methods available, in order of priority.

In the UK, recommendations on the safe use, installation and effective maintenance of safety nets on construction sites and other works where there is a risk of fall are given in BS 8411 (BSI, 2019a). Nets should have sufficient clearance below them to allow for deflection and to ensure that the person falling does not strike any object. It does not cover system V or tensioned access platform nets – for example, systems intended for persons to stand on, either for access, egress or work.

Proprietary safety net hooks are available to secure safety netting to the structure – for example, 'NetClaw' and 'Grippa'. These are capable of being installed and struck remotely using a pole (up to about 4 m in height), without the need for ladders. Advice has also been published by the Advisory Committee for Roofsafety (ACR) in its '*Blue Book*' (ACR, 2016a).

6.9. Tensioned access platforms

Tensioned access platforms provide a collective and passive working platform for light duty access. There is no European or British standard for the specification or use of these systems. It is therefore essential to follow the manufacturer's instructions, with designs proven by calculation and / or testing to demonstrate that they are suitable for the intended activity. Guidance has been published by FASET (2019a, 2019b). It recommends that suppliers / installers of such systems be able to provide at least the following technical information on a project-specific basis:

- material and component specifications, their use and lifespan
- forces from the components impinging on the structure from the installed assembly and the permissible 'live' working load, including people, tools, equipment and materials
- environmental loading, such as wind, snow, heat and cold
- durability and resistance to ultra-violet light, chemical attack, abrasion, hot works and so on
- design loads for the components and the system
- installer instructions
- user instructions, including loading limitations
- identification and traceability of components.

FASET (2019a, 2019b) recommends that access to a platform be by way of designated access and egress points. Those accessing a platform should be authorised to do so, with access controlled using a permit to work / access system. Before use, users should be aware of a platform's loading capability and limitations.

6.10. Roof over-netting

Working on a fragile roof is a high-risk activity. Accidents can occur because those carrying out the work – for example, inspection or repairs – are not competent to do so. Consequently, the work is carried out without the necessary planning or management and, often, without suitable equipment. There is no national standard for the installation or use of roof over-netting. Nor does the FASET safety net training syllabus and certification scheme prove that an installer has the required level of competence to perform this task safely. FASET has published guidance notes (FASET, 2018a, 2018b). These consider that the general principle should be:

> Roof Over-Netting should be viewed as the last resort for providing a fall protection system when over cladding a roof, installing solar panels etc. Efforts should be made to create access to the underside of the roof so traditional safety netting can be installed beneath the roof before Roof Over-Netting is considered. (FASET, 2018b, p. 1)
>
> This system of work should only be used where traditional Type S Safety Netting (rigged beneath the roof) is not feasible.' (FASET, 2018b, p. 4)

The safe system of work developed should consider at least the following:

- access to the roof structure
- loading of materials to the roof structure
- fall prevention / protection measures, such as perimeter scaffolds, edge protection, safety lines, harnesses and restraint lanyards, walk boards and so on
- weather conditions
- the method of placing the safety net on the roof.

6.10.1 Fragile roofs

Guidance on working on fragile roofs has been prepared by the ACR in its '*Red Book*' (ACR, 2019). The ACR has defined the UK's only test, recognised by the Health and Safety Executive (HSE), to determine the non-fragility status of roofing products. It has produced publications about safe working practices on both fragile and nonfragile surfaces, as well as other roof safety issues, including the '*Green Book*' (ACR, 2017) and the '*Black Book*' (ACR, 2016b). The *Red Book* describes a test standard that can be applied to any product which forms a roof or part of a roof and is intended to provide information about whether it can support the instantaneous loads imposed on it by a person stumbling or falling on it. Assemblies, when tested by a competent person, may be classified as 'nonfragile' – class A, class B or class C. If an assembly fails the test, then it must be classified as 'fragile'.

6.11. Stair towers

Stair towers are often installed to provide a safe means of access to a floor or roof structure. They may be tube and fitting or proprietary systems and should be tied to a supporting structure in accordance with the manufacturer's instructions (or suitably stabilised to a bespoke temporary works design). If fixing into an uncertain substrate, then pull testing should be carried out to justify the capacity of the fixings and substrate.

Ties are fundamental to the stability of a tower and typically are tied using tube and fitting components. Towers should be designed – for example, proprietary systems to BS EN 12810 (BSI, 2003a) – to a standard design or to TG20:21 (NASC, 2021) (a bespoke design). Installers should:

- ensure that the installation is carried out in accordance with the manufacturer's instructions
- tie the tower to the structure in accordance with a specific temporary works design
- be competent.

Guidance is available from FASET in bulletin ST01 (FASET, 2017) and NASC in SG25:20 (NASC, 2020).

6.12. Access platform decking

There are several lightweight proprietary access platform decking systems used to reduce / eliminate the risk of falls. Guidance has been published by FASET (2019c). Systems should have been load tested to BS EN 12811-1 (BSI, 2003b). It is also advised that systems, except those in steel, be tested to the ACR's *Red Book* (ACR, 2019). Where handrails are incorporated, these should meet the class A requirements of BS EN 13374 (BSI, 2013a) and manufacturers should provide evidence of load testing. Positional stability for free-standing access platform decking systems should be demonstrated as recommended in BS EN 12811-1 (BSI, 2003b), with the systems 'checked against sliding sideways, uplift and overturning'.

6.13. Personal fall protection

Personal fall protection equipment imposes a temporary load on its supporting structure and anchors should be tested. The testing of single user anchor devices is usually undertaken in accordance with BS EN 795 (BSI, 2012), which specifies the requirements for performance and associated test methods. For multiuser anchor devices, see CEN/TS 16415 (BSI, 2013b). The standard is not applicable to:

- anchor devices intended to allow more than one user to be attached at any one time
- anchor devices used in any sports or recreational activity
- equipment designed to conform to EN 516 or EN 517 (BSI, 2006a and BSI, 2006b)
- elements or parts of structures which were installed for use other than as anchor points or anchor devices – for example, beams, girders
- structural anchors.

Note: the design philosophy adopted in fall protection standards is not limit state. BS 8437:2022 (BSI, 2022), clause 16.2.3 provides some advice to designers using limit state methods but its veracity has been questioned with BSI.

6.13.1 Types of anchor device

Five types of anchor device are covered:

- type A – with one or more stationary anchor points, while in use, and with the need for a structural anchor(s) or fixing element(s) to fix to the structure (e.g. push-in socket eyebolt or D-ring)
- type B – with one or more stationary anchor points, without the need for a structural anchor(s) or fixing element(s) to fix it to the structure (e.g. tripod, quadpod, sling, door anchor or beam anchor)
- type C – employing a flexible anchor line which deviates from the horizontal by not more than 15° when measured between the extremity and intermediate anchors at any point along its length (e.g. rope, webbing or wire-based horizontal lifeline)
- type D – employing a rigid anchor line which deviates from the horizontal by not more than 15° when measured between the extremity and intermediate anchors at any point along its length (e.g. temporary horizontal rail)
- type E – for use on surfaces up to 5° from the horizontal where the performance relies solely on mass and friction between itself and the surface (e.g. deadweight anchor).

BS 8610 (BSI, 2017) covers anchor systems (rather than anchor devices). These include the anchor device, its fixings and base material. The same basic coding is used as with types A, C and D but expanded to cover the type of uses – for example, work restraint, fall arrest, work positioning, rope access, rescue, evacuation and whether the anchor system is a load limiting or non-load limiting design.

The performance requirements specified for components of personal fall protection systems and equipment are based on the need to ensure that the impact force on the user in the arrest of a fall (a dynamic force) does not exceed 6 kN. When considering anchor loads (as opposed to component strength), a minimum static strength of 12 kN is required for single person use. In terms of the user, this provides a factor of safety of at least 2.

Note: BS EN 795 (BSI, 2012) specifies a minimum static strength of 10 kN for anchors for single person use. Accordingly, the minimum factor of safety is 1.67. Therefore, when using a single anchor point the date of the product and / or its installation should be noted. For rope access methods for industrial purposes, a minimum factor of safety of 2.5 is recommended for single person use – that is, a minimum static strength of 6 kN × 2.5 = 15 kN. Refer to BS 7985 (BSI, 2013c), clause 12.3.5.2, BS 8610 (BSI, 2017) and the Industrial Rope Access Trade Association's *International Code of Practice* (IRATA ICOP) (IRATA, 2013).

Note: the additional factor is intended to consider any additional loading result from rescue.

6.13.2 Anchor points for restraint systems

For single person use, the minimum static strength of the anchor point and the restraint system should be equivalent to at least three times the user's body mass in the direction in which the load is to be applied in service (see BS 8437:2022, BSI, 2022, clause 16.4.1). Restraint systems should not be used for fall arrest.

6.13.3 Horizontal lifelines

When using a temporary rigid or flexible anchor line, subject to catenary effects, reference should be made to the manufacturer's information for advice on the required intermediate and / or end anchor loads. These depend on several factors, including span, material of construction, the provision of energy absorption (or not), the use of intermediate supports, in-line tension, number of users and so on. A factor of safety of at least 2.0 is usually applied.

6.14. Rope access

Rope access may be suitable for tall structures, unusual shapes or structures with multiple overhangs, where other options would not be reasonably practicable. It requires the high-level anchor system to be 'unquestionably reliable'. The authoritative source of advice on anchors for rope access is the IRATA ICOP:

> The anchor system is of primary importance in the rope access system and should be unquestionably reliable (IRATA, 2013, clause 2.11.2.1).

Anchors are used by way of their anchor points for the attachment of anchor lines (i.e. the working line and the safety line) to the structure or natural feature and also for other purposes, such as repositioning the anchor lines to avoid abrasion.

The following points are of note.

- When selecting, positioning and using anchors, the principle of double protection applies and, therefore, at least two independent anchors – that is, at least one for the working line and at least one for the safety line should always be used (clause 2.11.2.2).
- The recommendation to use two independent anchors applies even when attachment is to be made to an anchorage (i.e. a structure or natural feature) where it is apparent that the anchorage has more than adequate strength – for example, a large steel beam (clause 2.11.2.3).
- The foreseeable directions of loading and anticipated potential loads should be established and considered when setting up the anchor system (clause 2.11.2.5).
- To determine the minimum anchor strength recommendation, this code of practice uses a safety factor of 2.5. The maximum impact load on the user in the event of a fall should not exceed 6 kN; therefore, as a rule, the static strength of anchors, except for some deviation anchors, should be at least 15 kN (clause 2.11.2.6). Note: single point anchors are likely to have been type-tested to 10 kN or 12 kN (see section 6.13.1).
- There is no requirement for designers (e.g. building designers) to add a further safety factor but, of course, the static strength may be increased if it is considered prudent or necessary to do so (clause 2.11.2.7).

IRATA ICOP (IRATA, 2013), annex F gives advice and other information that could be relevant to users of rope access methods. It considers that anchor devices fall into two broad categories: installed anchor devices and placed anchor devices. Both anchor types should be installed and inspected by competent persons to manufacturer's instructions and static load tests carried out to prove the reliability of the base materials.

The use of a single deadweight should be undertaken with extreme caution. Test information in the public domain is limited. Similar advice is provided in respect of counterweight anchors:

> After testing and/or risk assessment, a single counterweight anchor may be used if it is judged to have sufficient mass to provide an unquestionably reliable anchorage for both the working line and the safety line, and that there are appropriate attachment points for these anchor lines. Where the mass of one counterweight anchor is insufficient, two or more counterweight anchors may be used. Their mass should be confirmed as sufficient by testing and/or risk assessment. (IRATA, 2013, clause F.3.4.2)

In trolley systems, the lever arm is small and therefore sensitive to set up. Additional consideration should be given to access and egress to the rope access system and the forces imposed by rescue. Wet or icy conditions can significantly affect the frictional performance of deadweight or counterbalance anchor systems. Consideration should be given to the requirements for rescue, which might involve the weight of two persons.

Additional design considerations include:

- the overturning of davit arms and the attachment to its base material (and the need for proof testing), see BS 7883 (BSI 2019b)
- the loading applied by remotely operated rescue devices (the 'pulley effect'), see Figure 6.3

- the loading applied by rigging to handrails and / or parapets (and their structural integrity)
- the loading applied at rope access deviation anchors (where the direction of the loading is changed), see Figure 6.4
- load magnification – for example, y-hangs (sometimes referred to as 'catenary action'), see Figure 6.5.

Note: horizontal anchor lines with multi-users should follow the manufacturer's instructions.

Figure 6.3 Pulley effect based on factored fall arrest for one worker of 15 kN (editor's own)

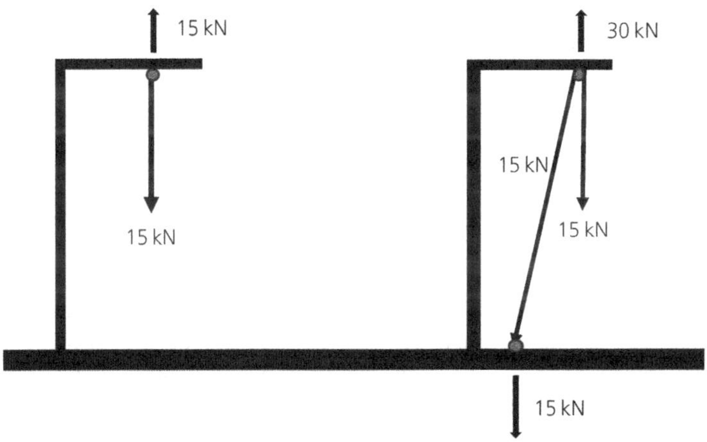

Figure 6.4 Example of how angle at a deviation anchor affects its loading (based on IRATA ICOP, IRATA, 2013) (editor's own)

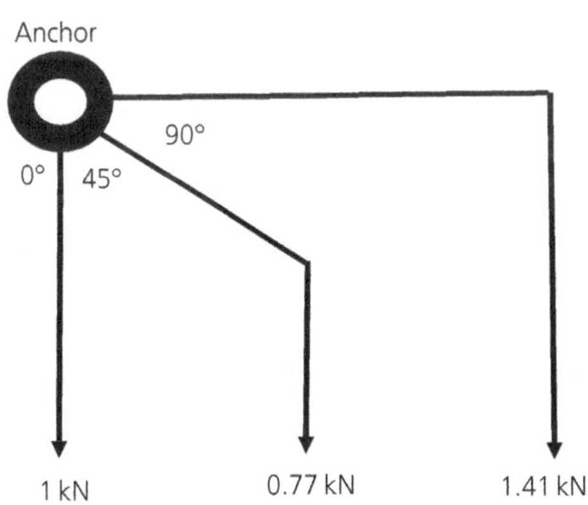

Figure 6.5 Example of increase in loading on anchors, anchor lines and anchor slings as 'Y' angle increases (based on IRATA ICOP, IRATA, 2013) (editor's own)

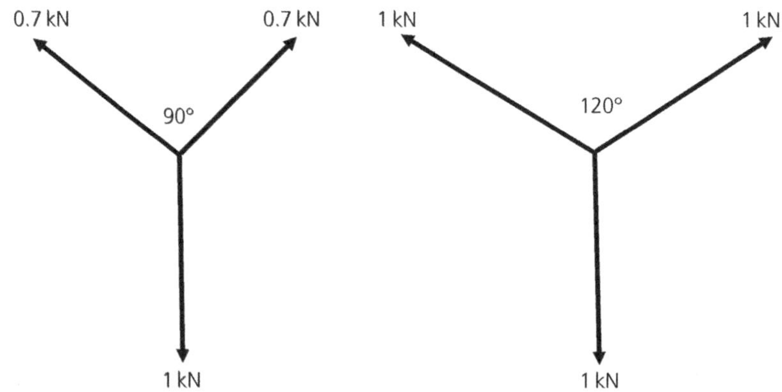

6.15. Suspended access

The safe use of suspended access equipment (SAE) includes:

- operating the SAE as specified in the equipment operating instructions
- never overloading the platform (a plate fixed to the platform should give the safe working load (sometimes referred to as working load limit or rated load) and the maximum number of users. Note: if the plate does not exist, place the platform 'Out of Service' by fitting a suitable sign and inform building management
- never assuming that the platform is safe to carry two people
- aways accessing or exiting the platform from a safe and approved place
- attaching the hook on the safety harness lanyard to a designated PPE attachment point in the platform, before manoeuvring the platform
- securing all loose items – for example, by tethering them to the platform.

6.16. Protecting holes and voids

Where possible, holes and voids should be avoided, as falling through an unprotected hole or void is a significant risk which should be assessed and managed properly. A similar risk exists if a roof, slab, deck, platform or floor is allowed to become overloaded with debris, tools, plant, materials and so on.

Protective measures could include:

- covers – made from plywood, timber, steel plate or similar which should be securely wedged or fixed in position; covers should be labelled 'void below – do not remove'
- guardrails with toe boards – scaffolding or proprietary systems which should be securely fixed
- proprietary covers – such as proprietary temporary manhole covers
- sliding gates – similar to those in use with passenger hoists, these can be used for lift shafts and risers where access is necessary.

The protective measure needs to consider potential loading – plant, operatives, impact and so on, and designed accordingly. Consideration should be given to how the protective measures are to be installed without exposing the operatives doing the installation (or removal) to risk (harnesses etc. may be necessary). Protective measures should be regularly inspected and maintained as necessary.

6.17. Remote methodologies

Where possible, work at height should be avoided. Modern technologies have been developed / adapted for this purpose.

- Remote digital scanners / sensors can 3D survey a building / structure to a great degree of accuracy from a remote location on the ground without the need to access heights for measurements.
- Drones can be used for survey / inspection purposes of buildings and structures without the need for scaffolding, powered access machinery or towers. They are very useful where access is challenging or where there are time constraints (major roads or railways) which may prohibit the construction and removal of scaffolding or towers.
- Remote controlled 'bots' can be used for inspection or cleaning purposes where access for personnel may be difficult or hazardous (e.g. cleaning inside high-level pipework). Magnetic climbing 'bots' are now increasingly used for similar purposes in high-risk environments, such as petrochemical or marine works.

REFERENCES

ACR (Advisory Committee for Roofsafety) (2016a) *ACR(CP)003:2016: Recommended Practice for Use of Safety Nets for Roofwork*, rev. 2. ACR, Whitchurch, UK.

ACR (2016b) *ACR(CP)005:2016: Guidance Note for Competence and General Fitness Requirements to Work on Roofs*, 3rd edn. ACR, Whitchurch, UK.

ACR (2017) *ACR(CP)002:2017): Safe Working on Fragile Roofs or Roofs with Fragile Elements*, 3rd edn. ACR, Whitchurch, UK.

ACR (2019) *ACR(M)001-2019: Test for Non-Fragility of Large Element Roofing Assemblies*, 6th edn. ACR, Whitchurch, UK.

BSI (2003a) BS EN 12810-1:2003: Facade scaffolds made of prefabricated components – Product specifications. BSI, London, UK.

BSI (2003b) BS EN 12811-1:2003: Temporary works equipment – Scaffolds. Performance requirements and general design. BSI, London, UK.

BSI (2005) BS EN 1991-1-4:2005+A1:2010: Eurocode 1: Actions on structures – General actions – Wind actions. BSI, London, UK.

BSI (2006a) BS EN 516:2006: Prefabricated accessories for roofing. Installations for roof access. Walkways, treads and steps. BSI, London, UK.

BSI (2006b) BS EN 517:2006: Prefabricated accessories for roofing. Roof safety hooks. BSI, London, UK.

BSI (2012) BS EN 795:2012: Personal fall protection equipment. Anchor devices. BSI, London, UK.

BSI (2013a) BS EN 13374:2013+A1:2018 (corrigenda January 2019 and August 2021): Temporary edge protection systems. Product specification. Test methods. BSI, London, UK.

BSI (2013b) CEN/TS 16415: Personal fall protection equipment. Anchor devices. Recommendations for anchor devices for use by more than one person simultaneously. BSI, London, UK.

BSI (2013c) BS 7985:2013: Code of practice for the use of rope access methods for industrial purposes. Recommendations and guidance supplementary to BS ISO 22846. BSI, London, UK.

BSI (2014a) BS EN 1263-1:2014: Temporary works equipment. Safety nets – Safety requirements. Test methods. BSI, London UK.

BSI (2014b) BS EN 1263-2:2014: Temporary works equipment. Safety nets – Safety requirements for the positioning limits. BSI, London, UK.

BSI (2017) BS 8610:2017: Personal fall protection equipment. Anchor systems. Specification. BSI, London, UK.

BSI (2019a) BS 8411:2019: Safety nets on construction sites and other works. Code of practice. BSI, London, UK.

BSI (2019b) BS 7883:2019: Personal fall protection equipment. Anchor systems. System design, installation and inspection. Code of practice. BSI, London, UK.

BSI (2020) BS EN 1004-1:2020: Mobile access and working towers made of prefabricated elements – Materials, dimensions, design loads, safety and performance requirements. BSI, London, UK.

BSI (2021) BS 13700:2021: Permanent counterweighted guardrail systems. Specification. BSI, London, UK.

BSI (2022) BS 8437: Code of practice for selection, use and maintenance of personal fall protection systems and equipment for use in the workplace. BSI, London, UK.

BSI (2024) BS 5975-2:2024: Temporary works – Falsework: design and implementation. Code of practice. BSI, London, UK.

EPF (Edge Protection Federation) (2014) *Code of Practice*. EPF, Shropshire, UK.

EPF/FASET (Edge Protection Federation/Fall Arrest Safety Equipment Training) (2022) *Good Practice Guide: Temporary Demarcation Systems*, rev. 1. EPF, Shropshire, UK.

FASET (Fall Arrest Safety Equipment Training) (2016) *Good Practice Guide: The Selection of Access Methods to Install and Dismantle Safety Netting*. FASET, Shropshire, UK.

FASET (2017) *ST01: Skill, Knowledge, Experience and Training for Temporary Stair Tower Installers*. FASET, Shropshire, UK.

FASET (2018a) *RON01: Advice for Clients and Contractors When Specifying, Tendering and Installing 'Roof Over-Netting' Solutions*, rev. 2. FASET, Shropshire, UK.

FASET (2018b) *Advice for Clients, Designers and Contractors When Specifying, Tendering and Installing 'Roof Over-Netting' Solutions*, rev. 3. FASET, Shropshire, UK.

FASET (2019a) *TAP01: Tensioned Access Platforms*, rev. 2. FASET, Shropshire, UK.

FASET (2019b) *Inspection Requirements for Tensioned Access Platforms*. FASET, Shropshire, UK.

FASET (2019c) *Guidance Note: Access Platform Decking*, rev. 2. FASET, Shropshire, UK.

HMG (Her Majesty's Government) (1998a) Lifting Operations and Lifting Equipment Regulations. The Stationery Office, London, UK.

HMG (1998b) Provision and Use of Work Equipment Regulations. The Stationery Office, London, UK.

HMG (2005) Work at Height Regulations (amended 2007). The Stationery Office, London, UK.

HMG (2015) Construction (Design and Management) Regulations. The Stationery Office, London, UK.

HSE (Health and Safety Executive (2014) *INDG 401: Working at Height, A Brief Guide*, rev. 2. HSE, London, UK.

HSE (2021) *HSG33: Health and Safety in Roof Work, Appendix 7*, 5th edn. HSE, London, UK.

IRATA (Industrial Rope Access Trade Association) (2013) *International Code of Practice: Safety Considerations When Installing or Placing Anchor Devices for Use in Rope Access*. IRATA, Ashford, UK.

NASC (National Access and Scaffolding Confederation) (2020) *SG25:20: Access and Egress from Scaffolds via Ladders and Stair Towers, etc*. NASC, London, UK.

NASC (2021) *TG20:21: Good Practice Guidance for Tube and Fitting Scaffolding*. NASC, London, UK.

NASC (2022a) *SG4:22: Preventing Falls in Scaffolding Operations*. NASC, London, UK.

NASC (2022b) *TG1:22: Temporary Edge Protection*. NASC, London, UK.

NASC (2022c) *SG27:22: Guidance on Temporary Edge Protection on Open Steelwork Roofs and Slab Edges (Class A and B Surfaces)*. NASC, London, UK.

Pallett PF and Filip RK (2019) *Temporary Works: Principles of Design and Construction*, 2nd edn. ICE Publishing, London, UK.

TWf (Temporary Works Forum) (2020) TWf2020:01: Effective Management of Scaffolding to BS 5975:2019. https://www.twforum.org.uk/viewdocument/effective-management-of-scaffolding (accessed 07/01/2025).

USEFUL WEBSITES

AIF (Access Industry Federation) – www.accessindustryforum.org (representing different sectors in the access industry: fostering and supporting codes of good practice, equipment standards, training, education and knowledge. Provides access to the Construction Industry Advisory Committee (CONIAC) ‘Safety Steps’)

Andun Engineering Consultants – www.andun.co.uk

ATLAS (Association of Technical Lightning and Access Specialists) – www.atlas.org.uk (represents the leading companies in the lightning protection and specialist access at height sectors)

BSI (British Standards Institution) – www.bsigroup.com

CFA (Construction Fixings Association) – www.the-cfa.co.uk (works to ensure best practice with fixings)

EPF (Edge Protection Federation) – www.epf-uk.org (acts as a representative voice and to ensure that high standards in manufacture, supply and application of edge protection systems are maintained and advanced)

FASET (Fall Arrest Safety Equipment Training) – www.faset.org.uk (the established trade association and training body for the safety net rigging and temporary safety systems industry)

IPAF (International Powered Access Federation) – www.ipaf.org/en-gb (promotes the safe and effective use of powered access equipment worldwide by providing technical advice and information; influencing and interpreting legislation and standards; and through safety initiatives and training programmes)

IRATA (Industrial Rope Access Trade Association) (formerly the Industrial Rope Access Trade Association) – www.irata.org (the world’s leading authority on industrial rope access, it directs and regulates, through its members, the training of all workers seeking its qualifications)

NASC (National Access and Scaffolding Confederation) – www.nasc.org.uk (national trade body for access and scaffolding with leading contracting firms, scaffolding suppliers and manufacturers, it has raising and maintaining safety standards at the heart of its objectives)

NFF (No Falls Foundation) – www.nofallsfoundation.org (the charity dedicated to preventing falls from height and helping people affected by the consequences of a fall).

PASMA (Prefabricated Access Suppliers' and Manufacturers' Association Ltd) – www.pasma.co.uk (the recognised focus and authority for mobile access towers, advancing safety, standards and best practice across a wide range of sectors and countries)

SAEMA (Specialist Access Engineering and Maintenance Association) – www.saema.org (national trade body for the permanent and temporary façade access equipment industry; committed to advancing safety, standards and best practice in the temporary and permanent suspended access industry)

The Ladder Association – www.ladderassociation.org.uk (not-for-profit lead industry body dedicated to promoting the safe use of portable ladders)

The Scaffolding Association – www.scaffoldingassociation.org (strives to continuously improve competence and minimise risk across the industry)

TWf (Temporary Works Forum) – www.twforum.org.uk (aims to encourage open discussion of any matter related to temporary works)

WAHSA (Work at Height Safety Association) – www.wahsa.org.uk (lead body representing interests of manufacturers, assemblers, installers, distributors and end users of PPE used for work at height and rescue)

FURTHER READING

BSI (2012) CEN/TS 16415: 2012: CEN Technical specification. BSI, London, UK.

BSI (2018) BS 8560:2012+A1:2018: Code of practice for the design of buildings incorporating safe work at height. BSI, London, UK.

BSI (2018) BS EN 363:2018: Personal fall protection equipment. Personal fall protection systems. BSI, London, UK.

HSE (Health and Safety Executive) (2014) *Information Sheet CIS58: The Selection, Management and Use of Mobile Elevated Work Platforms*. HSE, London, UK.

IPAF (International Powered Access Federation) (2014) *Management of MEWPs: Good Practice for Rental Companies and Contractors*. IPAF, Crooklands, UK.

O'Driscoll R (2019) Construction Engineering: A Guide to Temporary Works in Construction and BS 5975:2019. www.constructionengineering.uk (accessed 07/01/2025).

TWf (Temporary Works Forum) (2023) Information Sheet 18: Edge Protection and Demarcation: Types and Their Use. https://www.twforum.org.uk/blogs/david-thomas/2023/06/12/edge-protection-and-demarcation-types-and-uses (accessed 07/01/2025).

Ray Filip
ISBN 978-1-83608-525-6
https://doi.org/10.1108/978-1-83608-524-920251008

Chapter 7
Modern methods of construction

Adrian Bywaters
Mace, UK

Ray Filip
RKF Consult Ltd, UK

Martin Pike
Mace, UK

Kit Yardley
Laing O'Rourke, UK

7.1. Introduction

The construction industry and the construction process are constantly evolving and innovating. There has always been a drive for improved safety, speed and cost, with improved quality and sustainability (less wastage) – without this any organisation will stagnate and lose its competitive edge. This drive has led organisations to develop offsite (factory assembly) construction techniques as a means of mass production, with a focus on supply chain management as an alternative to traditional construction. This concept has been well established for many decades in other industries, such as manufacturing vehicles. In construction it has become known as 'modern methods of construction' (MMC) and has support from the government (especially for housing) due to the lack of traditional skills and perceived poor quality of the final product. The concept continues to evolve and develop into various sectors of the construction industry. MMC can take many forms, but at the centre of all such solutions is the drive to remove traditional construction activities from sites and deliver construction projects using elements manufactured off site, in essence changing the project site from a place of traditional construction to a place of assembly. An approach such as this will deliver many benefits but will also fundamentally change the construction process; these changes will apply as much to temporary works as permanent works.

Some examples of MMC could include:

- precasting reinforced concrete items off site and then delivering to site for assembly: tunnel lining segments, driven concrete piles, sewer pipes, stairs with landings, 'hollow core' planks, manhole rings and bridge beams (known as 'Y' beams) have been precast for several decades and, more recently, structural elements such as foundations, columns (with column shoes), walls (twin wall) and lattice planks (slabs), 'sandwich panels' (solid reinforced concrete with exterior insulation and a non-structural architectural face) for building systems

- prefabrication of mechanical, electrical and plumbing (MEP) installations or reinforcement cages (either on site or off site)
- prefabrication of building façade and cladding systems, which can include brickwork, glazing and guttering
- steel framing using structural sections to form a skeleton which can then be encased for fire and corrosion protection
- timber frame panels such as walls, floors and roof trusses; these could also include glue-laminated timber sections
- modular construction (also known as volumetric construction) of entire buildings / structures (modular buildings and bridges) or individual 'pods' such as bathrooms
- hybrid construction methods (semi-volumetric) which combine modular units and panel systems
- very recent innovations include bricklaying robots and 3D printing robots (may require temporary support frames) which can print components or entire concrete structures in their final position; they can be particularly useful for prototypes and complex shapes.

Also worth noting is 'design for manufacture and assembly' (DfMA), which is an initiative to standardise plans for large-scale projects, such as residential estates. Table 7.1 compares MMC with traditional construction.

Table 7.1 Comparison of MMC against traditional construction (continued on next page)

Advantages	Challenges
Improved overall predictability in terms of costs (bulk buying), which can also give greater certainty of product delivery, which in turn can improve construction time and potential for reduction in overall project risk.	The concept of MMC has been relatively slow to catch on with some clients, designers and contractors. There is a fear and reluctance in using innovative techniques without a long previous track record. It takes time and money to build factories and develop processes, which can lead to a lack of capacity within the industry to produce the necessary scale of production. There may be delays in approvals, regulatory complexity and inadequate certification.
Improved safety in a factory due to greater control over activities and increased repetition, with safety features being made permanent (rather than temporary on site). Less work at height on site.	Design responsibility for temporary works and permanent works needs to be clearly identified from an early stage, which can be different from a traditional responsibility split.
Improved quality and productivity in a factory as conditions can be controlled more easily, less affected by workmanship, not prone to the vagaries of the weather and urban traffic (ready mix concrete delivery to sites).	There is a desire to retain competition, which can often result in clients stating that they want an MMC solution but procuring a traditional design at an early stage. The contractor would then be expected to be responsible for converting the design to an MMC solution. A conversion design is always less than perfect.

Table 7.1 Continued

Advantages	Challenges
Repetition of the same process will improve speed of construction of elements. Robust moulds and frames can be created to be reused many times.	Reliance on logistics with inner-city traffic issues / restrictions. If the delivery and assembly schedules are not synchronised there will be issues with units having to be stored on site (if there is sufficient space) or standing time if the deliveries are delayed. Lorries may be returning from site empty.
Site assembly of pre-made elements is quicker and generally considered to be safer than traditional in situ construction. The construction speed can outweigh any additional production costs.	Consequences of any problems or mishaps can become serious due to the weight and size of the items being installed. Can take time to procure and deliver some replacements.
Greater potential to utilise recycled / reclaimed materials in a factory as there is more opportunity for innovation, bulk buying and often a greater area for materials storage.	Over-reliance on suppliers and significant issues for the project if a supplier becomes insolvent.
Reduction in wasted materials in factory conditions by enabling an optimisation of cutting patterns and enhanced procurement efficiency.	Requires people to understand the processes involved, take more technical responsibility for planning and coordination earlier in the process. It is no longer acceptable to assume that 'the site team will sort out problems' as the benefits of MMC will be lost (site will not keep up with delivery schedules). Additional awareness and training are required at all levels, from senior management to operatives.
	MMC components can be more expensive than traditional in situ works with an earlier commitment to cost. However, the increased cost is countered by a reduced delivery programme.
Less reliance on skilled trade operatives who are becoming increasingly scarce and costly (multi-skilled operatives are required).	Modular and repetitive elements can stifle architectural creativity.
Reduction in temporary works and elimination in some circumstances – e.g. some reinforced concrete slabs are designed to be self-supporting, thus eliminating the need for temporary falsework. Temporary works can also be integrated into the permanent elements, providing clear design responsibility is established (otherwise there could be a long and complex chain of responsibility).	Preventing damage during transportation, site storage (stacking) and lifting (induced temporary stresses and deformation of more flexible items). Items will often need to be turned from the horizontal plane (for transport) to the final vertical position. Lateral stability conditions can be the most severe cases and are often not considered early enough in the process and there is often a lack of coordination and responsibility.

Table 7.1 Continued

Advantages	Challenges
If sufficient planning and coordination is carried out, then site issues / problems will be minimised (especially when buildability, connection details and safety measures are incorporated).	More craneage time is required and the craneage limitations need to be fully understood. Items tend to be heavy so larger capacity cranes are likely, which require more space for rigging and operation. Environmental issues such as wind and poor visibility can limit crane usage.
'Just in time' delivery (used for many years by the automotive and retail sectors) will reduce / eliminate the need for any significant onsite storage areas. This can be important on inner city projects.	Moving heavy prefabricated items (mainly MEP) on trolleys across existing floors or slabs may require temporary works to prevent overloading.
Manufacturing and storing concrete components off site before delivery will mean that only full-strength items are delivered to site for immediate use and loading if necessary. This is particularly important as often 'low carbon' (environmental benefit) concrete is used in precast elements, which gains strength more slowly.	Any non-standard or infill concrete elements cast on site are likely to be loaded at an earlier stage in the construction process. This can cause problems with low carbon concretes which gain strength more slowly and are likely to require temporary works to prevent overloading.

Note: there have been some high-profile issues – for example, insufficient thought / planning given to storage of modular units prior to delivery to site (Construction Enquirer, 2023). For MMC and traditional construction, fire safety in the partially built condition requires careful consideration.

7.2. MMC concept

For the benefits of MMC to be fully exploited, there needs to be a change in traditional mind-set, roles and responsibilities.

The responsibility for planning and coordinating temporary works (including lifting) and permanent works needs to be identified at a very early stage and the two integrated together. It is generally beneficial to try to eliminate the need for temporary works where possible.

MMC has also flipped design responsibility, as more of the final design and detailing (MEP, fire safety, finishes etc.) must be substantially complete before construction starts, otherwise the efficiency benefits are lost by having to retrofit elements. This will require a redefinition of the planning and design process to ensure items such as MEP fixings are considered earlier. Principal designers (PDs) and permanent works designers (PWDs) need to take more responsibility for coordination of all elements, including sequencing and temporary works. This may involve engagement with specialist contractors, who will have their own perspective on what can be achieved from previous experience.

The design and detailing must also clearly consider the assembly sequence, including any necessary temporary works to provide stability, and make sufficient allowance for tolerance.

Figure 7.1 Temporary works influence map (reproduced with kind permission of Mace)

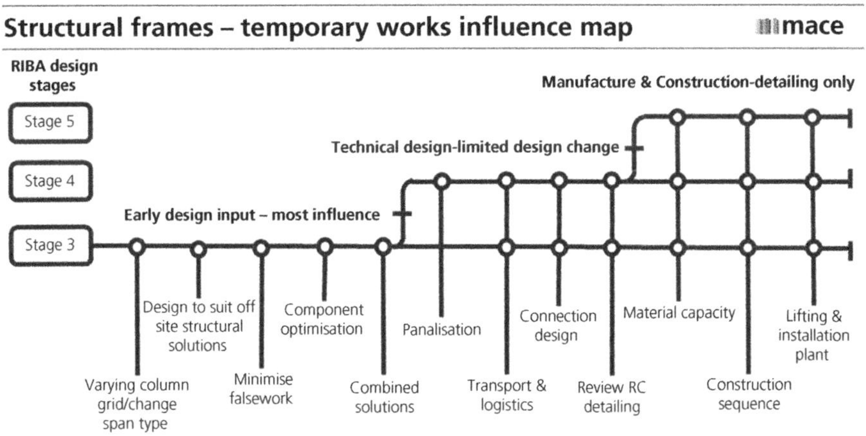

Careful detailing of connections needs to be undertaken as at present some of these details tend to have had little thought and are somewhat unsophisticated. From an early stage, clients, architects and designers must understand what can be produced, how it can be transported and how it can be assembled on site. All parties need to adapt to these new ideas, from foundation designers upwards, and clearly define who is responsible for what. The goal should be to identify a way (as early as possible) to standardise components and integrate elements, such as MEP, insulation and architectural finishes, into structural elements, without excessive complexity.

The use of 3D and 4D 'building information modelling' (BIM) can help maximise construction efficiency and coordination by identifying physical clashes between components and systems when multiple trades are involved. This should include potential clashes between temporary works items (such as propping) and permanent works. However, responsibility for this process should be established.

The Royal Institute of British Architects (RIBA, 2020) has produced guidance on how a construction project should be organised in a structured and time-efficient way. This guidance highlights eight stages from initial strategic definition to final use of the project. Part of this has been used to produce Figure 7.1 in relation to temporary works.

7.2.1 RIBA Stage 2 concept design and Stage 3 spatial coordination

From the concept stage of a project, a high-level approach to the project-specific challenges will ideally be put in place to ensure that the benefits offered by MMC can be maximised – for example:

- modular structural components
- cladding and wall components
- bathroom pods
- modular MEP installation.

To deliver these as efficiently as possible and to gain maximum benefit, an MMC strategy must be set from concept stage. Beyond this point, any MMC benefits will be diluted as it is likely it will become a conversion design, from traditional to MMC methods, and this will generally be imperfect with enforced constraints. In particular, the opportunity to deliver a multidisciplinary strategy will quickly be lost, which not only will compromise the benefits of MMC, but will also complicate the delivery of temporary works on the project. Examples of such multidisciplinary inclusion could include:

- precast structural components with cast-in fixings for MEP installations
- slab modules with pre-determined service openings
- slab modules with pre-fixed insulation
- pre-clad structural modules / components
- modular MEP risers with pre-installed fire compartmentation
- structural components with finishes applied or cast in.

This highlights that the implementation of an MMC strategy demands a complete re-think of the traditional design process, with architectural and MEP design required significantly earlier in the project development process. In addition, the permanent works engineers must not only provide a concept design but must have ongoing input and responsibility for the coordinated design, including temporary works design along with integration and detailing. Figure 7.1 shows decisions that are made throughout the design process that influence both MMC and temporary works proposals.

7.2.2 RIBA Stage 4 technical design and Stage 5 production information

At this stage, the detailed design of the MMC component / product is underway. However, there are still several factors that must be considered to fully define the temporary works impact on the design:

- manufacture
    - de-moulding strength requirements
    - lifting and turning elements, generally from horizontal to vertical orientation
- transport
    - orientation and stability of elements
    - supporting the elements – that is, the permanent state or at temporary support points
    - support frames and transport trestles (which in themselves are temporary works)
- installation
    - lifting points
    - connections to the permanent works, both temporary and permanent
    - temporary stability of individual elements and the structure as a whole
    - spatial coordination of temporary works to avoid clashes; coordination of other items of temporary works that are included in the element also at this stage (e.g. propping sockets, handrail posts etc.).

The minimisation (or elimination) of temporary works should be an aspiration and can be achieved by careful incorporation into the permanent works. The use of building information modelling (BIM) as a tool for coordinating permanent works is now standard practice and

Figure 7.2 Complex coordination of temporary propping (reproduced by kind permission of Mace and BG&E Consulting Engineers)

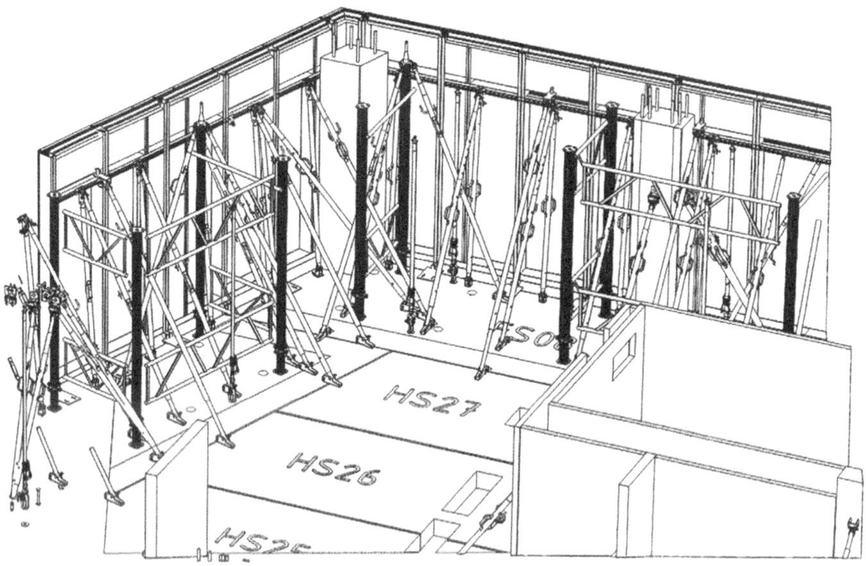

should also be extended to temporary works, which should routinely be incorporated into the federated model. In this way, accurate virtual build sequences can be developed, allowing coordination and clash detection not only between permanent works and other permanent works but also:

- temporary works to temporary works
- temporary works to permanent works
- temporary works to logistics and installation plans.

Figure 7.2 illustrates a complex scheme involving coordination of modular façade units, precast columns, precast walls and temporary inclined propping to walls and temporary vertical propping to allow the next precast floor to be added. Combined with this, the props were also spatially coordinated to allow the installation of bathroom pods.

7.2.3. RIBA Stage 5 manufacture and construction

The introduction of MMC solutions to a project may reduce the likelihood of an incident occurring by removing works and operatives from the workface. However, the consequences should a failure occur are now significantly higher and the temporary works solutions need to take this into account. The implementation risk category from BS 5975-1 (BSI, 2024), table 1 should be carefully considered at detailed design stage such that the design itself can be modified if necessary to design out risk. Additional precautions such as designed redundancy can also be included. An example is shown in Figure 7.3.

Figure 7.3 Precast column installation with temporary propping (editor's own)

- A precast column is installed in the centre of the ground floor by two experienced operatives using push–pull props to secure it in the temporary state. Risk of failure is low and consequences of failure will be low; hence, the implementation risk is 'low'.
- The same column is installed in the same position by the same operatives using the same method but at level 20 of the structure. Risk of failure remains low, but consequences of failure are now increased to medium due to increased height.
- The same column is now installed at level 20 by the same operatives, but now on the slab edge. Again, risk of failure is low, but consequences of failure are now high as it could fall off the edge of the building and potentially beyond the site boundary. Implementation risk is now high. Mitigation is needed and could include installing additional fixings into the column and slab to provide redundancy against failure (noting minimum spacing constraints for fixings). Alternatively, an additional prop could be installed perpendicular to the slab edge.

7.3. Manufacturing concrete elements

It is imperative that structural concrete mix specifiers and suppliers recognise the requirements of the manufacturing, transportation and site assembly process. They need to produce mixes that are sufficiently workable to flow into formwork, gain strength at a rate to suit the factory process, can be struck from formwork and then lifted and transported easily without damage.

Figure 7.4 View inside factory manufacturing precast units (reproduced with kind permission of Laing O'Rourke)

Factory-produced concrete elements have a relatively low temporary works interface (see Figure 7.4). The repetitive nature of prefabrication requires well-made, robust shutters capable of producing large quantities of units, with acceptable finishes and dimensional accuracy. These shutters should also be capable of adjustment / modification to allow the manufacture of non-standard elements. Attention must be paid to the methods selected to remove the elements from the shutters and subsequent lifting. In the interest of efficiency, the elements will be removed from the shutters as soon as possible, with the issue of lifting and transporting low-strength concrete having to be addressed. Tilting table shutters can be used to rotate beams and wall elements from horizontal to vertical prior to them being struck. It may also be necessary to employ lifting beams / frames to control the bending and shear stresses, and the pull-out capacity of lifting anchors in immature concrete needs to be established. Careful early planning and design is required to specify type and position of any fixings in the precast elements. These could be for lifting but also for site propping to ensure temporary stability. The suppliers of cast-in and post-fix anchors do not usually publish capacities of their products used in early age concrete; hence, testing will be necessary. Wherever possible, lifting used in the factory should be the same as used on site, with all lifting points and any propping positions clearly identified (why design multiple lifting points?), but this can be challenging due to reorientation of elements (from horizontal to vertical).

Columns will normally be constructed vertically, ensuring all visible faces will be formed surfaces. Casting vertically may pose difficulties with headroom for pouring the concrete and then lifting the column and the shutters. Pumping concrete from the base of the mould is sometimes employed and caution must be exercised to ensure the shutters are designed to take the increased concrete pressure. Locally to the pump attachment point, the pressure can be equivalent to the pump delivery pressure, which will be significantly higher than the full hydrostatic concrete pressure. Where columns and walls are poured horizontally (flat), the position and strength of demoulding lifters must be considered. Pouring and storing units flat also adds some additional complexity to the lifting stresses in the primary lifters as they are required to pitch the unit into vertical for placement. Failures have occurred with elements where lifters are stressed too early or are not sufficiently embedded for pitching.

Stair flights and landings are best poured on their sides (unless special finishing is required on the treads), again giving formed surfaces on all but one face, ideally the one which will not normally be visible, such as against a shaft wall.

Post-tensioned slabs and beams present further complications. Substantial foundations will be required to anchor the prestressing strands prior to pouring the concrete. The manufacture of 'twin-wall' panels requires one leaf of the wall to be concreted horizontally with the lattice girders embedded. When sufficient strength has been achieved, the first leaf is removed and rotated 180°. The second leaf is then poured when the first leaf, complete with lattice girders, is lowered and lattices embedded. The use of heating ovens will increase the production rate.

During the manufacture of the prefabricated item there will be a need for access (scaffolding, towers or mechanised access machinery), for attaching lifting accessories, making good and finishing.

Storing and transporting precast elements need careful consideration. Concrete elements can be very heavy and easily damaged (especially corners). Transportation issues will include, as a minimum, the additional loadings induced by braking, accelerating, cornering and 'bouncing'. Orientation of the elements needs to be considered to keep within the Department of Transport's 3.5 m width restriction. The police require at least 2 days' notice when loads overhang the sides of the vehicle and warning marker boards are required at each end of the vehicle. Wider loads are permitted but escort vehicles are required, which add to costs and complication. Those involved in transportation should become familiar with the various rules relating to transporting abnormal / overhanging loads. Guidance is available on the UK government website along with application / notification forms. There may be height restrictions to consider (low bridges enroute); the proposed route will need be to be investigated and planned to establish the maximum allowable loaded vehicle height. This may necessitate the elements being loaded in undesirable orientations utilising transport frames (which should be designed for the purpose) for unloading and placing. When unloading the elements it may be necessary to change the orientation. Rotating any concrete elements requires careful planning and execution. During rotation the corners of the elements are vulnerable to damage. The use of 'sand boxes' to rest the corners during rotation can reduce the potential for damage. Rotation in mid-air is possible using winches to pay out lifting cables to achieve the rotation.

Architects and detailers should be made aware that complex shapes with many recesses or protrusions will add to the complexity and cost of formwork. It may also make lifting these items and finding suitable positions for temporary propping more difficult.

It is imperative with all manufactured items that there is a robust quality control process to ensure quality standards are consistent and customer needs are met. This should include checking formwork, concrete mixes, reinforcement and lifting points.

Collaborative Reporting for Safer Structures (CROSS, 2012) describes some specific safety concerns over the temporary stability of precast wall units.

7.4. Manufacturing MEP and architectural elements

MEP elements are increasingly being pre-assembled off site and then transported to site (see Figure 7.5). This could be relatively simple items, such as ducting, or entire bathroom pods. As with concrete elements, planning and coordination is required to maximise the benefits. Lifting / rotating of elements (from horizontal to vertical and utilising temporary lifting frames) and element stacking during transportation to prevent damage are critical considerations. An example is large glazing units which will generally require a temporary grillage to allow them to be transported in the vertical position. The grillage should be designed to be sufficiently robust and adequately secured to prevent the units falling off the lorry due to acceleration, braking or cornering. Consideration should be given if heavy MEP is being moved on trolleys across suspended slabs / floors as back-propping beneath may be necessary. Lifting equipment / devices are also necessary to position MEP at the correct level (could be connected to ceilings).

Care needs to be exercised when installing MEP before façades are completed as wind loading would not have been allowed for and units can be damaged by exposure to the weather

Figure 7.5 Large MEP unit transported to site and being lifted into position (reproduced with kind permission of Mace)

Note: consideration has been given by the designers / manufacturers to lifting points.

(rainwater ingress, frost etc.). Wherever possible, MEP / architectural design and structural design should be incorporated so that MEP fixings / supports can be pre-formed. This creates an additional level of manufacturing complexity and transportation considerations.

7.5. Installation of elements

7.5.1 Logistical planning

Careful logistical planning is required as many projects will be in city centres with limited space for offloading and storage. Some projects will employ a 'just in time' delivery option, where elements are offloaded and lifted (possibly also rotated) directly into their final position (see Figures 7.6 and 7.7). All such operations require careful planning and control with inspection and test plans (ITPs), along with a permit system. Sequencing and temporary stability become critical considerations. Clients need to be made aware of these challenges to ensure adequate financial (or other) provisions are made.

A clear responsibility matrix should be established for the overall coordination, design, sequencing, lifting and temporary works requirements.

When elements arrive on site, they should be inspected for compliance and any damage. Lifting points and lifting frames / beams should be clearly marked and available with pre-planning as to how the chains / straps will be released from the load. Elements can be lifted into position with edge protection already attached to allow access for operatives. Alternatively, methods such as fall arrest systems or remote release shackles can be adopted. The

Figure 7.6 Careful planning of temporary works solutions for lifting and positioning (reproduced with kind permission of Mace and Permasteelisa)

Figure 7.7 The result of careful planning of temporary works solutions for lifting and positioning (reproduced with kind permission of Mace)

number and size of cranes (mobile, tower, crawler cranes or other specialist lifting equipment) are critical to assembly operations. Their positioning should consider overall site coverage and efficient working as 'crane hook time' is a premium. When multiple cranes are used, some degree of overlap of operating radii is desirable but careful coordination is required to prevent accidental clashes.

The operatives should be briefed on the assembly sequence with consideration given to contingency planning if there are delivery issues (late delivery or incorrect items) or elements arrive damaged. Construction tolerances need to be identified and allowed for, with slotted holes in steelwork, in situ concrete 'stitch-joints' (with formwork solutions), infill grouting and shims often used. Access for operatives to carry out the assembly is likely, with scaffolding, towers, fall arrest or mechanical access equipment being commonly used.

When the lifting attachment is released from the element, temporary stability can become a major issue. It is desirable to 'design out' the need for temporary propping where possible. For example, products such as column shoes are used to create a moment-resisting stiff column connection (can also be used on sandwich wall panels), which can eliminate propping requirements for stability. If temporary propping is unavoidable (e.g. for temporary stability and alignment of precast wall and column elements), the type of props, how many props, their connections (ensure correct anchors are being used and installed to manufacturer's specification), when and how they can be removed and potential clashes need to have been designed and considered at an earlier stage. Efficiency and safety can be improved if elements are lifted into position with fixings and props already partially installed. It is advisable to allow for some degree of redundancy when designing propping and fixings for the possibility of an accidental impact, unless the impact can be eliminated or controlled. It is more efficient to have pre-formed prop connections rather than 'drill and fix' connections once the elements are in position, but tolerance and adjustment need to be considered. Grouting connections are often critical to achieve temporary stability and load transfer in the permanent condition; in situ concrete infill (stitching) may be required for tolerance or for difficult shapes. The strength of the concrete and grout (especially in cold weather) and how quickly they can achieve the required strength to maximise efficiency of the overall assembly process need to be considered. Mechanical connections can be an alternative to grouting. Once elements are in position, they require further protection from potential accidental damage; timber or plywood packs and exclusion zones are common solutions. A contingency plan with access for operatives should be developed if there are issues or damage to deal with at a future point in the assembly process.

7.5.2 Specific assembly considerations

Precast concrete 'twin-wall' panels will have several areas where temporary works are required. Due to the manufacture of twin-wall panels, the poured concrete pressure exerted between the leaves needs to be restricted to 30 kN/m^2 (the industry standard); because the lattice girders are only emersed into the leaves of wet concrete, there is no mechanical connection. To achieve such a low concrete pressure, an extremely slow rate of rise needs to be enforced (unless other measures such as additional external propping are adopted). This may engender conflict with the construction team. At junctions between twin-wall panels there is a requirement for additional permanent works reinforcement to be placed in the void between the leaves. This impacts on the location of the lattice girders within the panel and results in undesirable cantilevers of the leaves at these junctions. At corners, this can be dealt with by adding external corner straps. At butt joints, small diameter ties can be passed through the gaps between panels. By adjusting the centres of the lattice girders, the strength of the panels can be enhanced, allowing greater concrete rates of rise to be achievable, but the projecting cantilevers at the junctions must also be addressed. Twin walls along with 'sandwich panels' and 'façade units' generally require propping for stability and plumbing. Shimming under the units is usually required to level them. The number and location of anchor points to attach 'push–pull' props need careful attention; large panels will attract a large wind force, so a degree of redundancy should be allowed for (e.g. additional bolts within prop base fixings).

Square and rectangular columns will need a minimum of three temporary push props – two to plumb the column in both axes and the third to control the orientation in plan, with shimming for level also required. Temporary bearing / shims should be designed for the purpose and spaced according to the design. These should be positioned away from the edge of the precast elements to prevent spalling and an unsightly finish. As mentioned in section 7.5.1, proprietary ‘bolted shoes’ such as ‘Peikko’ within the permanent works connection design may reduce or eliminate the need for temporary propping. The bolted connections may also be used to plumb the columns into position.

Lattice plank slabs are in effect soffit formwork that contribute to the final structural capacity of the permanent works slab. The concrete ‘biscuit’ takes the part of the formwork liner, the lattice girder is the secondary, so only the primary beams and props need to be provided. Problems arise because lattice planks are usually manufactured undersize to ensure easy installation. This results in gaps between the planks which will lead to grout loss. Some planks may also require temporary vertical support from proprietary falsework systems, which may also require consideration of additional lateral restraint at the head (see top-restrained falsework in Pallett and Filip, 2019, chapter 22 and TWf, 2023). The additional lateral restraint can be provided by additional bracing to the falsework, tying the falsework to the permanent works and tying the lattice girders to adjacent plank girders and column / wall starter bars. The optimum solution appears to be running neoprene seals (or a cheaper backer rod) between the planks, followed by grout over the seals and between the planks and the columns and walls adjacent to the pour (pre-grouting). The design of the falsework layout can be made more efficient if careful early intervention into the lattice plank design is undertaken. The lattice girders can be placed at various centres and their strength modified by using different sizes of reinforcement bar in their construction. Manipulating the structural capacity of the lattice planks will reduce the amount of falsework required, and more practical layout exploited. Some suppliers also provide prestressed lattice planks to further enhance capacity.

‘Hollow-core’ slabs may not require any temporary works at all; at most it will be one or two lines of propped primary beams to reduce the overall span of the slabs. Prefabricated stairs and landings again will require minimal temporary works, generally only propping at each end of the units while the permanent connections are made.

Bridge beams may require substantial supports at each end, which may incorporate some formwork to cast the in situ stitch, if required. The bridge deck soffit will also need addressing; prefabricated permanent formwork systems have been successfully utilised using reinforced concrete, glass-reinforced plastic and glass-reinforced concrete. Glass-reinforced plastic and concrete both have the advantage of either encapsulating the steel components in watertight plastic or having no steel components, therefore not compromising the cover to any steel reinforcement.

Abutment shells, manhole rings, culvert units and vertical concrete barriers generally only require lifting beams or frames.

Bathroom pods are usually installed after the building frame and need to be ‘fed’ into the partially completed building. To facilitate this, a loading platform designed to receive the units from the crane at the correct level should be provided. As an alternative, cantilevered counterbalanced lifting frames have been used to insert the pods. Before the pods are inside the building, a check must be made of the ability of the permanent works slab to take the load of the pod and its transportation equipment. This must be done for the route the pod will take to its final location. Hopefully the structure will have been designed for the pods in their final location.

For maximum efficiency, MEP elements can be incorporated into structural elements. However, where this is not achieved, MEP elements will still require some temporary works consideration, such as transportation (even spread of loading for stacking, securing the load and additional loads to prevent damage), safe lifting and fixing into their final position. The frames that MEP components are fixed to are also items of temporary works that need to be suitably designed for lifting, turning and installation. Often these elements might be installed at height (fixed to ceilings), so safe access and working platforms need to be allowed for. There have been issues when heavy MEP elements have been wheeled across suspended slabs (e.g. by pallet trucks) without adequate consideration given to the imposed loading on the slab and the possible need for back-propping or load spread.

Note: this issue can equally apply to moving precast elements or pods across slabs.

Temporary Works Forum has produced a guide to precast concrete (TWf, 2019). There is also useful information on prefabricated reinforcement cages in Temporary Works Forum guidance documents (TWf, 2020, 2021).

7.6. Temporary works design considerations

MMC construction relies on considerable coordination between parties through several project stages. Many variables must be considered, including weight and size of elements, types of connections, site constraints and temporary works support. Elements are often subject to many different load combinations (including striking from moulds, transit, reorientation, lifting, placement) and are subject to loads which often no single party is responsible for. For example, the PWD provides a design for the element in the final location (as part of the final structure) but may not necessarily consider the loads the element is subject to in reaching the final position. Experience and expertise of those involved should be assessed and the use of a responsibility matrix early in the adoption of MMC solutions will support managing these interfaces and set clarity for all parties.

The main items to be considered include:

- integration of various components from structure to architectural finishes; clash avoidance is imperative
- concrete strength for striking from mould
- factory lifting and handling
- stacking and storage (prevention of damage and protection from weather)
- additional loading and prevention of damage from transportation (prevention of racking failure due to insufficient bracing)
- lifting and handling / reorientation on site (especially when lifting arrangements are different from factory)
- temporary stability until permanently stable
- strengths required for removal of temporary support.

Further complexity is added to this when the temporary support (e.g. bolted support) is either partially or fully designed into the permanent works. In this case, the PWD should ideally take responsibility for any temporary stability.

Falsework decks are not required with full floorplate precast construction; they are erected without extended working platforms beyond the slab edge. In traditional construction, falsework and climbing screens provide protection against falling elements, whereas precast construction needs to consider how this risk is reduced without adding unnecessary elements into the build. Consideration may be given to pre-installing edge beams / slabs with edge protection or, similar to steel construction, providing fans / netting to protect against the risk of falling bolts / plates. Leading UK contractor Mace utilises a 'high-rise solution' (HRS) approach, whereby the façade itself is installed as an assembly with the floorplate, providing leading edge protection and preventing falling objects.

All parties involved with MMC should realise that cranes are the most critical item of machinery on site. When compared to traditional build, more cranes will be necessary to achieve the required efficiency. The cranes are also likely to be larger, with larger foundations (due to larger imposed loads). Temporary offloading gantries, protective fans / screens and bespoke lifting frames are also common. Traffic management issues need to be resolved with the relevant authorities and permissions obtained for any temporary works items that may overhang or impact roads, pavements and services. These issues require early identification so that adequate space can be provided and responsibility apportioned so that a design and checking process can be undertaken.

Designers need to give particular attention to alignment, stability and placement tolerance for individual elements as well as understand the limitations of each element (e.g. potential for local failure within twin-wall corners). They should also consider secondary loads (e.g. wind, eccentricity and impact) and possible out of balance in the temporary condition, which can induce temporary stresses, strains and torsion, especially on primary support elements and connections (e.g. installation of stairs / landing or platforms while walls are propped in the temporary support condition).

7.7. Principal safety considerations and risks

The traditional approach to construction is one that tends to be labour intensive; the typical hazards are high in frequency yet low in severity of accidents (slips and trips, manual handling injuries etc.). This does not suggest that higher severity accidents do not occur, however, as the number of hazards is often greater. In turning to offsite manufacture, construction projects require operatives to be more highly skilled, often carrying out multiple functions. The workers have reduced exposure to lower severity hazards (e.g. related to site housekeeping) and much greater exposure to lower frequency / higher severity accidents – for example, falls from height, falling objects, crushing injuries, swinging precast elements and structural stability. Exclusion zones beneath work at height should be enforced where possible; otherwise, measures to prevent falling objects hitting operatives or the public below should be instigated (also see chapters 5 and 6).

The use of products manufactured off site is not simply different products, but a different organisational approach which changes roles, responsibilities and risk exposure throughout the process. Incidents in MMC are often related to failures in lifting anchors, propping sockets and movement or collapse of elements during lifting. Ensuring quality throughout the manufacturing process, designing the product for pitching and placement loads and building robustness into temporary support reduces many of these risks. The consequences of failure of these larger, heavier elements can be more severe and take longer (and are more expensive) to resolve, with significant programme / element delivery implications.

In terms of worker behaviour, key areas of focus should be training operatives in workplace positioning (staying clear of suspended loads) and working at height while removing products from the crane hook. In addition to these basics, through having a thorough understanding of the installation, the elements are installed faster and the workforce exposed to the risks for a shorter time. The teams should therefore fully develop the method and sequence of assembly and erection, including any temporary works (where possible pre-installing fixings), erection tolerances and complexities with lifting arrangements.

Other issues

- Sequencing, construction / manufacturing tolerances and connection details need to be understood so that interface details between different materials and elements can be managed. In volumetric modular construction, attention should be given to the 'eighth module' problem (how you land the eighth module on the fourth below and the third adjacent at the same level to complete the cuboid – and still get into the central corner), as well as how to close the intermodular void to control fire spread.
- Freeze–thaw can occur when you have well voids.
- Construction / delivery programmes need to allow for the possibility of bad weather (high winds and poor visibility) which will limit crane use on site.
- How are repairs and retro-fitting carried out?
- Embodied carbon – high cement replacement equates to lower early strength.
- Temporary works failures (e.g. props or fixings) can be catastrophic – could be caused by poor installation or impact from mechanical equipment.

7.8. Future de-construction and re-purposing

To truly meet the future demands of the construction industry, the whole lifecycle of a structure needs to be considered. This includes the consideration of maintenance, reuse and de-construction / demolition. To facilitate this, temporary works should be considered in a similar way to the permanent works, including:

- consideration of design life and durability
- accessible lifting points for easy dis-assembly
- details of temporary works and dis-assembly methods / sequence included in the project health and safety file.

REFERENCES

BSI (2024) BS 5975-1:2024: Management procedures for the control of temporary works. Code of practice. BSI, London, UK.

Construction Enquirer (2023) Storage issues. *Construction Enquirer*, 9 March.

CROSS (Collaborative Reporting for Safer Structures) (2012) Report 402, April, CROSS, UK.

Pallett PF and Filip RK (2019) *Temporary Works: Principles of Design and Construction*, 2nd edn. ICE Publishing, London, UK.

RIBA (Royal Institute of British Architects) (2020) *Plan of Work 2020 Overview*. RIBA, London, UK (available from www.ribaplanofwork.com).

TWf (Temporary Works Forum) (2019) *TWf 2019:01: Precast Concrete: Good Practice and Common Issues in Temporary Works*. https://www.twforum.org.uk/viewdocument/pre-cast-concrete-good-practice-and (accessed 07/01/2025).

TWf (2020) *TWf 2020:03: Temporary Condition of Reinforcement Cages Prior to Concreting: Part 1 – Management Guidance*. https://www.twforum.org.uk/viewdocument/rebar-guide-part-one-management (accessed 07/01/2025).

TWf (2021) *TWf 2021:01: Temporary Condition of Reinforcement Cages Prior to Concreting: Part 2 – Technical Guidance*. https://www.twforum.org.uk/viewdocument/rebar-guide-part-two-technical (accessed 07/01/2025).

TWf (2023) *TWf Information Sheet 17: Lateral Stability: Top-Restrained Falsework (or not)?* https://www.twforum.org.uk/viewdocument/twf-information-sheet-no-17-late (accessed 07/01/2025).

USEFUL WEBSITES

BSI (British Standards Institution) – www.bsigroup.com
CC (The Concrete Centre) – www.concretecentre.com
CIRIA (Construction Industry Research and Information Association) – www.ciria.org
CROSS (Collaborative Reporting for Safer Structures) – www.cross-safety.org
NAO (National Audit Office) – www.nao.org.uk
NHBC (National House Building Council Foundation) – www.nhbc.co.uk
RIBA (Royal Institute of British Architects) – www.architecture.com
Simply Engineering – www.simply.engineering
TWf (Temporary Works Forum) – www.twforum.org.uk

FURTHER READING

CIRIA (Construction Industry Research and Information Association) (2023) *X538F: Offsite Construction: Concept Design and Delivery*. CIRIA, London, UK.

Farmer M (2016) *The Farmer Review of the UK Construction Labour Model*. Construction Leadership Council, London, UK.

HMG (Her Majesty's Government) (2022) *Modern Methods of Construction Strategy (MMC)*. The Stationery Office, London, UK.

NAO (National Audit Office) (2005) *Using Modern Methods of Construction to Build Homes More Quickly and Efficiently*. NAO, London, UK.

NHBC (National House Building Council Foundation) (2018) *Modern Methods of Construction, Who's Doing What?* NHBC, Milton Keynes, UK.

Oakley M (2017) *WPI Economics Report for Heathrow: The Value of Off-Site Construction to UK Productivity and Growth*. https://wpieconomics.com/publications/off-site-construction/ (accessed 07/01/2025).

The Concrete Centre (2019) *TCC/03/064: Offsite Concrete Construction*. The Concrete Centre, London, UK.

Ray Filip
ISBN 978-1-83608-525-6
https://doi.org/10.1108/978-1-83608-524-920251009

Chapter 8
High-rise construction

Adrian Bywaters
Mace, UK

Ray Filip
RKF Consult Ltd, UK

Martin Pike
Mace, UK

Kit Yardley
Laing O'Rourke, UK

8.1. Introduction

High-rise construction includes residential / commercial buildings and civil engineering structures, such as industrial chimneys, cooling towers for power plants, wind turbines, bridge piers and dams. They require advanced engineering solutions to withstand the forces of nature (wind, snow, earthquakes etc.) so that they remain stable and perform as intended. Such structures require massive foundations and often their shape can be complex to limit lateral forces and sway which may cause structural damage or collapse. Choice of materials (most often steel and high-strength concrete), construction sequence, logistics and temporary stability are critical considerations. The associated temporary works to construct these buildings and structures will also be complex and have significant associated risks.

Temporary works at height follow the same considerations and design process as for any other temporary works. However, the loadings and conditions experienced at height can be more challenging and more significant due to increased environmental considerations (e.g. wind – see section 8.3).

High-rise construction poses some specific challenges.

- As the building or structure progresses, it takes increasingly longer to get personnel, plant, equipment and materials to the 'workface' from ground level, which has programme implications.
- There are capacity and capability limitations associated with cranes, concrete pumps, scaffolding and edge protection systems.
- Careful consideration should be given to how items of temporary works, such as concrete placing booms, cranes, edge protection, falsework, structural propping, access platforms and so on are installed and, perhaps even more challenging, how they will be

subsequently removed. Temporary laydown areas may be required at ground level for dismantling / storage before items are loaded onto lorries for transport. This can be a challenge on city centre projects where space is at a premium.

- Emergency procedures for rescue, fire and evacuation must be established and maintained.
- As a building or structure progresses upwards, the ever-increasing height will increase the 'implementation risk' (BS 5975: 2024, BSI, 2024a, table 1) of all temporary works. The main risks will be associated with increased environmental loading (see section 8.3) and falling objects / debris / personnel (also likely to travel further horizontally the further they fall vertically), as the consequences of failure are likely to be severe.

High-rise buildings and structures, with a significant degree of repetition, may be particularly suitable for offsite manufactured components which are transported to site and assembled (see chapter 7).

8.2. Logistics

High-rise construction presents unique challenges with respect to logistics. Construction teams must consider how personnel, construction materials and temporary works (including false-work, formwork, back-propping, edge protection etc.) are moved up the building or structure (or recycled from lower floors / levels) to the uppermost working level. Often this may rely on the use of temporary platforms within cores, or externally supported loading platforms or cantilever decks (see section 8.7) that place load on the partially constructed structure which may not have achieved full strength. Where the structural core is extended beyond the working level, access to the core itself may require the use of a hoist to ensure efficient construction.

Temporary works items, such as formwork / falsework / back-propping / edge protection and so on, need to be considered in the wider context of the building (similarly for civil engineering structures), especially where follow-on trades are installed in parallel with the structural frame. An example of this may be the lifting in of bathroom pods onto a floorplate while the project is 'blue-sky' (during the construction of each floor) rather than bringing them in using a hoist later. The placement of falsework and any additional bracing for the temporary support of the structure must also consider access routes, emergency escape routes and the potential for obstruction.

High-rise buildings are mainly constructed in city centres and available space for offloading / loading vehicles and storage can be a challenge. Ties for tower cranes, hoists, access stairs, scaffolding and run-off areas must be carefully planned to avoid bottlenecks on the floorplates which could lead to a slower building fit out. In the case of incomplete structures, items of temporary works may remain within the structure for a significant time (such as crane supports) and may not be removed until very late in the programme. Project teams must therefore consider how these items are removed and the permanent works completed, while not disrupting the ongoing internal fit out.

A further challenge presented in high-rise construction is the provision of water and control of ingress of rain which may cause damage to internal finishes. Temporary protection may need to be installed at intermediate floor levels to allow lower floors of the building to proceed with fit out. Hole protection, which is installed at the time of the structural frame, may therefore need to serve multiple purposes: to support temporary loads and protect falling items but also to prevent the passing of water through the building. On some projects, water bunds are required.

8.3. Environmental loading

Temporary works associated with high-rise structures will attract higher wind loads (BSI, 2024b, clause 6.4.1). Wind increases with elevation: the wind force at 100 m height can be around 50% greater than that at 20 m height. Tall buildings will not benefit from shielding as they will be higher than the surrounding structures. At lower heights wind speeds generally increase along a gradient as height increases, but wind pressure varies as the square of the wind speed. A small increase in wind speed can have a significant effect on wind pressure, especially at extreme heights. Some proprietary temporary edge protection systems may not be suitable for use at significant heights as they may not be able to withstand the increased wind loads. The effects of this increased wind loading should also be considered for machinery such as tower cranes and hoists. The American National Standards Institute (ANSI, 2016) identified 1125 worldwide tower crane accidents between 2000 and 2010, resulting in 780 fatalities. In 23% of these cases, it was identified that exposure to wind loading was the main issue and a contributing factor in many other cases.

BS EN 1991-1-6 (BSI, 2005) and BS 5975-2:2024 (BSI, 2024b) allow for a reduction in wind loading due to the length of exposure. However, some items of temporary works (such as climbing screens) may be in use for prolonged periods on very tall structures and this should be carefully considered when designing for wind loading. Also, care should be taken when installing equipment in buildings before the envelope is airtight; their fastenings may not have been designed for wind loads in the temporary condition of the structure. Temporary works designers are reminded that careful consideration should be given to localised wind effects around tall buildings and structures and the effect on temporary structures such as hoardings and falsework (see BSI, 2024b, clauses 6.4.1.9 and 6.4.1.10; CROSS, 2016; Structural Safety, 2015).

At great heights, in cold weather with a wind chill, snow and ice can accumulate, thus increasing the loads on items of temporary works and the permanent works structure. Designers should allow for these additional loads, with inspection and protection measures also being used on site.

The taller the structure, the greater the chance of lightning strikes. Consideration must be given to earth bonding items of temporary works, such as scaffold, climbing frames and jump forms. Guidance documents are available from NASC (2014) for scaffolding and from CPA (2021) for tower cranes.

Specialist advice should be sought on this issue as it is generally outside the experience and competence of temporary works engineers.

8.4. Strapping down

It is essential that sites have robust safety procedures and designs for restraining / strapping down items from being blown away or blown over. Incidents in windy conditions have occurred involving items such as temporary works equipment (formwork and falsework), sheeting, advertising banners, tools and rubbish bins (blown off buildings), which can cause significant damage. Stacking materials such as sheeting is commonplace. A stack of plywood sheets will be sufficiently heavy; however, the top sheet may be blown off if not strapped down or weighted. Generally, these can be treated as category 0 or category 1 design checks, unless the items are very large sails.

8.5. Placing concrete at great heights

Concrete strength, particularly early strength development, is critical in high-rise construction as it directly impacts the programming and safety of temporary works items, such as formwork and falsework. It is especially important in high-rise construction due to the criticality of efficient floor construction cycles on overall structural build times. Early strength gains allow for faster removal and reuse of temporary supports. In addition, many anchor fixings which may be installed after a slab is poured (e.g. for column or wall propping) rely on minimum concrete strengths for use. Insufficient concrete strength (perhaps due to cold temperatures or through use of lower carbon concretes) can therefore delay sequences, increase costs and, in cases of high-rise, where teams can become complacent with the repetitive nature of the work, can lead to unsafe actions. Concrete strengths for unloading props, re-propping, lifting jump forms, loading out, placing fixings and so on must all be considered, carefully planned and monitored to ensure the works are carried out in a controlled manner.

Concrete can be lifted to great heights in skips by very large mobile cranes, tower cranes or a cable crane. The capacity of the crane limits how much concrete can be placed at any one time and batching capacity can limit the total amount of concrete to be placed in a single pour. Crane operators tend to be more cautious when lifting concrete, as a fluid will behave differently from a solid and spillages can occur. Productivity reduces as construction continues upwards as it takes progressively longer to lift and place the concrete.

Advances in concrete mix design and pumping technology mean it is now common to place large volumes of concrete using pumps and placing systems such as booms which offer greater efficiency. Concrete needs to be pumped vertically and horizontally to great heights with precision, minimal wastage and speed to ensure the structural performance of the permanent works. High-pressure concrete pumping technology has advanced significantly and allows for great efficiency in concrete placement. The Putzmeister pumping system, used at Burj Khalifa in Dubai, achieved a pumping height of over 600 m with a measured concrete pressure of 200 bar and an output of 30 m^3/hour.

It is essential to consider access and space requirements for the concrete delivery and pumping system, along with suitable temporary foundations for the machinery. Wind speeds should be monitored to ensure safe working wind speeds are not exceeded. Pumping distances are affected by rheology and workability of the concrete mix. If the mixture is too thick or of low workability it is unlikely to flow smoothly through the pump pipeline and thereby will reduce the range. It may also cause blockages, as may concrete with a large aggregate size. Admixtures can improve concrete workability and ‘pumpability’. Adjusting pumping pressures according to distance and height can ensure a smooth, continuous flow of concrete which will ensure mix stability, mix ‘homogeneity’ and prevent ‘cold joints’. In very hot climates it is common to place concrete at night when it is cooler and there is likely to be less traffic to cause delays to concrete deliveries, so appropriate temporary lighting systems will be necessary. Good planning is essential to ensure a constant and steady supply of concrete, thus avoiding delays which may lead to blockages in the pipeline.

When pumping concrete great distances, the concrete gang will initially utilise a crane to lay out the steel pipeline between the pump and the area to be concreted, before manually attaching each of the line sections to create one continuous line between the end of the static / mobile pump and the pour zone. The sections of pipe are connected using a coupling and coupling seal (either Victaulic or male / female type), with an 'R' clip safety pin. The sections of pipe nearest to the pump will see the highest pressure and the greatest wear and accordingly these sections should be thick-walled and should be installed in 'like new' condition. The maximum concrete pressure of the pump should be the primary factor used to determine pipe wall thickness and types of ends required (in the case of a blockage, the maximum pressure of the pump will be exerted). The wall thickness of the pipe should be checked regularly using an ultrasonic digital thickness gauge and, if found to be less than 2.2 mm, the section should be replaced. Where the pipeline crosses walkways, temporary ramps or steps will be installed to prevent trip hazards.

The steel delivery pipeline is designed to be suitable for the pressure. Careful planning is required to position, support and restrain this pipeline. Often it is preferable to position the pipeline in a riser (ready-made hole through the building). If this is not possible (due to other work occurring in the riser or for civil engineering structures), then the pipeline may be positioned on the outside of the building (or structure) or may be located within the building, passing through temporary holes in slabs which will have to be filled afterwards. The reinforcement should be appropriately detailed around any such temporary holes. The pipeline and couplings are not designed to be freely suspended and require temporary supports to secure the line. They are generally fixed horizontally with steel brackets to floors or walls. Both types typically utilise 20 mm-diameter threaded bar with 200 mm-long anchors fixed with resin and lock nuts for support and restraint. Vertical lines are anchored in a similar way with wall brackets at regular intervals (at each pipe section). All changes in direction of a pipeline should also be secured with brackets and fixings which prevent movement. Where the pipeline cannot be secured directly to the permanent works, then a suitable temporary thrust block should be provided. BS 5975-2:2024 (BSI, 2024b), clause 6.3.3.5 provides further guidance. Fixing positions will be chosen such that they do not coincide with fair-faced concrete work and their location should be checked with the structural engineer and, if necessary, the client and architect. Manufacturers should provide the relevant loading information; many contractors have subsequently developed their own temporary works 'standard details' for the support / restraint systems. The support / restraint systems generally rely on the permanent works (or temporary works systems such as falsework) and structural checks should be carried out to ensure they are suitable for the applied loads.

When the pumping limits of a pump are reached, then concrete can be multi-stage pumped (i.e. pumping from one pump to another to increase range). When pumping vertically, subsequent pumps will be positioned on the structure at pre-planned locations. These locations should be designed / checked for the applied loading and may require strengthening or temporary back-propping.

8.6. Scaffolding

Erection of access / working scaffolds to significant heights is limited by the capacity of the standards to take the axial compression (standards buckling under applied vertical loading), with an effective length governed by the working lift height (normally 2.0 m), vertical spacing of the ties and bay length. Taking data from *TG20:21* (NASC, 2021) for tube and fitting, access scaffolds with single standards have an absolute maximum height of 56 lifts (i.e. 112 m high; see Figure 8.1). This limitation would be for a three-board wide scaffold, with a 1.2 m bay and with one boarded level loaded to 0.75 kN/m^2; the scaffold would need to be tied into the adjacent structure at every lift.

More typically, a 5 + 1 scaffold (five boards wide with one inside board), with 2.1 m-long bays, with two boarded lifts (loaded at 2.0 kN/m^2 and 1.0 kN/m^2) tied at every other lift would have a maximum allowable height of 17 lifts (i.e. 34 m high).

Doubling the standards can be beneficial but, at extreme heights, the self-weight of the tube plus the additional fittings required will take a very large proportion of the additional capacity provided.

When the limitations of scaffolding are reached, temporary steelwork frames / supporting brackets and / or proprietary propping systems can be used at lower levels, with scaffolding constructed above to increase the achievable height; however, building very high scaffolds is a time-consuming and expensive endeavour. Increased wind loading with height needs to be considered on high-rise scaffolding, especially when netted or sheeted. Type and positioning of ties should be considered to avoid clashes and the capacity of ties (and the structure the ties rely on) should be established.

Figure 8.1 Scaffolding at the limits of height, Victoria Tower, Houses of Parliament, London, (reproduced with kind permission of K Yardley)

8.7. Tower cranes, hoists, mast-climbers and canti-decks

High-rise buildings and tall civil engineering structures will often require tall tower cranes. Tower cranes can free stand to a significant height; however, at great height the crane mast will require additional restraint by tying to the building / structure under construction. The crane supplier / manufacturer should provide the tie arrangement (generally three pin jointed struts and a braced collar around the mast) and the loading information. A temporary works design (and design check) is then required for the anchors / connections of the struts to the structure. The structure itself should also be checked for the reactions from the anchors / connections. It is essential that responsibility for these designs / checks is clearly defined and there is coordination between temporary works and permanent works. The crane supplier / manufacturer will advise on the positioning of the ties and careful thought should be given to these positions to avoid clashes which can cause construction programme delays and site difficulties.

Tower cranes can be positioned inside a tall building (often in a lift shaft) or externally and can be 'climbed' (jacked upwards) as the construction progresses and then back down once the crane is no longer required (finally removed by mobile crane at lower level). The crane is supported by temporary collars at two levels (with a third support provided during climbing) and these collars transfer the crane forces to the supporting structure by way of a steel grillage. As with tower crane ties, responsibility for designing / checking all temporary works and checking the structure should be clearly defined.

Figure 8.2 shows a project in central London where a tower crane could not be positioned on a conventional foundation (see Pallett and Filip, 2019, chapter 4) at ground level. The temporary steel frame allowed the tower crane to be to be positioned at a higher level, allowing the road and pavement below to be kept clear. This was a complex design which also relied on the permanent structure for support, so careful coordination between temporary works and permanent works was necessary. The cantilever nature of these crane grillages transferred significant horizontal forces into the permanent structure, which in turn required significant amounts of additional reinforcement and Macalloy bars to transfer those forces back to the central core.

There is a major challenge in providing access for operatives (and materials) to the workface at high levels, as scaffolding or proprietary staircases can be impractical. Hoists and mast-climbers can be provided to transport operatives and tools / materials to a high level relatively quickly. The number of operatives and size / weight of materials that can be transported will be stipulated in the operating instructions. The mast of the hoist / mast-climber is tied to the structure at regular intervals. The ties are connected to the structure and, as with tower crane ties, the supplier / manufacturer should supply the loading information. A temporary works design is required for the fixings and a check carried out on the structure itself and, as with tower cranes, coordination between permanent and temporary works design is essential.

Tower cranes, hoists and mast-climbers are all subject to control measures in windy conditions which can be experienced at exposed, high locations. Anemometers will measure wind speeds and operation will be halted if permitted limits are exceeded.

Figure 8.2 Complex tower crane supports for high-rise inner-city project (reproduced with kind permission of MACE)

Note: the front cover of this publication shows the tower cranes tied to the structure as the works progressed.

For further information see

- Construction Industry Research and Information Association (CIRIA) (CIRIA, 2006)
- Construction Plant-hire Association (CPA) (CPA, 2011, 2019)
- There are a numerous 'YouTube' videos showing these processes.

Temporary cantilever loading bays (often known as 'canti-decks') can be used to allow a crane to set down a load safely outside the building before it is moved inside (perhaps by pallet truck). These loadings bays will rely on support from the permanent works, with temporary fixings and back-propping often necessary for support. The load on the outer (leading) edge of the permanent works slab will be approximately twice the weight of the platform, plus the item placed on the platform. The slab above will also require a check as it is unlikely to be designed in the permanent condition for the upward force from the prop (if used instead of fixings to the slab below). Kentledge can be placed on the slab above or 'up-propping'. Figure 8.3 shows a typical arrangement.

Figure 8.3 Temporary support arrangement to cantilever deck (editor's own)

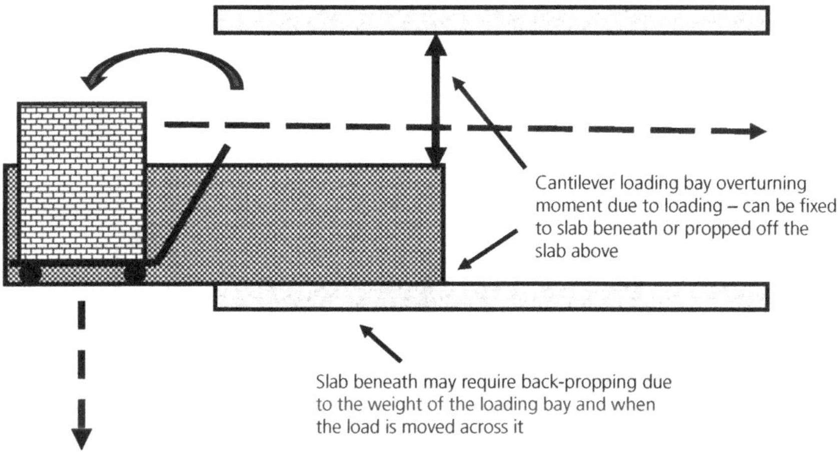

8.8. Temporary stability

Determining the construction sequence and most onerous temporary conditions for the permanent works are critical considerations and should be identified during the planning / design phase. The cores of tall buildings (lift shafts, stairs etc.) will often provide lateral stability to the overall building. The core will be built ahead of the floor plates using jump formwork or slip formwork (see Pallett and Filip, 2019, chapter 25), with temporary formers for door openings. The floor slabs are subsequently tied into the cores as construction progresses upwards. It is critical to consider the temporary stability of the exposed cores above the floor plates (especially as the concrete may not be fully cured), as the core itself may potentially become unstable. The permanent works designer (PWD) should provide guidance on how the structure can be constructed safely (see chapter 2 and HMG, 2015). Additional temporary bracing may need to be installed as the core is constructed to ensure stability and to assist the overall construction programme by maintaining progress of the core under construction. A similar issue can be encountered with other isolated elements, such as very tall piers for suspension bridges. Figure 8.4 demonstrates part of the construction sequence for rectangular reinforced concrete suspension bridge piers. The diaphragms perform a function in both the temporary and permanent conditions (once the deck is completed). In the temporary condition, they can provide restraint to the piers (in one direction); if the piers are steel, then cross-bracing can be used. It is essential that the PWD considers the most onerous structural condition (for certain elements this may be in the temporary condition), buildability and temporary stability and develops suitable solutions and sequences with the contractor.

8.9. Temporary climbing screens

Temporary climbing screens allow work to take place behind them by 'sealing off' the inner area against the effects of wind, thereby improving safety and efficiency. Proprietary systems are available from most formwork / falsework suppliers. Externally, they comprise a framing

Figure 8.4 Staged construction of rectangular reinforced concrete piers for suspension bridge (editor's own)

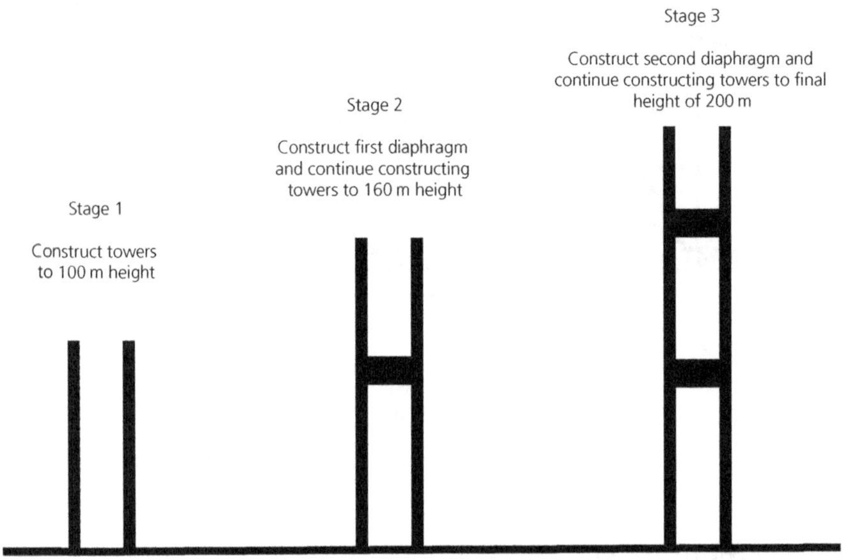

system and, to reduce weight, the infill can be polycarbonate sheeting, providing an effective 'sealed' inner area. Internally, they incorporate integrated working platforms and access. The screens are hydraulically jacked upwards on a rail-guided climbing system or can be crane-lifted upwards as the work progresses; jacking is generally more efficient as craneage may not always be available when required. The screens should be designed for the applied wind loading and induced forces during jacking / lifting, with the screen being designed to maximise the free height above the structure. A limitation during lifting will be the safe permitted operating wind speed of the crane and the size of the screen section (sail area). It is critical to consider the gap between the structure and the screen and how this will be covered when the screen is in position and when lifting. Consideration should be given to more severe environmental loading with height, as described in section 8.3.

8.10. Edge protection

As described in section 8.3, wind loads increase with height, which can affect edge protection systems. Checks should be carried out to ensure the edge protection system and its connections are sufficient for the increased loading, especially at corners. Additional issues are created by adding netting to these systems and the possibility of impact loading from debris, which can lead to failure. High-rise construction relies on consistently working at heights and working close to the edge of buildings or structures. Typically, climbing screens are used to support these works which shelter operatives from exposure but, in some cases (particularly using prefabricated methods such as steel or precast construction), the use of proprietary edge protection is necessary.

Although minimum standards of edge protection (BSI, 2013) require 950 mm of height (minimum guardrail height with toe-board), many contractors have set higher standards, requiring more than 1.5 m, with close mesh panels and / or netting. Where the height of the edge protection is increased (normally with increased risk associated with falling items), the load applied to the edge protection can increase due to the applied moment at the top of the panel. Wind loads also increase at height, which could significantly impact the capacities of proprietary systems and connections to the structure. These should be checked against specialist load cases and may require bespoke design to account for discrepancies. Installation must be carried out by trained personnel to prevent gaps or weak points and consider all potential items that may fall, including during the installation of the edge protection itself. Coordination of the edge protection is crucial to ensure that it does not clash with follow-on operations (such as the installation of cladding) and therefore fixing points should be planned and accommodated within design at an early stage. Where prefabricated systems are used, this may mean casting in channels or providing slotted holes in beams to receive bolted connections. Where plant / machinery is being used at the edge of the building, additional protection measures should be implemented, suitable for the equipment being used – for example, stop blocks.

8.11. Top-down construction

A tall building with multiple levels of basements can be constructed using ‘top-down’ methodology, which can have programme benefits which allow both the sub- and superstructures to be constructed simultaneously. However, this form of construction can also create challenges regarding construction sequence, stability, access, temporary loading and waterproofing. Specific considerations regarding temporary works associated with top-down construction include:

- plunge column design
- ground floor slab design to support temporary storage loading, construction vehicles and so on
- support grillage (steel or concrete) for core construction.

An outline of a more conventional construction methodology is as follows.

- Prepare the site and install a temporary (which may become permanent) retaining wall (sheet piles, contiguous piles, secant piles or diaphragm wall) around the basement perimeter.
- Progressively excavate downwards, installing temporary props / horizontal frames / anchorage systems and so on to support the retaining wall (which is unlikely to be able to stand as an embedded cantilever).
- Continue excavating and installing temporary supports to formation level.
- Construct foundations such as piles, pile caps or raft and so on.
- Construct the lowest slab. Install temporary formwork to construct the columns (with other load-bearing elements such as walls and cores) which support the slab above. The slab will also be connected to the perimeter retaining wall.
- Progressively construct the below-ground slabs and columns, working upwards and removing any internal temporary supports as the slabs are constructed.
- Construct the ground floor slab and start the above-ground construction.

Below are some differences associated with top-down construction.

- Prepare the site and install a temporary (which may become permanent) retaining wall (sheet piles, contiguous piles, secant piles or diaphragm wall) around the basement perimeter.
- Piled foundations can be constructed from ground level (open bore above pile cut-off level) and full-height (below-ground) columns can be 'plunged' into position from ground level as the retaining wall is constructed.
- Construct the ground floor slab (formwork and falsework may be necessary although there are examples where the ground below the slab can be used to support the soffit formwork, thereby omitting the falsework). The slab will have large holes (known as 'moling holes').
- Use the moling holes to excavate under the ground floor slab to the first basement level.
- The first basement slab can be constructed, again with moling holes. The slab will be connected to the load-bearing elements, such as columns and perimeter retaining wall.
- As the below-ground works continues, the superstructure works can commence simultaneously.
- The below-ground sequence continues until the lowest slab is constructed. The moling holes are then progressively filled in. The superstructure works continues.

8.12. Emergencies

Sites need to plan for the possibility of an emergency. This may include having to rescue a tower crane operator (the fire brigade undertake training exercises), fire prevention procedures and emergency evacuation. Temporary walkways, staircases, lighting, egress points and signage will need to be provided. All escape routes should be regularly inspected with a permit system. Progressively install fire compartmentation as the works progress. Particular attention should be given to maximum permitted travel distances in case of fire, number of escape stairs required and ensuring there are alternative means of escape.

There are two useful documents:

- *Fire Prevention on Construction Sites* (Construction Confederation, 2009)
- *Fire Safety in Construction* (HSE, 2010).

Fire prevention is key and can be helped by eliminating the need for hot works, the correct storage of flammable materials (if it is not possible to avoid their use) and choice of construction machinery (avoiding combustible fuels). Construction methodology, such as offsite construction (see chapter 7: Modern Methods of Construction) can assist with this.

8.13. Case study

Figures 8.5 and 8.6 show a very high-level scaffold that was required for a refurbishment project at the top of the British Telecommunications (BT) tower in central London. Part of the project involved carrying out extensive work to the revolving restaurant at the top of the tower and refurbishing / replacing the glazing and external cladding at a height of over 150 m above pavement level below. The length of all components (including scaffolding) that could be taken to the top of the tower was limited, as they all had to fit into the lift inside the tower.

Figures 8.5 and 8.6 Two views of the partially constructed scaffold around the top of the BT tower (editor's own)

The scaffolding was designed by an experienced scaffold designer and category 3 checked (see BS 5975-1:2024, BSI, 2024a, table 1), with detailed sequences developed for the installation, inspection, maintenance and subsequent removal. At the lowest level of the revolving tower section, scaffolding beams cantilevered out through the windows and were then supported off the concrete slab below and 'punched' off the slab above to resist the cantilever of the beams. At concrete roof level, scaffolding beams were anchored to the concrete and cantilevered out. The lower-level scaffold was built upwards from the beams and the upper-level was built downwards until they met. The entire scaffold was sheeted for further protection.

The risk of falling objects was a major consideration because it was not possible to create an effective exclusion zone at pavement / road level beneath the works. Netting and crash decks were also rejected as they would pose their own risk during construction, there were limited places where they could be positioned (mainly at much lower levels) and they would not be suitable to protect from objects falling significant distances. The scaffold was built and inspected by a team of very experienced scaffolders, who ensured they had no loose items in their pockets (phones, coins, cigarette lighters etc.) and all small items (scaffold clips, tools) were tethered to their wrists or belts. Anchors were drilled into a concrete frame, so ropes could be attached, to handle larger objects, such as scaffold tubes and scaffold boards. The ropes were only removed once the tube or board was positioned. Once positioned, scaffold boards were lashed in position to prevent them getting blown off.

REFERENCES

ANSI (American National Standards Institute) (2016) Wind Effects on Tower Cranes. Blog 7 July 2016, blog.ansi.org (accessed 21/10/2024).

BSI (2005) BS EN 1991-1-6: 2005: Eurocode 1: Actions on Structures – General Actions. Actions During Execution. BSI, London, UK.

BSI (2013) BS EN 13374:2013+A1:2018, Incorporating corrigenda January 2019 and August 2021: Temporary edge protection systems. Product specification. Test methods. BSI, London, UK.

BSI (2024a) BS 5975-1:2024: Management procedures for the control of temporary works. Code of practice. BSI, London, UK.

BSI (2024b) BS 5975-2:2024: Temporary Works – Falsework: design and implementation. Code of practice. BSI, London, UK.

CIRIA (Construction Industry Research and Information Association) (2006) *C761: Tower Crane Stability*. CIRIA, London, UK.

Construction Confederation/Fire Protection Association (2009) *Fire Prevention on Construction Sites: The Joint Code of Practice on the Protection from Fire of Construction Sites and Buildings Undergoing Renovation*, 7th edn (with support of Association of British Insurers, Chief Fire Officers Association, London Fire Brigade). Construction Industry Publications, Bedford, UK.

CPA (Construction Plant-hire Association) (2011) *TCIG 1101: The Climbing of Tower Cranes*. CPA, London, UK.

CPA (2019) *CHIG 1901: Tying Construction Hoists and Mast Climbing Work Platforms to Supporting Structures*. CPA, London, UK.

CPA (2021) *TIN 036: Tower Crane Electrical Supplies, Earthing Arrangements, Emergency Escape, Lightning and Lightning Protection*. CPA, London, UK.

CROSS (Collaborative Reporting for Safer Structures) (2016) Newsletter 43, July 2016, Report 548. https://www.cross-safety.org/uk/news-and-events/cross-uk-newsletter-43 (accessed 07/01/2025).

HMG (Her Majesty's Government) (2015) Construction (Design and Management) Regulations. The Stationery Office, London, UK.

HSE (Health and Safety Executive) (2010) *HSG168: Fire Safety in Construction*. HSE, London, UK.

NASC (National Access and Scaffolding Confederation) (2014) *SG3:14: Earthing of Scaffolding Structures*. NASC, London, UK.

NASC (2021) *TG20:21: Good Practice Guidance for Tube and Fitting Scaffolding*. NASC, London, UK.

Pallett PF and Filip RK (2019) *Temporary Works: Principles of Design and Construction*, 2nd edn. ICE Publishing, London, UK.

Structural Safety (2015) *Wind Adjacent to Tall Buildings*. SCOSS Alert, December 2015. https://www.cross-safety.org/sites/default/files/2015-12/wind-adjacent-tall-buildings.pdf (accessed 07/01/2025).

USEFUL WEBSITES

Alimak – www.alimak.com

BSI (British Standards Institution) – www.bsigroup.com

CPA (Construction Plant-hire Association) – www.cpa.uk.net

CROSS (Collaborative Reporting for Safer Structures) – www.structural-safety.org
Erect Group – www.erectgroup.com
HAS (Health and Safety Authority) – www.has.ie
Holcim – www.holcim.com
Liebherr – www.liebherr.com
London Hoist – www.londonhoists.co.uk
Putzmeister Pumps – www.putzmeister.com
Select Plant Hire – www.selectplanthire.com
Structural Safety – www.structural-safety.org
TWf (Temporary Works Forum) – www.twforum.org.uk

FURTHER READING

HAS (Health and Safety Authority) (2018) *HSA0479: Guidance on Safety in Concrete Pumping*. HAS, Dublin, Eire.

Holcim Technology Ltd (2015) *Concrete Pumping Handbook*, rev. 1. https://www.holcim.be/sites/belgium/files/documents/concrete_pumping_handbook_v7.2_rev1_-_en.pdf (accessed 07/01/2025).

The Concrete Centre (2014) *Tall Buildings*. MPA/The Concrete Centre, London, UK.

Ray Filip
ISBN 978-1-83608-525-6
https://doi.org/10.1108/978-1-83608-524-920251010

Chapter 9
Tunnelling

Prof Colin Eddie
CECL Global, UK

Dr Henry Pairaudeau
CECL Global, UK

9.1. Introduction

Tunnels form an essential part of our infrastructure networks, with diverse applications including transportation, utilities, mining and underground storage. By using tunnels, it is possible to free up space on the surface, particularly important in increasingly congested cities, as well as avoiding geographical obstacles and reducing environmental impacts.

It is crucial to understand the ground conditions when planning for tunnel construction. Coupled with an understanding of the client's requirements, this allows the selection of an appropriate tunnel construction methodology. Several tunnelling methodologies are available.

- Bored tunnelling – a tunnel boring machine (TBM) is used to temporarily support the ground and a permanent lining is installed behind the machine.
- Mined tunnelling – an open-face excavation is made either by hand or with machinery (e.g. a backhoe excavator). Temporary support can be provided with timber frames, sprayed concrete and / or rock bolts, depending on the ground conditions.
- Drill and blast tunnelling – explosives are used in hard rock geologies to create the tunnel profile.
- Micro-tunnelling / pipejacking – a small diameter TBM is used for excavation with a series of lining pipe sections, usually for utility tunnels.
- Cut and cover tunnelling – similar methodology to basement construction.
- Jacked box tunnelling – precast concrete box sections are sequentially jacked into place, with a shield at the front providing temporary support and protection for excavation. Usually used for short drives under rail / highway embankments.
- Immersed tube tunnelling – used for underwater tunnels whereby prefabricated sections of tunnel are floated into position and lowered into a dredged trench. The sections are sealed and drained with backfilling on top to prevent uplift.

In the design of tunnel supports, the boundary between temporary and permanent works is often unclear and ideally should be undertaken in a holistic manner. In some situations, good permanent works design will eliminate the need for temporary works entirely. Transient load cases may exist which represent a worst case for the lining. Conversely, long-term loadings may represent an onerous condition against which it would be unnecessary and uneconomical to design temporary support measures. Robust design management is required, with a competent tunnelling engineer assigned overall responsibility (the controlling mind). Where temporary works and permanent works design responsibilities are undertaken by separate designers, good communication is essential to ensure a suitable design is in place for all stages of tunnel construction and use.

9.2. General considerations

9.2.1 Temporary tunnel stability

The very act of advancing a tunnel creates a temporary state of stress in the ground. Unloading of the pre-existing, in situ stresses results in a redistribution, which can lead to collapse if an equilibrium state is not found. The consequences of such collapses can be severe, threatening the lives of operatives in the tunnel and the public at the surface. Tunnel instability also has significant economic consequences through project delays, damaged equipment and costly remediation measures. Stability considerations are key to selecting an appropriate tunnelling methodology.

In hard ground (i.e. rock) tunnelling, a commonly used, empirical approach to understanding the stability and support requirements is the use of a rock mass classification system. There are several in common use today, including:

- rock mass rating (RMR) (Bieniawski, 1973, 1989)
- Q-system (Barton, 2002; Barton *et al.*, 1974)
- geological strength index (GSI) (Hoek and Brown, 2019; Marinos and Hoek, 2000).

Probably the most widely used and most accessible to the wider engineering community is the Q-system (Equation 9.1). In this system, a numerical Q-value is calculated to give a description of stability for underground openings in jointed rock masses:

$$Q = \frac{RQD}{J_\mathrm{n}} \times \frac{J_\mathrm{r}}{J_\mathrm{a}} \times \frac{J_\mathrm{w}}{SRF} \tag{9.1}$$

where:

RQD = rock quality designation (i.e. degree of jointing)
J_n = joint set number
J_r = joint roughness number
J_a = joint alteration number
J_w = joint water reduction factor
SRF = stress reduction factor.

Guidance on the derivation of these individual parameters that aggregate to give the Q-value can be found in the Q-handbook from the Norwegian Geotechnical Institute (NGI, 2022).

With knowledge of the Q-value, the tunnel excavation dimensions and the excavation support ratio (ESR – a value related to the use category of the tunnel), an estimate of the support measures required can be obtained from the chart. However, caution must be exercised when using this approach. The Q-system is most applicable for initial estimates of stability and support requirements in jointed rock masses where instability is governed by falling rock blocks. Extrapolation of the approach to other rock mass behaviours is not recommended.

Tunnel stability is a particularly pertinent consideration for soft ground tunnelling, where the ground may have limited or no ability to support an open face. In cohesive ground, a stability number 'N' (Equation 9.2) can be readily calculated which can be used to assess the stability in the short-term undrained condition:

$$N = \frac{\sigma_s + \gamma z - \sigma_T}{C_u} \tag{9.2}$$

where:

σ_s = surcharge loading (if present)
γ = bulk density of the ground
z = depth to tunnel axis
σ_T = face support pressure
C_u is a representative value of the undrained shear strength of the ground.

As a general rule, for $N \leq 3$, the stability of the tunnel face is usually ensured; for $3 < N \leq 6$ special consideration of the ground settlement risk must be made and for $N > 6$, instability is highly likely. For an explicit understanding of the critical stability number, N_c, charts based on geotechnical centrifuge testing of tunnels with various geometries can be used. Given the ratios of cover to the tunnel crown divided by excavated diameter (C/D) and the unsupported tunnel length to excavated diameter (P/D), the charts (Figures 9.1 and 9.2) can readily be used to evaluate N_c. The ratio of N_c/N provides a factor of safety on tunnel instability. Considering the requirements of Eurocode 7 (BSI, 2013), tunnel stability would be considered as a geotechnical (GEO) load case and, as such, a minimum factor of safety of 1.4 would be compliant, considering the partial factors Design Approach 1, Combination 2 in the UK (the critical case). It is essential to note, however, that the lower the factor of safety, the higher the ground movements. It is hence prudent to opt for a higher factor of safety where control of ground movements is essential (e.g. in an urban environment), which could involve providing a face support pressure (e.g. using an earth pressure balance TBM), even where stable conditions are predicted.

When tunnelling through granular (coarse-grained) soils (e.g. sands), stability analysis is more complex. Broadly, failure is governed by the presence of water, the frictional properties of the soil and the tunnel diameter. Face support when tunnelling in such ground conditions is essential, with a positive pressure provided which slightly exceeds the prevailing groundwater pressure.

9.2.2 Assessment of ground loading on tunnels in temporary states

A classic methodology for estimating the loading conditions and support requirements for a tunnel in rock is the convergence–confinement or interaction analysis (Panet and Sulem, 2023).

Figure 9.1 General chart for critical stability number variation with tunnel heading geometry (reproduced with kind permission of Mair and Taylor, 1997)

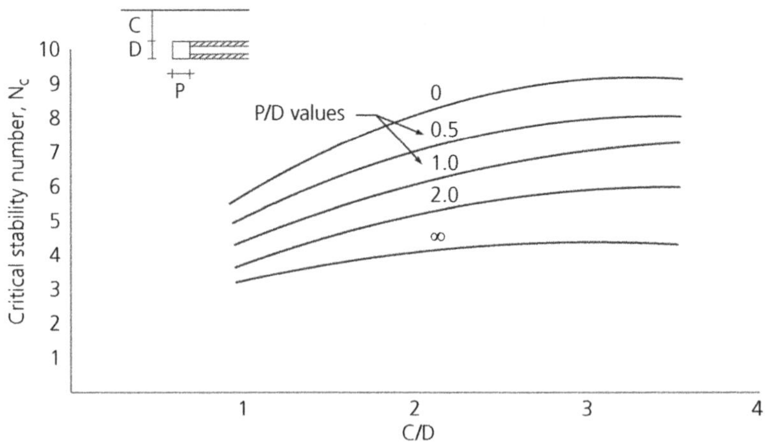

Figure 9.2 Detailed chart for critical stability number variation for lined tunnel headings (reproduced with kind permission of Mair and Taylor, 1997)

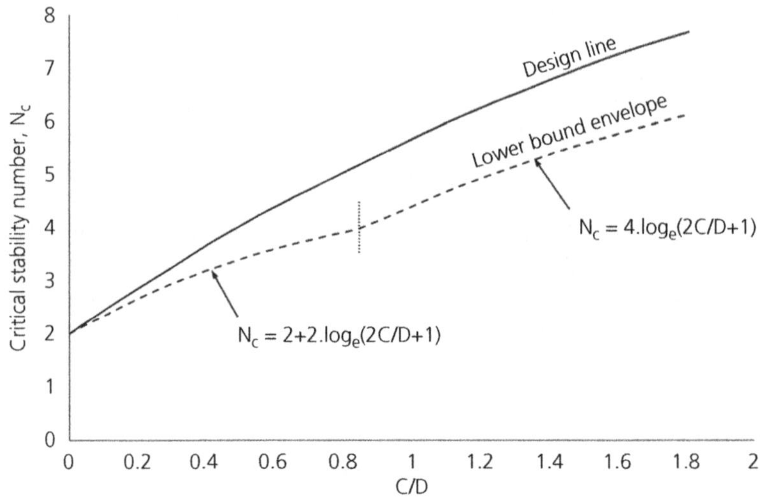

By considering a cross-section for an advancing tunnel heading, a ground reaction curve can be plotted of the radial support pressure (y-axis) which is required to limit the radial displacement at the tunnel periphery (x-axis) to a given value. This ground reaction curve can be generated through analytical solutions (assuming a particular ground behaviour) or by means of numerical analysis and specialist software. On the same graph, the load–displacement capacity of the support can be plotted, with the intersection point providing an estimate for equilibrium of the excavation (Figure 9.3). Interaction analysis highlights a key feature of tunnelling in rock, where delaying the installation of support measures will reduce the loading on the support, but at the expense of increased deformation. Engineering judgement is required as to the best approach given the ground conditions and the performance requirements for the tunnel.

In soft ground tunnelling, it is not uncommon for the lining to be designed for a full overburden loading condition in the long term. However, for temporary conditions, the designer may consider benefit from arching within the ground, reducing the loads within the lining. Empirical formulations based on the work of Terzaghi (1946) in Table 9.1 can be used to estimate a reduced loading on tunnels considering soil arching. In granular (coarse-grained) soils above the water table, the arching effect is dependent on the density of the soil, with denser sands / gravels / soils showing increased benefit (Figures 9.4 and 9.5). The benefits of ground arching cannot be relied on for shallow tunnels and full overburden load should be considered when the depth of cover is less than 1.5 $(B + H)$. In cohesive ground, the undrained shear strength of the soil can be used to calculate a reduced overburden load. Note that, in some cases, this may give a very small or even negative value for the load. In such cases, it is prudent to consider a minimum of 30% of the full overburden load for temporary works design purposes.

Adoption of loads less than full overburden requires the soil to form a natural arch above the tunnel. In areas where the natural ground may have been disturbed by previous activities (service trench etc.) or which is subject to vibration from traffic, as shown in Figure 9.6, it is recommended that full overburden loading is adopted.

Figure 9.3 Concept of tunnel interaction analysis (reproduced with kind permission of CECL Global)

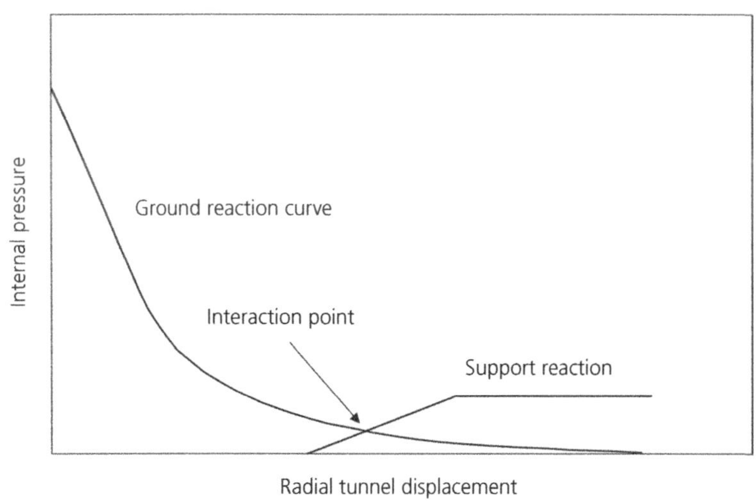

Table 9.1 Temporary loads acting on tunnels (after Terzaghi, 1946)

Temporary loads acting on tunnels		
Type of ground	Ground condition	Design load (m of overburden)
Granular (coarse-grained) material above the water table (Notes 1 and 2)	Loose	$0.5\ (B + H)$
	Medium	$0.4\ (B + H)$
	Dense	$0.3\ (B + H)$
Cohesive (fine-grained) material (Notes 2 and 3)	Homogenous	$\left(C + \frac{H}{2}\right) - \frac{2C_u\left(C + \frac{H}{2}\right)}{\gamma(B + H)}$

where:

C = depth of cover: m

B = tunnel width: m

H = tunnel height: m

C_u = undrained shear strength of ground (cohesive/fine-grained): kN/m^2

γ = bulk unit weight of ground: kN/m^3

Note 1: where the depth of cover is less than 1.5 $(B + H)$ full overburden depth should be used.

Note 2: a minimum design load of 0.3 $(C + H/2)$ should be considered in cohesive ground (m of overburden).

Note 3: the authors recommend that $H = B$ for circular tunnels.

Figure 9.4 Estimation of loads on tunnels considering arching in granular (coarse-grained) soils (reproduced with kind permission of CECL Global)

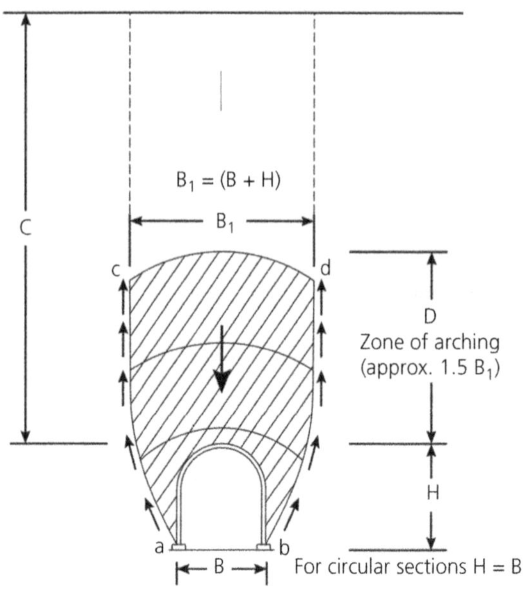

Figure 9.5 Estimation of loads on tunnels considering arching in cohesive (fine-grained) soils (reproduced with kind permission of CECL Global)

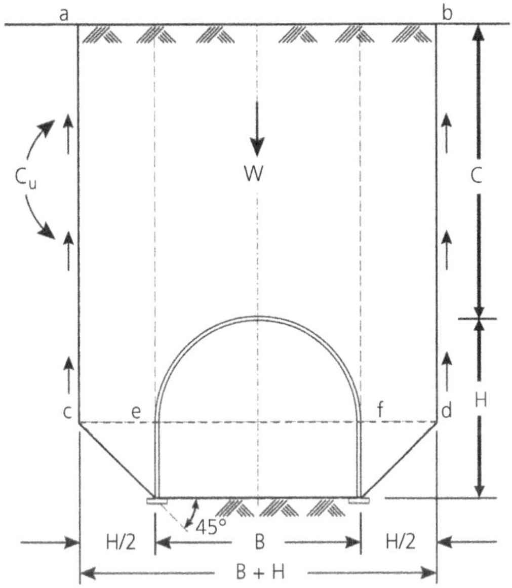

Assessing loads on tunnel supports can also be undertaken with numerical analysis methods. Full numerical analysis should be considered if the tunnel design situation is complex or structurally critical in nature. Techniques that fall under this category include finite element analysis; Mar (2002) has provided guidelines on its responsible application in practice. Finite element analysis is powerful and complex, and it is critical that it is performed correctly. It can provide detailed understanding and insight of factors that are important in design such as:

- stability
- deformation
- soil–structure interaction behaviour.

Tunnelling projects deal with nonlinear behaviour of both the ground and the structures in contact with it, so it is important to apply methods such as finite element analysis to complement the simplified and hence limited assessment that traditional / empirical methods can deliver. Full numerical analysis enables important soil–structure interaction effects to be explored, such as:

- soil property variation
- structural property variation (including complex geometries)
- groundwater conditions

Figure 9.6 Ground disturbance due to services and vibration from the road (reproduced with kind permission of CECL Global)

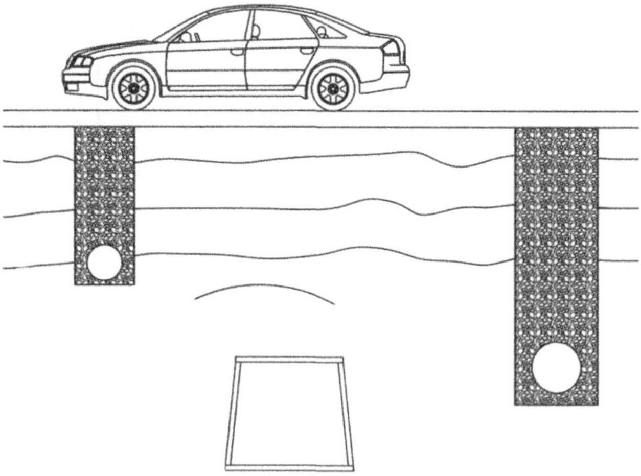

- external loading effects
- short- and long-term effects of the ground and the structures
- evaluation of stress and strain distribution in the ground and the structures to better understand load-transfer effects.

Numerical analysis should always be undertaken by an experienced engineer and the results must be critically reviewed.

9.3. Safety considerations

Due to the uncertainty and variability of the ground within which tunnels are built, and the fact that they are confined spaces, several specific risks exist for tunnel construction. These are particularly relevant for temporary works in tunnelling due to the dynamic nature of the tunnel construction process. BS 6164 (BSI, 2019) provides useful guidance for health and safety practices for tunnel construction. Key points to factor into temporary works design for tunnels are as follows.

- Access and egress – the nature of tunnel construction is such that the site layout is dynamic. Designers, both for permanent and temporary works, should ensure that suitable space is provided within the tunnel construction envelope for the necessary plant, equipment and materials, while ensuring safe access and egress for construction operatives. Digital modelling of the tunnel construction operation can assist with such space-proofing considerations.
- Manual handling – in the confined space environment of a tunnel, manual handling can be unavoidable. Attention should be given to the sequence of construction for any temporary works and limiting the loads which need to be manually handled. The Health

and Safety Executive (HSE) manual handling assessment charts (MAC) process can be used to understand the risks of manual handling (HSE, 2019).

- Ventilation – a minimum fresh air supply of 0.3 m^3/min per person is required and additional ventilation is needed to account for construction plant and equipment. Air quality is important, with oxygen levels needing to be kept above 19%. Where dust is generated (e.g. in sprayed concrete application), dust suppression requires forced ventilation for both fresh air blowing and dust extraction, along with filtration systems.
- Fire – fire in a tunnel can spread rapidly and smoke generated can quickly impede egress. It is therefore important to limit the use of combustible materials within tunnelling temporary works. Activities such as welding or cutting steel with an oxyacetylene torch should be avoided within a tunnel where possible. Timber members used in timber heading construction should be treated with fire-retardant before being used underground.
- Control of water – tunnelling may often take place below the water table and the design and sequencing of tunnel construction must make suitable provision to control water ingress. High-quality ground investigation is required to identify where risk of inundation during tunnelling may occur and to inform the design of appropriate control measures.

9.4. Mechanised tunnelling using tunnel boring machines

Recent decades have seen a rapid expansion in the mechanisation of tunnelling using tunnel boring machines (TBMs) which have the capacity to excavate the ground, temporarily support the tunnel and construct the tunnel supports. While these machines have the capability to fulfil several temporary works requirements for building a tunnel (e.g. using a shield to provide temporary support), the temporary works engineer is nonetheless required to support the process. Temporary load cases on tunnel lining segments, TBM launch, reception and turnaround, forming tunnel cross-passages and spoil conditioning / handing all require input from a temporary works perspective and are addressed in this section.

9.4.1 Temporary load cases for segmental tunnel linings

A common solution for the permanent support of bored tunnels is a lining formed from precast concrete segments. It is notable that the most onerous load cases for precast tunnel lining segments are often the temporary load cases associated with handling and erection, with damage to the segments during these phases potentially causing unacceptable durability issues for the tunnel lining. A precasting facility used for the manufacture of tunnel lining segments should aim to produce consistent, defect-free segments while maintaining a production rate which ensures an appropriate supply of segments to the tunnelling site. For productivity, a minimum characteristic compressive strength of the concrete at the time of demoulding should be discussed and agreed; a value of between 11 MPa and 15 MPa would be usual. A check should be undertaken to consider the required strength for the demoulding of segments using a vacuum lifting device (Figure 9.7), considering the geometry of the segment and the size of the vacuum lifting plate, with the self-weight load acting as a cantilever at each end of the segment.

Figure 9.7 Removing a tunnel lining segment from the mould with a vacuum lifter following casting (reproduced with kind permission of CECL Global)

The incorporation of cast-in lifting anchors into the circumferential joint of the segments can assist with safe handling in the production facility and on site. The loading on such anchors must consider:

- the segment weight, with attention paid to the curved segment geometry
- the number of anchors used; two per segment are usually provided on the circumferential joint edge
- dynamic loading – controlled lifting of segments at early age is recommended to minimise damage
- angle spread factor – greater load per anchor is considered when the angle on any lifting slings is increased. Specifying a vertical lift for segments at early age is prudent since this avoids shear loading of the anchors with a potential for corner damage.

Noting the curved geometry of tunnel lining segments, the anchors should be positioned in line with the centre of mass of the segment, ensuring that the segment hangs vertically when lifted.

Manufacturer's guidance can be used to specify suitable anchors. The guidance in BS EN 1992-4 (BSI, 2018) can also be used to explicitly calculate the anchor resistance for various failure modes. Segments will often be stacked following manufacture prior to transporting to site and / or stacked at the site prior to delivery to the TBM. Segments can be stacked either:

- on their 'edges' (i.e. on the circumferential joints)
- on their 'backs' (i.e. on the outside face of the segment).

In the case of the former, suitable packers should be placed on the loaded edges of the segments to avoid damage to the gaskets and segment corners. In the latter case, segments comprising one full ring are generally placed in a single stack. For this load case, the lowermost

Figure 9.8 Eccentricity to be considered when stacking segments on their back (reproduced with kind permission of CECL Global)

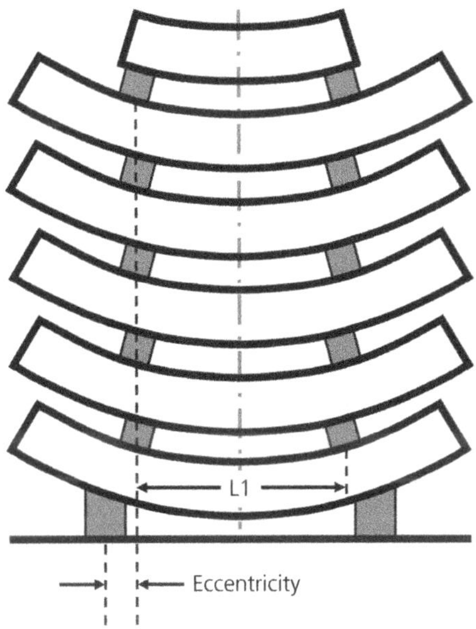

segment can be analysed as a simply supported beam with self-weight and the load of the upper segments taken as point loads with a maximum eccentricity of 100 mm. A peak tensile stress from such loading can be derived from elastic analysis and compared to the tensile strength of the concrete at the time of stacking (Figure 9.8).

Often the most onerous load case for a precast concrete segment comes during the shoving of the TBM off the last built ring. TBMs can exert substantial thrust forces and any build imperfections or dimensional tolerances within the precast rings themselves can result in substantial tensile stresses being developed. This condition needs to be carefully considered on a case-by-case basis; the designer will need to decide whether additional reinforcement may be required to prevent excessive cracking. For larger tunnel segments, a hybrid solution of steel fibre reinforced concrete (SFRC) together with glass fibre reinforced polymer (GFRP) bars on the perimeter of the segment can offer a robust and cost-effective solution.

9.4.2 TBM launch

The TBM requires a thrust reaction so the tunnelling process can advance. At the start of tunnelling, where no lining has been built, temporary works arrangements are required to provide the necessary thrust reaction as well as ensuring that the first tunnel rings can be built accurately, without in-plane misalignments that could cause damage to the segments and / or cause TBM steering difficulties at launch. Determination of the thrust force for design of the launch temporary works requires careful consideration. The full thrust capacity of the TBM shove rams would yield a worst-case upper bound for the total design load; however, launch structure design using such loads would likely lead to an uneconomic design. Realistic thrust forces at launch may likely be reduced compared to those which the TBM hydraulics have been designed for due to several factors.

- Geological conditions – understanding of the hydrogeological conditions at tunnel launch and the presence of any ground treatment will influence the estimation of the launch thrust loads.
- TBM backup gantries – TBMs may be launched in a 'short' mode (i.e. without their full complement of backup gantries). Reduced thrust may be warranted in such cases since the TBM does not need to initially tow the gantries.
- Skin friction on the shield – this is not present initially at launch but will develop as the TBM progresses.
- Skin friction on the initial grouted rings – as the TBM advances further from the tunnel portal, the thrust load will be carried by skin friction between the ground and the grouted tunnel rings. As an estimate, it could take around 30 rings before the thrust loads can be carried by the lining skin friction.

The distribution of loading is also of importance when designing TBM launch arrangements. In soft ground tunnelling, higher forces may be concentrated in the lowermost rams of the TBM to counteract the tendency for the machine to dip during advance. It is important to discuss the launch loads and the TBM launch protocol with the site team to ensure that assumptions about loads and load distribution used for temporary works design are reflected in the launch operation.

A critical factor in TBM launch temporary works design is the control of deflections. If the stiffness of the arrangement used is insufficient, then the initial rings that are built may be damaged due to misalignments and consequential unfavourable vectoring of forces. As a guide, the deflections should be limited to maintain planarity of the initial rings that are built and reduce lipping and stepping of adjacent segments. Monitoring during launch is prudent to provide information to the TBM operators, with verification that the movements are within the design envelope.

A common design solution for TBM launch is the use of a steel thrust frame (Figure 9.9). There is a wide variety in the geometric configuration of thrust frames used, with the form dependent on the particular constraints of the launch site. Temporary tunnel rings are built against the thrust frame which must be supported and then dismantled once the TBM advance has progressed sufficiently. Steel members in the thrust frame transfer the thrust loads into the launch shaft / box structure and temporary works designers should seek to understand the constraints of the permanent works with respect to how the launch loads can be transferred from the thrust frame and into the permanent structure (e.g. bearing capacity limitations). Consideration must also be given to how the thrust frame is to be assembled and disassembled. Decisions must be made as to the number of structural joints; fewer joints would result in a stiffer structure; however, member sizes and frame sections may be large in such cases and the logistics arrangements at the launch site may not be able to accommodate these.

Figure 9.9 Typical TBM launch frame arrangement (reproduced with kind permission of CECL Global)

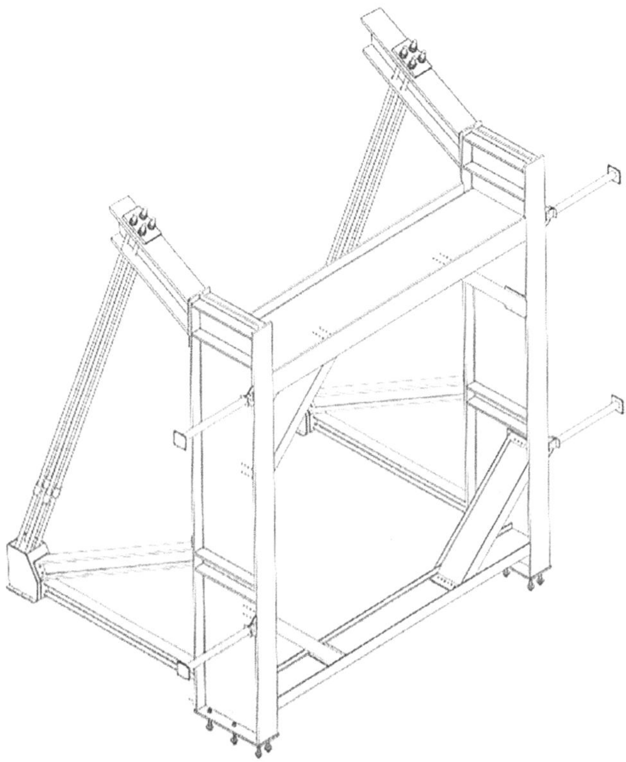

An alternative to a steel thrust frame is the fast or 'flying' launch system (Figure 9.10). Here, the launch thrust is provided by external hydraulic hollow-centre jacks which pull on high tensile steel bars connecting a fixed steel frame at the tunnel portal with a moving steel thrust ring at the rear of the TBM. The TBM's own thrust rams are locked off and all thrust control at launch is undertaken by the computer-controlled hollow-centre jacks. Once the TBM has fully entered the portal, the thrust ring is locked in position and the permanent tunnel ring building can commence. The system has some advantages over the use of a conventional thrust frame, with no temporary rings to build and support, allowing a potentially quicker launch. Stiffness is again a key issue for the temporary works designer to address in the fast launch system, with distortion of the thrust ring and the fixed frame needing to be limited. With the steel bars under high-tension loads, the factor of safety should be appropriate to prevent tensile failure. Robust quality assurance should exist for the high tensile bars and the load capacity of any couplers should be verified to ensure no local reduction in tensile capacity.

9.4.3 TBM portals at launch and reception

In shafts or box structures which are created for launching and receiving TBMs, allowance for the penetration of the TBM is made by incorporating a 'soft eye' into the wall. In

Figure 9.10 'Flying' launch arrangement (reproduced with kind permission of CECL Global)

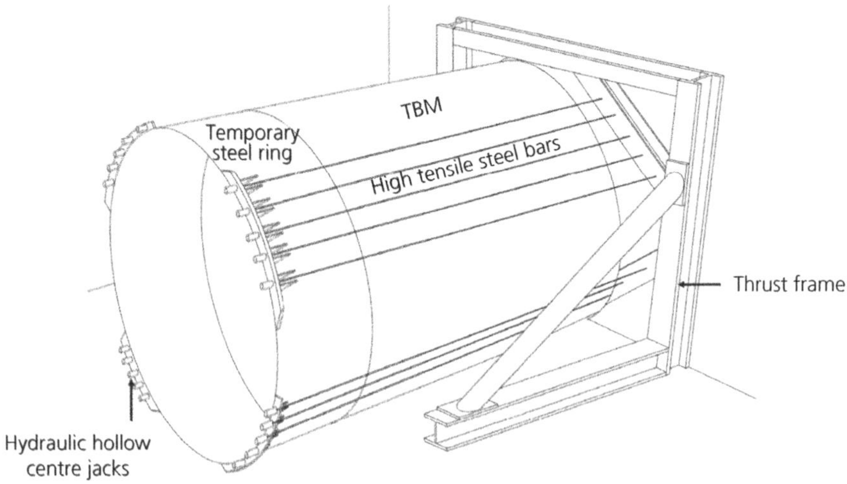

concrete-lined shafts, since steel reinforcing bars would cause significant damage to the TBM cutterhead and spoil handling systems, no steel is incorporated at soft eye locations. The tunnel eye can be locally unreinforced to allow the TBM to break through, with structural capacity of the shaft lining to be checked around the portal due to localised stress concentrations caused by the opening. Where reinforcement is structurally required in the shaft lining, this can be provided by glass fibre reinforced polymer (GFRP) rebar locally in the tunnel eye, which will disintegrate when excavated and avoid damage to the TBM systems. Designing concrete structures with GFRP requires caution since GFRP is less stiff and does not have the same ductility as steel rebar. Designers should note that while GFRP bars often have high tensile strengths (higher than steel), high factors of safety are often used in design to account for the lack of ductility and avoid failure. Given the reduced stiffness, deflections will be higher than for an equivalent structure with steel reinforcement. However, as the concrete in the portal eye is temporary, additional deflection and cracking in this location may be tolerable. Design guidance for designing with GFRP can be found in ACI (2015).

In water-bearing ground, the tunnel portal requires a sealing arrangement to be incorporated to provide a seal between the ground and the TBM shield during launch and reception. Rubber seals are incorporated into a circular steel ring which is either connected or cast into the portal. Control of deflections is critical for the sealing ring to ensure that it remains circular during installation.

To support the TBM at the invert during launch and reception, a cradle is provided. The cradle is commonly a temporary structure made from structural steel sections, but a curved invert section can be formed in concrete as part of the shaft base structure. The bearing capacity and ground movements beneath the TBM support need to be assessed, particularly where there is no base slab structure in the launch / reception pit.

Figure 9.11 Nitrogen skates supporting a TBM on a reception cradle during turnaround operations (reproduced with kind permission of PHL Hydraulics and Mammoet)

For twin tunnel drives, programme and cost may require the same TBM to be received and then re-launched from within the shaft. For this purpose, a nitrogen skate system may be employed (Figure 9.11). This system avoids risks associated with manoeuvring the TBM with a crane in a confined space and can allow for a quick turnaround of operations. The system uses a TBM cradle fitted with skates consisting of hydraulic feet with rubber skirts. The TBM is then raised by the hydraulic feet and compressed nitrogen (an inert gas) fills the base to create a cushion. The hydraulic supports are then switched off, with the TBM resting on the compressed nitrogen cushion. Steel plates placed on the base of the shaft facilitate a low-friction interface. A series of chain pulls anchored to the shaft wall are then used to manoeuvre the TBM into position for re-launching.

9.4.4 Spoil management

Bored tunnelling processes generate a significant quantity of spoil very quickly. It is important that this spoil is managed appropriately, from the point of excavation through to disposal, to ensure the efficiency of tunnelling operations and environmentally responsible handling and disposal. Hard ground TBMs will produce rock fragments which are relatively straightforward to handle and are often re-purposed as fill material (following characterisation and processing to ensure compliance with the relevant specifications). Bored tunnelling in soft ground, however, can present challenges due to the methods used for providing face support. Where coarse-grained soils are encountered, the face support is provided by a controlled slurry pressure, with the spoil being transported away from the face hydraulically in pipelines. The slurry used is generally a bentonite suspension where the bentonite supports the formation of a 'filter cake' on the excavated face, preventing the loss of slurry pressure. The spoil and slurry mixtures are then pumped to a slurry treatment plant (STP). The STP is required to remove soil particles from the slurry, through a combination of screens, cyclones and / or centrifuges, and rejuvenate the slurry before pumping it back to the tunnel face as part of the slurry circuit. The smooth operation of the STP is hence essential to productivity of the overall tunnelling process. Significant temporary works may be associated with slurry handling and the STP, with pipework supports, STP foundations, superstructure and processed spoil storage requiring careful design, particularly for constrained sites. Where earth pressure balance TBMs are used (for tunnelling in fine-grained soils), the aim is to create a plug of excavated material within the front chamber of the TBM which balances the ground and groundwater pressure. This pressure is controlled by an Archimedes screw which discharges the excavated material at a controlled rate such that the appropriate pressure is maintained. Soil on its own is generally not able to form a consistent plug of material so conditioning agents are added, either at the face or in the screw conveyor. Water, foams and polymers are added to reduce the stickiness of the soil, modify its flowability and reduce abrasion on the TBM's moving parts. Selection of the appropriate conditioning agents requires an understanding of the geotechnical parameters of the soil, particularly the particle size distribution, plasticity indices and natural water content.

9.4.5 Compressed air for ground support

When tunnelling below the water table, inflow of water into the excavation in weak or non-cohesive soils will rapidly create instability and collapse. Since the 19th century, engineers have employed compressed air to balance the water pressure to prevent infiltration of water. The technique was first used to construct bridge piers (e.g. the Golden Gate Bridge in San Francisco) and soon after was in regular use in tunnels around the world. While the early techniques employed were successful, workers suffered (and in many cases died) from what was then termed caisson disease, which is known today as decompression sickness. Stringent regulations are now in place to regulate exposure periods of compressed air workers and decompression procedures.

The use of compressed air and the design of the associated temporary works is highly specialist in its nature. Even at relatively low pressures, the stored energy in compressed air systems is enormous and the greatest care must be taken when designing components to these systems, including temporary bulkheads and air decks. For this reason, stringent regulations exist to ensure that the health and safety risks of compressed air workers are kept as low as reasonably practicable (ALARP). The utilisation of compressed air and the design of temporary works associated with compressed air working should only be undertaken by specialist contractors experienced in this work and should be in full compliance with standards and guidance published by the Health and Safety Executive (HSE) and the British Tunnelling Society (BTS). The authors consider that it would be imprudent to provide any technical guidance in this publication as this would be inevitably incomplete and therefore potentially dangerous.

9.4.6 Junctions and cross-passages

For cross-passages and junctions in bored tunnels, openings in the segmental lining need to be formed. Cutting an opening in a segmental lining involves locally breaking the ring structure that carries the ground and groundwater loading; therefore, another means of carrying the load must be found. A common approach is to implement a temporary steel frame arrangement

Figure 9.12 Opening set with cill beam connected to the segmental lining with shear dowels (reproduced with kind permission of CECL Global)

which can transfer the loads around the opening. Cill beams (Figure 9.12) are fixed with shear dowels above and below the opening, with soldier members between them to carry the load either side. Regarding design loading on the frame, ground arching means it is unlikely that the full loading from the segmental ring is transferred to the steel frame. Undertaking 3D finite element analysis where the construction sequence is modelled and reviewing against case history data will provide a more robust approach to determining the design loads. The shear dowels are critical components of this approach and hence their performance must be carefully assessed and verified. The capacity of the dowels should be checked for:

- steel failure (shearing of the dowel)
- edge failure (shearing of the concrete near a segment edge)
- pry-out failure (of the concrete).

BS EN 1992-4 (BSI, 2000) provides guidance on the design for the above conditions. In addition to design calculations, it is prudent to undertake testing on the shear dowel capacity, with a load test providing verification of the design. Where several cross-passages are required, the temporary steel frames have the potential to be reused.

Along with support for the segmental lining, consideration must be given to ensuring stability of the cross-passage excavation itself. It is very unusual for cross-passages to be excavated with a TBM (although there are a few case studies of this being done). Depending on the ground and groundwater conditions, it may be necessary to undertake ground improvement measures at the cross-passage location to allow local excavation and support of the ground using mined tunnelling techniques. Measures that could be used are:

- grouting
- ground freezing
- dewatering / depressurisation.

The creation of a cross-passage opening will inevitably cause disturbance to the gasket seals in the segmental tunnel lining. As this is the principal waterproofing measure in a segmental tunnel lining, the design of the cross-passage must incorporate appropriate waterproofing details. Communication between the temporary works designers and the permanent works designers is essential to ensure that such detailing is well thought out and incorporated into the sequence of works for cross-passage construction.

9.5. Mined tunnelling

9.5.1 Sprayed concrete linings (SCL)

- Sprayed concrete lining or SCL is commonly used to describe any construction methodology which uses sprayed concrete as a tunnel lining support. It has its origins in the New Austrian Tunnelling Method (NATM) which was developed as an observational excavation and support process for mined tunnelling in rock. SCL tunnelling methods are particularly suited for situations where more flexibility in geometry is required, such as tunnelling for metro stations, cross-passage construction and non-circular tunnels. In the UK, investigations following the high-profile collapse of an SCL tunnel at Heathrow Airport in 1994 led to the development of more robust processes being developed and implemented in practice. In modern SCL tunnelling design and construction, several principles should be applied.

- A risk-based approach to design and management is required.
- Design for each element of work should be fully developed before construction commences.
- Observation should be used to verify a design; it should not be used as a design method in its own right.
- An integrated approach should be taken for the design of permanent and temporary works.
- Every stage of excavation should be designed.
- The design should determine the monitoring regime and contingency actions.
- The achievement of quality is essential if an SCL project is to be successful.
- Buildability (ease of construction) should be considered in design and construction planning.
- Contingency plans and emergency procedures are required to deal with adverse events.
- Competence of the SCL team is crucial and should be assessed.

Traditional practice considers the sprayed concrete primary lining as temporary works, with a secondary lining (permanent works) designed for the full long-term loading and installed later. However, this has been recognised as inefficient and it is usual practice for modern SCL tunnels to be designed such that the primary lining contributes to the long-term load-carrying capacity. It is essential that the design philosophy and responsibilities between design teams are clarified at the outset to ensure that the design is coordinated, noting that the separation between temporary and permanent works design is particularly blurred for SCL construction. To ensure stability and limit ground movements, subdivision of the excavation face is undertaken in SCL construction, particularly for larger tunnels. This subdivision also reduces the work to be done in terms of excavation and support of each advance increment, thus providing earlier support and invert closure. Possible sequences for subdividing the excavation face include:

- top heading (crown), bench and invert
- sidewall drift and enlargement
- twin sidewall drifts with central core for enlargement
- pilot tunnel with enlargement (Figure 9.13).

The choice of sequence may be driven by programme, geometry and contractor's preference. There is evidence to suggest that the pilot-enlarge sequence is favourable in terms of reducing ground movement associated with SCL tunnelling, with the pilot tunnel acting like a large face dowel to reinforce the ground ahead of the enlargement. A pilot-enlarge sequence also will obviate the need for construction joints between sections of lining, with such joints being challenging to design and construct.

Historically, SCL construction has used lattice girders and steel mesh to provide control of the tunnel profile. However, this exposes operatives to unacceptable hazards from falling ground on unsupported excavation faces. Modern SCL techniques (such as LaserShell) use a combination of features to deliver safety and quality:

- real-time and remote profile control through specialist geospatial survey systems (e.g. TunnelBeamer)
- high-quality, steel fibre reinforced sprayed concrete
- a domed and inclined face providing protection for the operatives.

In design, it is essential to explicitly analyse each stage of the construction sequence, noting that some temporary stages of the sequence could be more onerous for the design than

Figure 9.13 Excavation of enlargement tunnel using the Lasershell approach for SCL construction (reproduced with kind permission of CECL Global)

the final condition. Prior to undertaking the design, a comprehensive design plan should be prepared that clearly defines the geotechnical parameters, lining material parameters, external loading and analysis methodology used. Structural analysis can then be undertaken with closed-form analytical solutions or with numerical analysis with assumptions. A comprehensive design analysis should incorporate a parametric study of key ground and structural parameters such that the design can be assessed against a range of conditions that could be encountered in the field. The design should incorporate appropriate factors of safety at each stage, with partial factors applied to unfavourable loads (or load effects) and partial factors on the sprayed concrete compressive strength. Design drawings must show all dimensions required to construct the works and explicitly detail every phase of construction and advance step. An appropriate allowance for construction tolerance must be included in the design. This tolerance value will be project-specific, depending on the nature and size of work. An allowance of between 50 mm and 100 mm on the radius dimension is not uncommon.

Prior to and during construction of SCL works, testing and monitoring is essential to validate the design. This is to ensure that behaviour is within the expected performance envelope and, where deviations are observed, to identify appropriate actions that must be taken.

Testing and inspection should include, but not be limited to:

- sprayed concrete testing regarding flexural strength, compressive strength and permeability; such materials testing should take place prior to commencement and during execution of the works
- as-built profile and thickness data
- geological face-logging
- surface and sub-surface ground movement monitoring data; this should include monitoring of the built environment which could be impacted by tunnelling operations
- lining deformation monitoring data.

The most reliable method of monitoring an SCL tunnel is by accurately measuring displacement of the lining at arrays along the length of a tunnel. Depending on the size of the tunnel and the urban setting, these arrays would normally consist of a minimum of five optical targets set at 5 m to 15 m intervals. The establishment of trigger levels is necessary for the interpretation of the monitoring results and effective control of the works. A clear distinction should be made between trigger levels set for the SCL tunnel deformations and trigger levels that may be set for the protection of the existing built environment. Trigger levels for the SCL are required to ensure the safety of the new construction and to reduce to as low as reasonably practicable the chances of instability. Trigger levels associated with the existing built environment are normally set at levels that are designed to prevent commercial loss (e.g. the prevention of cracking). In establishing appropriate trigger levels for the lining, the SCL designer must recognise that some delay in obtaining the first reading is inevitable. Following excavation and the application of the initial safety layer, the optical target support pins will be set and sprayed in. Following this the optical targets themselves will be set and baseline readings will be taken. It is inevitable that some small amount of movement will have occurred in the SCL lining before the base reading is taken and this should be taken into account in the establishment of appropriate target levels.

Traditionally, trigger levels have been in the form of absolute displacement (X, Y, Z) of a tunnel lining, irrespective of when or how the displacement occurred. More recently, however, it has become commonplace to set triggers which are based on strains in the tunnel lining and which also take account of the maturity of the lining. Shortly after construction, the effective stiffness of the sprayed concrete is low and thus the lining can tolerate more movement than a stiffer, mature lining. That is to say, a 1 mm displacement on a newly sprayed lining will produce far less in the way of stress than 1 mm displacement on a fully matured lining. The age at which the monitoring data is obtained is therefore important. Simple compressive strain values for each trigger horizon (Table 9.2) are suggested as a conservative, yet safe and efficient means of interpreting monitoring data in real time on a busy construction site. If the designer expects tensile strains to develop, appropriate triggers shall also be set.

During the implementation of an SCL design, it is important that a structured and rigorous approach to the review of all available data is undertaken. This normally takes the form of a daily review meeting at which key personnel from the project review all monitoring and quality data and agree on the required excavation and support sequence (RESS) for the next day's (or shift's) work. This review process ensures that further construction work is undertaken on an informed basis of the actual conditions encountered during construction.

9.5.2 Timbering in soft ground

The use of timber to support underground excavations has been common practice for hundreds, if not thousands, of years. Traditionally the 'design' of the timberwork would be undertaken by experienced lead miners and engineers and used reflective practice (i.e. no design calculations were undertaken). While the authors would strongly recommend consultation with miners and engineers with previous relevant experience, it is also nowadays necessary to undertake

Table 9.2 Trigger levels for SCL tunnelling

Trigger level	Factor of safety*	Compressive strain: %	Response
Green (predicted)	2.1	0.18	Do nothing – lining is behaving within the expected performance envelope in the design
Amber (trigger)	1.8	0.25	Increase the frequency of monitoring. Inform key personnel, including the designer
Red (action)	1.5	0.35	Suspend further excavation. Implement mitigations to arrest further movement (e.g. increase sprayed concrete thickness). Hold an emergency design review
Black (evacuate)	1.1	0.40	Implement emergency preparedness plan and evacuate all tunnel personnel and at-risk members of the public

*The design global factor of safety of 2.1 comprises a partial factor on ground loading of 1.4 and a partial factor on concrete compressive strength of 1.5.

calculations to demonstrate the adequacy of the proposed support for the prevailing conditions. Traditional timberwork solutions would in most cases not comply with modern timber design codes. This is why, when the BTS published *Traditional Timbering in Soft Ground* (BTS, 2014), they did not describe this as a design guide but 'a historical review'. Notwithstanding, this publication should be the starting point for any engineer tasked with designing a timber solution.

Clearly, the structural integrity of the timber is vital and only structural stress-graded timber to the stress classes in accordance with BS EN 338 (BSI, 2016) (structural timber) should be used. The minimum timber grade for use in tunnelling temporary works should be grade C24. Structural design of the timber elements should be undertaken in accordance with BS 5268-2 (BSI, 2002). Further useful guidance is provided in the Timber Research and Development Association (TRADA) publication *Timber in Excavations* (TRADA, 1990). While typical design values are provided for each grade of timber, a number of modification factors have to be used, depending on the application, environment and loading duration. For most timber headings, the authors recommend the assumption of wet exposure and short-term loading. Depth correction factors should also be applied.

The design process can be summarised as follows:

- Determine loading conditions from the ground (using Terzaghi (1946) or similar); see Figure 9.14.
- Reduce (by 50%) the uniform distributed loading of flexural members to allow for arching; see Figure 9.15. No reduction should, however, be taken when considering axial stresses.
- Calculate the permissible stress for the timber components (including modification factors).
- Check that the combined bending and axial stresses are within the permissible values.

Figure 9.14 Short-term loading action on a timber heading (reproduced with kind permission of CECL Global)

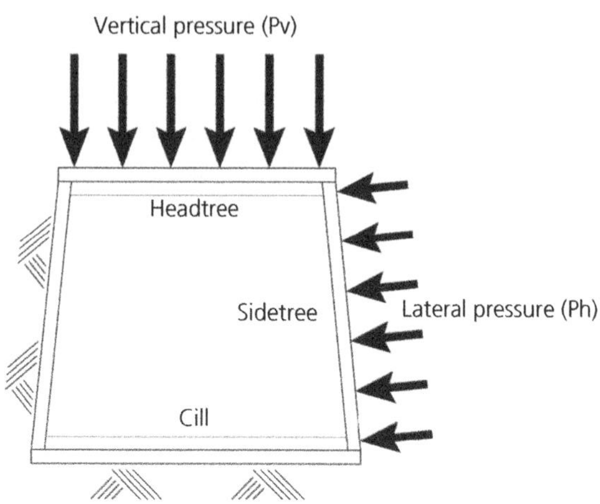

Figure 9.15 Load reduction on flexural members due to arching (reproduced with kind permission of CECL Global)

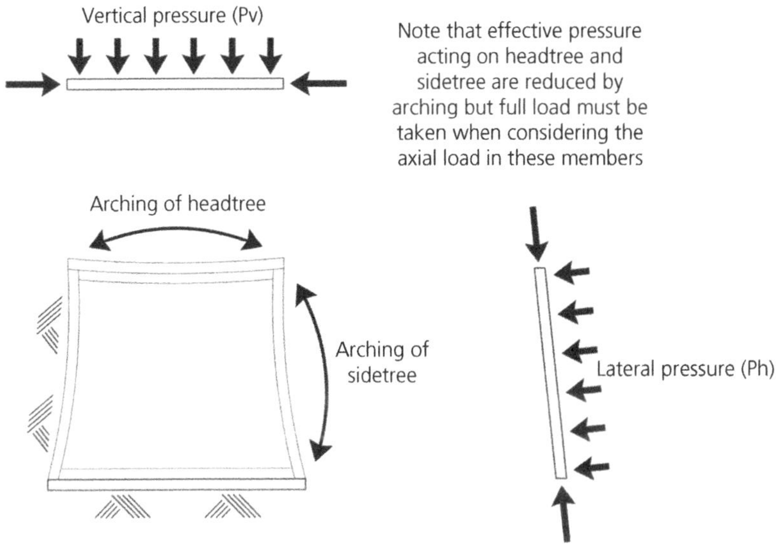

9.5.3 Use of ground improvement techniques

In certain ground conditions, there may be a need to undertake ground improvement measures before mined tunnelling methods can be used. This may be due to the ground having a low stand-up time and / or being too permeable for open-faced tunnelling to progress safely. The three main ground-improvement methods to support tunnel construction are:

- grouting
- ground freezing
- dewatering.

9.5.3.1 Grouting

Grouting is a permanent means of ground improvement whereby an engineered grout material is injected into the ground to improve its strength / stiffness and / or reduce its permeability. Using a Tube-a-Manchette (TAM), grout can be delivered to a specific location in the soil body, making grouting a more targeted technique compared to other ground improvement and ground treatment techniques. In a TAM, inflatable packers are used to isolate specific grout port locations, allowing a precise location to be grouted.

Permeation grouting is used in coarse-grained, permeable soils (e.g. sand and gravels) where the grout (usually a microfine cement-based grout) is injected to fill void spaces in the ground (reducing permeability) and cement the soil together (increasing strength). This technique can also be used as a pre-treatment option to improve outcomes for ground freezing operations.

To mitigate the impact of tunnel-induced settlements in urban areas, compensation grouting can be used. Here, arrays of TAMs are drilled and installed below sensitive buildings or assets which are being tunnelled underneath, with grout injected at specific locations to compensate for the stress relief caused by tunnelling. The grouting undertaken by this method is displacement grouting whereby primarily one of two processes is used, depending on the ground conditions.

- Compaction grouting – in coarse-grained soils a thick, low-slump cementitious grout is used to expand a bulb of grout around the injection point.
- Fracture grouting – in fine-grained soils (e.g. clays) a low-viscosity grout is injected at a pressure which is enough to fissure the ground and create lenses of grout. Once hardened, it is possible to make successive injections at the same grouting location.

TAMs for compensation grouting arrays can be drilled and installed from nearby shaft / box excavations or tunnels. For a successful compensation grouting operation, pre-treatment grouting fills voids and reverses any ground loosening due to the TAM installation. Grouting is then undertaken concurrently with the tunnelling operation. This requires understanding of the predicted tunnelling ground movements and optimisation of the grouting to effectively compensate for this movement. For calculating the required grout volume, a grout efficiency factor (GEF) is used:

$$\text{GEF} = \text{heave volume} / \text{injected volume} \tag{9.3}$$

The GEF for stiff clays is between 0.3 and 0.5; for coarse-grained soils, it is around 0.1. The pre-treatment stage allows an evaluation of actual GEF for each grouting point. Concurrent compensation grouting with tunnelling also requires a grouting exclusion zone (which moves as the tunnel is advanced) to be enforced to mitigate the risk of stability issues from grouting operations. For SCL tunnels, a suitable exclusion zone will also avoid overstressing the immature sprayed concrete near the tunnel face. Throughout grouting and tunnelling operations, real-time monitoring of the ground movement data is necessary to control the grouting operations and inform tunnel construction. While it is preferable to undertake compensation grouting concurrently, sometimes an additional corrective grouting (or grout jacking) phase is required following the passage of the tunnel to ensure that movements are within the required limits.

9.5.3.2 Ground freezing

Ground freezing is a temporary means of ground treatment where heat is extracted from the ground to turn the groundwater into ice. This increases the strength of the ground and immobilises the groundwater such that mined tunnelling can be undertaken safely. The method is more expensive than grouting but is a useful alternative where grouting is unsuitable for the ground conditions (e.g. in sandy ground below the water table).

The principle of ground freezing for tunnelling is to drill a series of freeze pipes around the intended excavation and then circulate coolant in the freeze pipes to create a continuous ring of frozen ground within which the tunnel can be excavated. Two options exist for freezing systems.

- Brine – refrigerated brine (a solution of water and calcium chloride salt) is circulated down an inner tube in the freeze pipes which then removes heat from the ground as it returns back through the annulus between the freeze pipe casing and the inner pipe. A refrigeration plant then re-chills the brine before recirculating it.

- Nitrogen – liquid nitrogen is delivered to site and stored in tanks. This is sent down a copper inner tube in the freeze pipe and then returned in the outer annulus. As it is returned, the liquid nitrogen changes state into nitrogen gas which can then be vented to the atmosphere.

The decision to use freezing and which option to use is dependent on the ground conditions. Loose soils with voids, high porosity and a high water content will take much longer to freeze and require more energy. In some cases, pre-grouting can be undertaken to reduce the porosity of the soil to improve the efficacy of the freezing process. If freezing is required rapidly (e.g. for emergency stabilisation measures), then nitrogen systems will be quicker. However, there are safety risks when using nitrogen in confined spaces and atmospheric monitoring and alarm systems must be used to ensure that nitrogen levels in the tunnel atmosphere are not too high. Where the groundwater at the location to be frozen is moving, this can reduce the efficiency of freezing systems, with brine systems able to freeze where the groundwater velocity is less than 2 m/day and nitrogen systems able to freeze at higher velocities (less than 15 m/day).

In many cases, freezing and maintaining the temperature of the ground to below –10°C will provide a reliable increase in the compressive strength of the ground. By modelling this zone of increased strength (e.g. in numerical analysis), the tunnel can be designed in the temporary state. Testing to ascertain the strength of the frozen ground should be undertaken to facilitate this. Note that where the groundwater is saline, the freezing point will be lower and a lower temperature will be necessary to ensure competency of the freeze wall. Again, testing of the ground will inform this. During construction, monitoring of the ground temperatures is required to check that the freeze is being adequately maintained.

To calculate the energy requirements for freezing, the volume of frozen ground needs to be estimated along with the water content. Most of the energy (around 60%) required for freezing is needed to change the groundwater from a liquid into ice (latent energy). The latent heat of water is 334 MJ/m^3; hence, by knowing the volume of water in the soil which is to be frozen and dividing this by 0.6, a reasonable estimate of the energy required for creating the freeze can be obtained. The power output of the brine refrigeration units used can then give an estimate of the time to freeze in a brine system. For a nitrogen system, one cubic metre of liquid nitrogen can approximately remove 250 MJ of heat from the ground, allowing estimates of the volume required. Additional energy will be required to maintain the freeze during construction. Frost heave may be an issue during ground freezing operations, with ice occupying approximately 9% more volume than water. A first estimate of the volume increase, ΔV, associated with freezing can be found as follows:

$$\Delta V = 0.09 \times n \times V_{\text{frozen}} \tag{9.4}$$

where n is the porosity of the ground and V_{frozen} is the volume of the frozen soil.

Around a tunnel, this volume increase gives a 'reverse' volume loss which can then be used with the Gaussian solution for tunnel settlements (O'Reilly and New, 1982) to estimate the heave effects. Frost heave is more problematic in fine-grained soils (in the silt region) since coarse-gained soils have lower frost susceptibility due to drainage.

9.5.3.3 Dewatering

Where groundwater pressures and flows are significant, dewatering techniques may be employed to temporarily reduce these to enable construction to take place safely. In tunnelling, this may be useful for cross-passage construction, shaft sinking and during TBM / launch reception. Sump pumping is one of the simplest forms of dewatering, where groundwater is allowed to flow into an excavation with a submersible pump installed in a sump at the base of the excavation to remove water that has flowed in. Sump pumps must always be submerged when in operation. Where inflows are high, fines from the soil may be pumped which may cause instability of the excavation base.

Deep-well dewatering systems are used where drawdowns of over 6 m are required and are generally the optimum choice for dewatering of deep, thick aquifers with a permeability $K > 10^{-6}$ m/s. Multi-stage submersible pumps are installed in lined boreholes, with a permeable filter pack and well-screen installed in the target aquifer. A cone of depression around the well is formed as water is pumped out of the deep-well; the intersecting cones around a shaft excavation can permit shaft construction and tunnel breakouts in a dry condition. Depending on the depth and ground conditions, deep-wells can cause drawdown over a large area and consideration must be given to ground movements which may arise from water extraction.

Wellpoint systems use short boreholes connected to suction pumps to provide drawdowns of less than 6 m. They are relatively quick and inexpensive to install and, due to their compact nature, they are highly suitable for in-tunnel installations. Ejector systems are used where the ground has lower permeability (down to $K = 10^{-7}$ m/s). In these systems, high-pressure water is used to create a vacuum in the well (by way of the Venturi effect). These systems have the advantage of being able to work with low volumes and can pump air and water mixtures. However, they are more expensive than wellpoint systems.

9.6. Secondary linings for tunnels

Following excavation and primary support, there may be a requirement to build a secondary tunnel lining. Such linings are usually constructed using cast in situ concrete and may be required for long-term ground and groundwater support or for specific client requirements, such as tunnel profile or long-term durability. A reuseable, travelling formwork arrangement is generally used to cast the secondary lining in sections along the tunnel. This may be a two-stage process, where a concrete section in the invert is initially cast and then followed by an arch section (Figure 9.16), or a single-stage process, where the secondary lining is cast as a full-round process (Figure 9.17). The pressure of fresh concrete drives the load conditions on the shutter and hence the design of the travelling shutter is dependent on a combination of several factors:

- geometry of pour
- rate of pour
- concrete mix design
- ambient temperature.

CIRIA report 108 (CIRIA, 1985) provides guidance on design formwork pressures. For secondary lining shutters, uplift is a unique load case which must be checked in the design. In a

Figure 9.16 Travelling shutter for secondary lining casting above a pre-formed invert (reproduced with kind permission of CECL Global)

Figure 9.17 Full-round secondary lining shutter (reproduced with kind permission of CECL Global)

circular tunnel, pressure from fresh concrete acting below the axis level of the shutter could lead to flotation. Checking of this design condition should include conservative estimates of the shutter self-weight (the stabilising load) and the concrete pressure (the destabilising load). The travelling shutter design may need to incorporate uplift restraints (e.g. spud bars) to resist flotation. Where self-compacting concrete (SCC) is used, such restraints may need to be left in place until the concrete has cured, leaving voids in the secondary lining which would require a further operation to fill and repair. This should be avoided and the temporary works designer should specify the maximum acceptable rate of rise within the shutter such that uplift pressure is controlled. The use of load cells on the spud bars in the crown is also recommended to monitor the uplift forces in real time to inform the control of the concreting process. The locations of concrete pumping port positions and integrated formwork vibrators should ensure good coverage and allow even distribution of concrete during the pour, with the aim of avoiding out-of-balance loading.

To achieve optimum quality of the formed secondary lining, consideration must be given to minimising the effects of restrained cracking. The striking times of the shutter should be optimised to control the temperature of the concrete and limit thermally induced cracking. To minimise cracking from drying shrinkage of the concrete, consideration should be given to an appropriate curing arrangement, possible by means of a curing gantry following the shutter, facilitating control of the temperature and humidity at the cast surface.

9.7. Shaft sinking for tunnels

There are many ways in which vertical shafts can be formed, including:

- segmental precast concrete rings built by underpinning
- segmental precast concrete rings built by caisson sinking
- in situ caisson
- sheet-piled cofferdam
- secant piles
- contiguous piles
- diaphragm wall.

Guidance on the piling methods and diaphragm wall method can be found in Pallett and Filip (2019, chapters 13 and 14). This section shall concentrate on caisson sinking methods as this has related design issues for which there exists little guidance.

Caisson shafts can be constructed from precast concrete segments or, more rarely, can be of in situ reinforced concrete construction. Typically, they range from 3 m to 25 m internal diameter, but much larger (up to 65 m diameter) have been constructed for applications such as bridge piers. While the technique is robust, careful attention to detail is required to ensure that the shaft is constructed to the full depth, within acceptable construction tolerances and without damage. Temporary works plays a vital role in the success of any caisson operation and practitioners have found to their cost that trying to cut back on these measures can prove to be a false economy. Caissons can be sunk ‘dry’ or ‘wet’ or indeed a combination of the two where the ground conditions change with depth. ‘Dry’ means that the shaft formation level is always accessible and groundwater inflows are either slight due to the ground conditions

Figure 9.18 Typical caisson set-up (dry) (reproduced with kind permission of CECL Global)

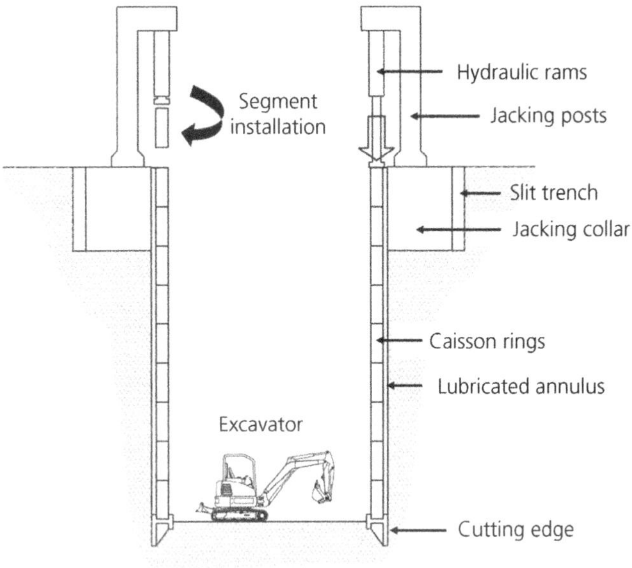

or are controlled by sump pumping or external dewatering (Figure 9.18). Where it is neither practical nor desirable to attempt to control groundwater, the shaft can be maintained in a flooded state such that the level of the water inside the shaft at least balances the water level outside. In this 'wet' condition, it is important that the external level does not exceed the level inside the shaft as this will result in flow into the shaft which can cause ground loss and settlement of the surrounding ground or, at worst, total collapse. In wet caisson construction, excavation is normally performed using a grab (either crane-mounted or pole-mounted on a long-reach excavator). If the caisson is still in a wet condition at the base level, it is advisable that divers are utilised to ensure that the formation is clear of debris before the concrete base can be tremied (not placed). Groundwater inflows can also be controlled using compressed air. The three key aspects for success of a caisson sinking operation are:

- cutting edge
- lubrication system
- thrust and steering system.

9.7.1 Cutting edge

To maintain the shape of the caisson during sinking and to enable harder ground to be penetrated, it is strongly advised to use a robust, purpose-designed steel cutting edge. Precast concrete cutting edges are provided by some of the manufacturers, but these should only be considered for very small and shallow caissons as they generally lack the strength and stiffness required to maintain the shape of the shaft. Typically, the cutting edges will be manufactured off site in segments from steel plates ranging from 30 mm to 75 mm. The steel segments

Figure 9.19 Typical steel cutting edge arrangement (reproduced with kind permission of CECL Global)

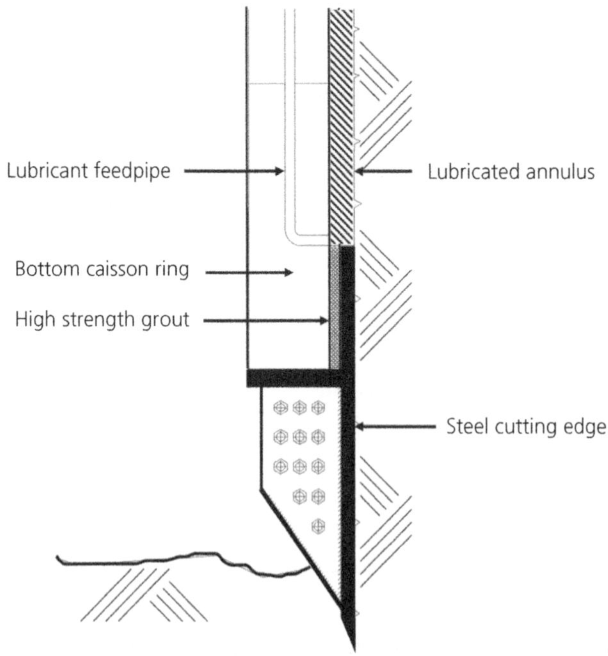

are then assembled on site by bolting together and then preferably finally welded together for added rigidity. Welding of the segments does, however, make the subsequent removal more difficult and, in these circumstances, consideration can be given to sacrificing the cutting edge by burying it beneath the base slab. If the cutting edge is to be removed, fabricated pockets should be provided to enable it to be jacked apart from the shaft above. During sinking, it is highly desirable to keep the knife of the cutting edge always buried to prevent the lubricant from escaping from the annulus into the shaft. Such a loss leads to a significant increase in the sinking effort and failure to reach full depth. A typical cutting edge (an example from an 11 m ID shaft, 55 m deep) is shown in Figure 9.19.

9.7.2 Lubrication system

The lubrication system is vital to limit drag (friction and / or adhesion) with the ground. An annulus of between 50 mm and 100 mm is created by the cutting edge around the caisson and this annulus much be kept continuously full and pressurised with a suitable lubricant (normally a bentonite slurry). It is vital that the lubricant is supplied just above the cutting edge. Simply trying to top the system up (essentially using gravity) is highly ineffectual and has led to the failure of many caissons. The prevailing ground conditions will inform the optimum choice of lubricant, with many products now available on the market. For many caissons, a relatively simple 5% bentonite lubricant will be sufficient. In other cases, it may be necessary to use materials suited to the prevailing environmental conditions (e.g. pure Wyoming bentonite is suitable for use in most aquifers) or more problematic (open) ground conditions. Like the engineering considerations given to a slurry wall or a D-wall panel excavation, it is important that the lubricant is not lost to the ground. In very coarse granular conditions, for instance, it may be necessary to increase the dosage of bentonite or use specialist polymers to improve viscosity (Milligan, 2000). It is also important that bentonite is properly hydrated before it is placed. This will require it to be mixed several hours before and stored in suitable holding tanks.

9.7.3 Jacking force

Inadequate lubrication systems and under-designed cutting edges can be blamed for many failures of caissons. If, however, the advice given above is followed, then jacking forces should be acceptable and should allow caissons to be sunk to a depth of over 50 m. Jacking force data from a deep caisson is provided in Figure 9.20. The caisson had an external diameter of 11.7 m (350 mm-thick segments) and was sunk entirely in London clay. The caisson was fitted with a robust steel cutting edge (Figure 9.19) and was continuously lubricated with a bentonite feed pipe down to the cutting edge. When corrected to account for the beneficial self-weight of the concrete caisson rings, the effective jacking effort was a modest 1.6 kN/m^2.

Figure 9.20 Measured jacking force (11 m internal diameter caisson in London clay) (reproduced with kind permission of CECL Global)

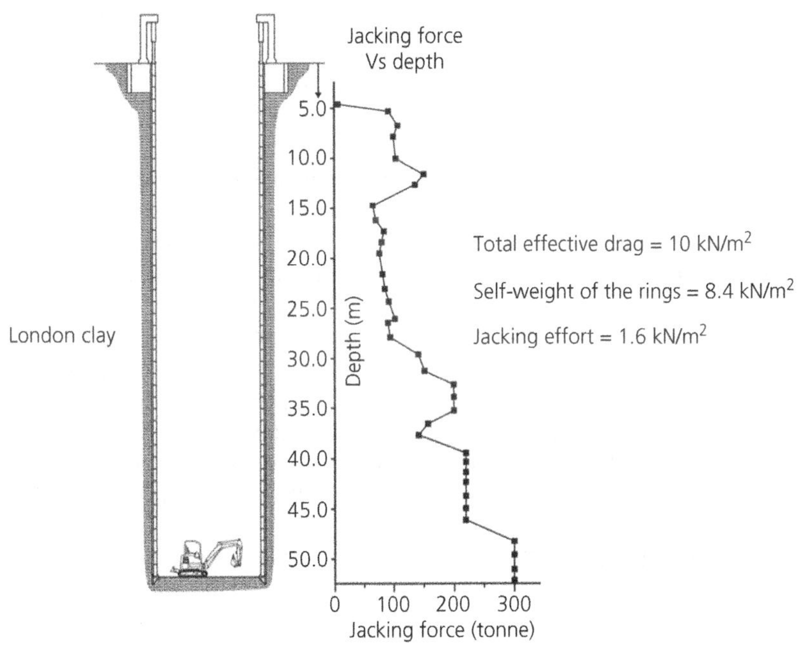

REFERENCES

ACI (American Concrete Institute) (2015) *ACI 440.1R-15: Guide for the Design and Construction of Structural Concrete Reinforced with Fibre-reinforced Polymer (FRP) Bars*. ACI, Farmington Hills, MI, USA.

Barton N (2002) Some new Q-value correlations to assist in site characterization and tunnel design. *International Journal of Rock Mechanics and Mining Sciences* **39**: 185–216.

Barton N, Lien R and Lunde J (1974) Engineering classification of rock masses for the design of tunnel support. *Rock Mechanics and Rock Engineering* **6(4)**: 189–236.

Bieniawski ZT (1973) Engineering classification of jointed rock masses. *Transaction of the South African Institution of Civil Engineers* **15**: 335–344.

Bieniawski ZT (1989) *Engineering Rock Mass Classifications*. Wiley, New York, NY, USA.

BSI (2002) BS 5268-2: 2002: Structural use of timber – Code of practice for permissible stress design, materials and workmanship. BSI, London, UK.

BSI (2013) BS EN 1997-2: Eurocode 7: Geotechnical design. Ground investigation and testing. BSI, London, UK.

BSI (2016) BS EN 338: 2016-TC: Structural timber. Strength classes. BSI, London, UK.

BSI (2018) BS EN 1992-4: 2018: Eurocode 2: Design of concrete structures – Design of fastenings for use in concrete. BSI, London, UK.

BSI (2019) BS 6164: 2019: Health and safety in tunnelling in the construction industry. Code of practice. BSI, London, UK.

BTS (British Tunnelling Society) (2014) *Traditional Timbering in Soft Ground Tunnelling – A Historical Review*. BTS, London, UK.

CIRIA (Construction Industry Research and Information Association) (1985) *R108: Concrete Pressures on Formwork*. CIRIA, London, UK.

Hoek E and Brown ET (2019) The Hoek–Brown failure criterion and GSI–2018 edition. *Journal of Rock Mechanics and Geotechnical Engineering* **11(3)**: 445–463.

HSE (Health and Safety Executive) (2019) Manual handling assessment charts (the MAC tool). https://www.hse.gov.uk/pubns/indg383.pdf (accessed 28/08/2024).

Mar A (2002) *How to Undertake Finite Element Based Geotechnical Analysis*. NAFEMS, London, UK.

Marinos P and Hoek E (2000) GSI: A geological friendly tool for rock mass strength estimation. *Proceedings of the GeoEng 2000 at the International Conference on Geotechnical and Geological Engineering, Melbourne, Australia*, pp. 1422–1446.

Milligan G (2000) Lubrication and soil conditioning in tunnelling, pipe jacking and micro-tunnelling – A state of the art review. Geotechnical Consulting Group (GCG). https://pipejacking.org/assets/pj/uploads/docs/Report_Milligan_2000.pdf (accessed 08/07/2024).

NGI (Norwegian Geotechnical Institute) (2022) Using the Q-system. Rock mass classification and Support Design. https://www.ngi.no/en/research-and-consulting/infrastructure-container/tunnels-and-the-q-system/ (accessed 08/07/2024).

O'Reilly MP and New BM (1982) Settlements above tunnels in the United Kingdom – their magnitude and prediction. *Proceedings of the Tunnelling 1982 Symposium, Institution of Mining and Metallurgy, London, UK*, pp. 173–181.

Pallett PF and Filip RK (2019) *Temporary Works: Principles of Design and Construction*, 2nd edn. ICE Publishing, London, UK.

Panet M and Sulem J (2023) *Convergence-Confinement Method for Tunnel Design*. Springer, Cham, Switzerland.

Terzaghi K (1946) Rock defects and loads on tunnel supports. In *Rock Tunnelling with Steel Supports* (RV Proctor and TL White (eds)). Commercial Shearing and Stamping, Youngstown, OH, USA, pp. 17–99.

TRADA (Timber Research and Development Association) (1990) *Timber in Excavations*, 3rd edn. Thomas Telford, London, UK.

USEFUL WEBSITES

ACI (American Concrete Institute) – www.concrete.org

BSI (British Standards Institution) – www.bsigroup.com

BTS (British Tunnelling Society) – www.britishtunnelling.com

CIRIA (Construction Industry Research and Information Association) – www.ciria.org

EFNARC (European Federation of Producers and Applicators of Specialist Products for Structures) – www.efnarc.org

GSL (Geological Society of London) – www.geolsoc.org.uk

HSE (Health and Safety Executive) – www.hse.gov.uk

ITA (International Tunnelling Association) – www.ita-aites.org

NGI (Norwegian Geotechnical Institute) – www.ngi.no

TWf (Temporary Works Forum) – www.twforum.org.uk

FURTHER READING

British Tunnelling Society (2011) *Monitoring Underground Construction: A Best Practice Guide*. ICE Publishing, London, UK.

Brown DA (2004) Hull wastewater flow transfer tunnel: Recovery of tunnel collapse by ground freezing. *Proceedings of the Institution of Civil Engineers – Geotechnical Engineering* **157(GE2)**: 77–83.

BTS (British Tunnelling Society) (2005) *Closed-Face Tunnelling Machines and Ground Stability: A Guideline for Best Practice*. ICE Publishing, London, UK.

BTS (2021) *Guidance on Good Practice for Work in Compressed Air*. British Tunnelling Society Compressed Air Working Group. https://britishtunnelling.com/pages/work-in-compressed-air (accessed 07/01/2025)

BTS (2024) *Specification for Tunnelling*, 4th edn. ICE Publishing, London, UK.

Cashman PM and Preene M (2020) *Groundwater Lowering in Construction*, 3rd edn. CRC Press, London, UK.

Chapman DN, Metje N and Stärk A (2017) *Introduction to Tunnel Construction*, 2nd edn. CRC Press, London, UK.

Eddie C and Neumann C (2003) LaserShell™ leads the way for SCL tunnels. *Tunnels and Tunnelling*, June 2003, pp. 38–42.

EFNARC (European Federation of Producers and Applicators of Specialist Products for Structures) (1996) *European Specification for Sprayed Concrete*. EFNARC, Farnham, UK.

HSE (Health and Safety Executive) (2000) *The Collapse of NATM tunnels at Heathrow Airport*. HSE Books, Sudbury, UK.

ICE (Institution of Civil Engineers) (1996) *ICE Design and Practice Guide: Sprayed Concrete Linings (NATM) for Tunnels in Soft Ground*. Thomas Telford, London, UK.

ICE (2022) *UK Specification for Ground Investigation*, 3rd edn. ICE Publishing, London, UK.

ITA (International Tunnelling Association) (2015) ITA-AITES Report Number 10: Guidelines for Good Working Practice in High Pressure Compressed Air. https://about.ita-aites.org/publications/wg-publications/1187/guidelines-for-good-working-practice-in-high-pressure-compressed-air-version-2015 (accessed 07/01/2025).

Maidl B, Herrenknecht M, Maidl U and Wehrmeyer G (2012) *Mechanised Shield Tunnelling*. Ernst & Sohn, Berlin, Germany.

Mair RJ and Taylor RN (1997). Theme lecture: Bored tunnelling in the urban environment. *Proceedings of the 14th International Conference on Soil Mechanics and Foundation Engineering, Hamburg, Germany*, **4**: 2353–2385.

Meda A, Rinaldi Z, Spagnuolo S and Eddie E (2021) Behaviour of FRC segments with GFRP cage under TBM thrust in presence of GAPs. *Tunnelling and Underground Space Technology* **107**.

Preene M, Roberts TOL and Powrie W (2016) *C750: Groundwater Control: Design and Practice*, 2nd edn. CIRIA, London, UK.

Puller M and Puller D (2016) *Deep Excavations: A Practical Manual*, 3rd edn. ICE Publishing, London, UK.

Ray Filip
ISBN 978-1-83608-525-6
https://doi.org/10.1108/978-1-83608-524-920251011

Chapter 10
Nuclear projects

Ronan O'Driscoll
EDF Energy Ltd, Hinkley Point C Project, UK

10.1. Introduction

A nation requires a mix of energy sources to serve the population and industry. Fossil fuels are not an inexhaustible resource and cause pollution. Currently, renewable sources such as wind, wave and solar are not totally reliable and may not be able to meet peak demands on their own without major lifestyle changes to reduce consumption. Fossil fuel power stations using coal or gas provide a base load of electricity for the national power grid, and renewable forms of energy cannot currently do this. Other low-carbon energy sources, such as nuclear, can provide this base load of power supply and remain one of the cleanest options for a base load on the grid for the time being, especially with potential for future technological development. Hinkley Point C (HPC) nuclear power station is under construction in Southwest England, and the UK currently has plans to build several other plants. These massive projects require extensive and complex use of temporary works, with management systems appropriate to the nature of the work. Safety (both conventional and nuclear), security and assurance are paramount considerations for these projects.

Logistics can be particularly challenging as nuclear plants tend to be in remote locations and transportation along country lanes can be problematic. Assurance of the finished product is key, with additional levels of testing, supervision and inspection. Further challenges involve providing facilities for workers and security, including vetting of those involved in the project from design to construction. These issues would be over and above what would be expected for a conventional large civil engineering project.

Existing power stations require regular maintenance, upgrading and eventual decommissioning or demolition, and these operations also require extensive and complex temporary works.

Temporary works are described in BS 5975-1:2024 (BSI, 2024), clause 4.1.1.1 as providing an 'engineered solution' that is used to:

- support or protect either an existing structure or the permanent works during construction
- support an item of plant or equipment or the vertical sides or side-slopes of an excavation during construction operations on site
- provide access
- control stability, strength, deflection, fatigue, geotechnical effects and hydraulic effects within defined limits.

The 'engineering' refers to civil, structural or mechanical design.

This definition is well known in the civil engineering and building sectors; where the term 'temporary works' originates from is not described here – further explanation can be found in O'Driscoll (2022). The question often arises: are temporary works different in the nuclear industry? The answer is that they are not different, but the 'defined limits' mentioned above may have specific requirements. Temporary works are temporary works wherever you go; the definition does not change. Regarding any specific contractual requirements in the nuclear industry, BS 5975-1:2024 (BSI, 2024), clauses 10.2.3.1 and 10.3.3(o) state that the temporary works coordinator's (TWC) duties include confirming or verifying that the relevant client and / or third-party approvals required by the contract have been received. Understanding the defined limits and ensuring the required approvals are in place are already highlighted in BS 5975 (BSI, 2024) as standard. From this, anyone suitably experienced and competent in temporary works and following BS 5975 (BSI, 2019, 2024) will be able to coordinate temporary works in the nuclear industry. Temporary works are an 'engineered solution' for:

- construction
- maintenance / upgrading
- decommissioning / demolition.

The 'engineered solution' is to be suitably coordinated and controlled on site to minimise any risks, as far as reasonably practicable. The complexity often lies in the terminology 'temporary works'.

One of the biggest challenges in the nuclear industry is clarifying what temporary works are and what the definition of temporary works covers. This is due to several reasons:

- the large mechanical element of the nuclear industry, as it is not purely civil / structural-based
- the multinational element of the nuclear industry: having individuals and contractors from outside the UK who may not be familiar with the term temporary works, as it is not commonly used across Europe or many other countries.

For example, if a specialist mechanical contractor from outside the UK is working on a construction site in the UK, they may not be familiar with the terminology. They need to be educated as to what is meant by temporary works and how they should be managed to comply with UK legislation, regulations and best practices for the nuclear industry. Working in the nuclear industry in temporary works requires a clear understanding of what temporary works are, where the term originated from, and the ability to explain this to others who do not understand this terminology.

Adding the fact that a construction site in the nuclear industry is, by its geometry, congested and can be technically and logistically complex, with many specialist contractors working in close proximity to each other, the understanding of temporary works becomes more critical.

10.2. HPC temporary works management structure

HPC is a very large and complex project involving multiple disciplines, multiple interfaces and multiple contractors from different countries. An important initial consideration at HPC was which organisation(s) would undertake the roles of principal designer (PD) and principal contractor (PC) and how the multitude of varied and complex temporary works utilised by a variety of specialist contractors should be managed. The client for HPC is jointly EDF Energy and China General Nuclear Power Group (CGN), who created a separate subsidiary called Nuclear New Build Generation Company (NNB) to build and operate the power station. They undertake the role of client, PD and PC for the project. The project is overseen by the Office of Nuclear Regulation (ONR), which was set up under the Energy Act (HMG, 2013) with legal authority to regulate safety, security and ensure high construction standards. The project is then split into a series of large packages, which are managed by UK and international Tier 1 contractors, often working in joint ventures due to the scale and complexity of each package. In effect, the Tier 1 contractors are operating as subcontractors on this project and are managing their own temporary works using their own procedures.

Previous revisions of BS 5975 did not clearly define this complex type of situation, and BS 5975:2019 (BSI, 2019), clause 9.3.2.5, repeated in BS 5975-1:2024 (BSI, 2024), was introduced to reflect how the PC's temporary works team manages temporary works at the HPC construction project. This was following discussions and input from the project team; clause 9.3.2.5 (BSI, 2019) is as follows:

> Where a particularly technically or logistically complex project can be split into more than one well-defined area, then a PC's TWC may be appointed for each area. Each individual PC's TWC should be responsible for all the temporary works in the area allocated, irrespective of which organization carries out the temporary works in the area. Only one PC's TWC should be responsible for a specific area. One of these individuals should be appointed as lead PC's TWC to manage the interfaces between the areas.

The split of the overall site must consider the variety of different contractors from different backgrounds and different countries. This can be challenging, but it can also make these projects unique for temporary works coordination and very rewarding when done correctly.

The PC's temporary works team at HPC is the first team to have done this, and this would not be possible without working in partnership with the contractor's temporary works teams on site. The team has split the areas of responsibility for temporary works on the main site into clear geographical areas to match the construction layout. There are other areas, but the three main areas are designated:

- nuclear island
- conventional island
- heat sink.

The PC's temporary works team at the HPC construction project have developed and grown their understanding of how to manage temporary works on this technically and logistically complex project since the official start of the construction phase in 2016. This development on site is ever-changing and evolving as the project moves into more of a mechanical phase (from the civils phase). The PC's temporary works team works in close partnership with the very experienced and competent contractors' temporary works teams on site and, as a combined team, they have successfully managed temporary works on site for over 8 years. Figure 10.1 shows the organogram at HPC.

The PC's temporary works coordinator (PC's TWC) for each of these areas ensures that contractors who manage temporary works have the following.

- A suitable temporary works procedure is in place, with the company designated individual (DI) clearly identified.
- The company TWC is suitably appointed and identified on site. The skills, knowledge, experience and behaviours (SKEB) of the TWC are verified.
- If required, any temporary works supervisors (TWS) are appointed and identified on site. The SKEB of the TWS are also verified.
- The contractor's TWC ensures their temporary works register is maintained.

Figure 10.1 Hinkley Point C temporary works organogram (reproduced with kind permission of Hinkley Point C)

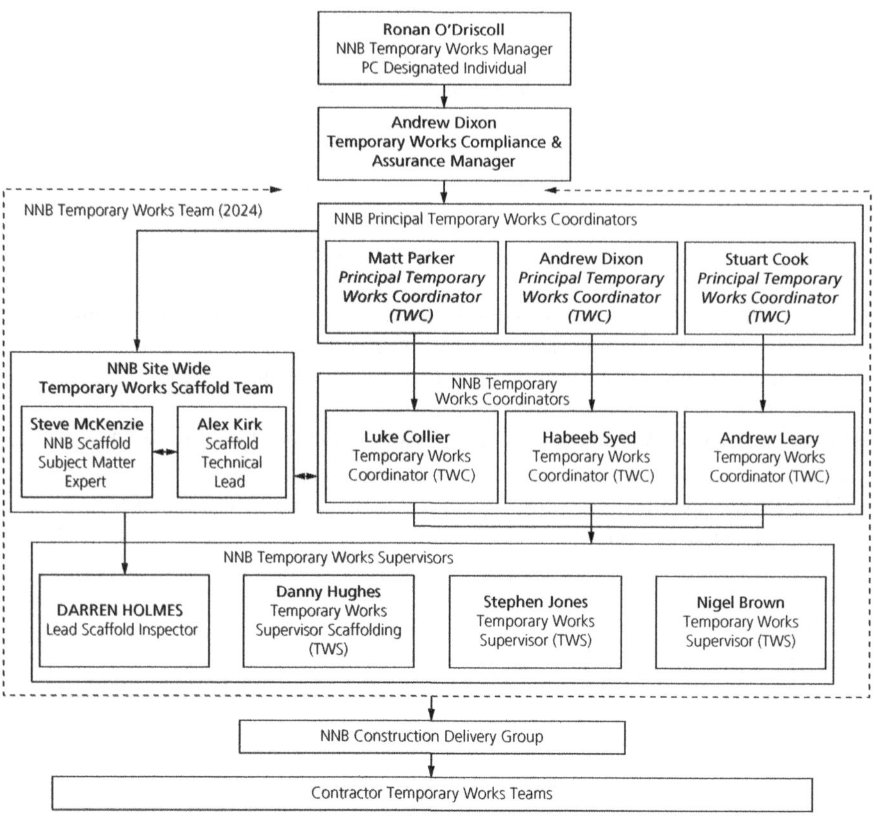

- Temporary works designs are checked to the level required in table 2 of BS 5975-1:2024 (BSI, 2024) and further approved where required by the PC, who ensures overall coordination of the various design elements.
- Temporary works designer's (TWD) residual risks are clearly communicated on the design drawings used on site.

The PC's TWC and TWS undertake regular audits and site inspections of each contractor's temporary works to ensure compliance to the agreed standards. The PC's temporary works team has regular meetings with the other temporary works teams on site and chairs a regular temporary works interface meeting to ensure temporary works interfaces between contractors are suitably highlighted and managed on site. There is a very collaborative and supportive relationship between the PC and each contractor's temporary works team to ensure the successful delivery of temporary works (and the overall project) on this scale of project.

10.3. Planning

With all temporary works, the earlier they are identified and considered (or eliminated if possible), the greater the benefits to the construction project. There can be improvements / benefits in terms of managing the risks, understanding the required construction programme, and identifying potential changes that could improve or reduce risk, improve safety, improve efficiency, improve quality, reduce costs and reduce the overall construction programme.

The early identification and careful forward planning of temporary works could also have longer-term benefits with respect to maintenance of structures and eventual decommissioning / demolition, which can have a huge impact on nuclear safety issues during these phases of work. For example, for the construction phase, the permanent works design must consider the temporary works required for the civils construction, which are then to be removed as early as possible to facilitate mechanical installation. The temporary works required for the civils construction are not required to consider specific loading conditions from the mechanical phase unless the same temporary works arrangement is required for both. Otherwise, the temporary works planning must consider both the civils and mechanical load cases and their specific requirements, such as defined limits for allowable deflections of constructed elements.

Permanent works designers (PWDs) must consider temporary works requirements for maintenance / outage works where equipment will be maintained or possibly replaced during the life of the nuclear facility. The possible impact on the reduction of risk by suitably considering this at the early planning stage is huge for any standard construction project and is increased when nuclear safety issues are considered. Similarly, PWDs must consider temporary works requirements for safe and efficient decommissioning once the nuclear facility has fulfilled its design life (a requirement for any nuclear facility). Reducing and simplifying temporary works and ensuring safety and stability should be the goals. The decommissioning process may take many decades, and the temporary works requirements should consider this timescale. If a PWD does not have the competency to fully determine the temporary works required for any of these stages of their design, they should consult with a temporary works specialist who has the required experience and competence in similar projects in the nuclear sector. PWDs should be aware of their duties under CDM Regulations (HMG, 2015) and these are discussed in chapter 2.

Some items of temporary works (such as haul / access roads, fencing, temporary batters and working platforms for tracked machines) will be in use for many years. This should be recognised by designers who should consider the use of materials for longevity. Onsite robust and regular inspection and maintenance regimes should be in place to ensure that items of temporary works remain in a fit condition for continued use.

10.4. Overview of nuclear sector temporary works

Nuclear new build projects such as Hinkley Point C are among the largest and most technically complex construction projects, with every conceivable type of temporary works on a huge scale (see Figure 10.2). The management team and procedures are required to manage interfaces between multiple different items and disciplines.

During the civils construction phase, typical items of temporary works include:

- site establishment and security measures
- tunnelling (onshore and offshore)
- marine works for temporary material jetties
- temporary slopes, earth bunds, excavations, sheet piling and controlling water
- formwork and falsework for reinforced concrete (including stability of reinforcement cages)
- foundations for machinery and lifting frames
- scaffolding, edge protection and temporary workshops.

Figure 10.2 Variety of temporary works in use at Hinkley Point C (reproduced with kind permission of R O'Driscoll)

During the mechanical construction phase, there is an increased requirement for access and the associated temporary works, such as scaffolding, edge protection and staircases, to facilitate this. The temporary works required to support lifting and moving of plant and equipment are also more prominent and there is much closer interaction with the site lifting teams, who work to the Lifting Operations and Lifting Equipment Regulations (LOLER) (HMG, 1998). Temporary works loading capacities from permanent structures are increased at this stage of the construction. There is a requirement for large temporary works structures to support lifting and manoeuvring of equipment. These structures will load the permanent works in unique ways that the permanent works designer must have considered at the design stage. During the commissioning phase, there will be further access requirements and temporary restraint systems for pressure testing of pipelines. There may also be some temporary works requirements to support plant and equipment used for the commissioning operation. When the structures and facilities used for the construction and commissioning phase are no longer required, there will be a range of temporary works (such as crane foundations and working platforms; see Figure 10.3) required for the dismantling / demolition of these structures. The project can boast what is believed to be the most powerful land-based crane in the world, with a capacity of 5000 tonnes at a lifting radius of 40 m and can be configured with a jib length of 250 m.

Figure 10.3 Extensive use of craneage at Hinkley Point C (reproduced with kind permission of R O'Driscoll)

10.5. Civil, structural and mechanical elements

Projects such as HPC are required to manage and coordinate major interfaces between civils / structural elements and mechanical elements for construction, maintenance and decommissioning / demolition. Temporary works are often a key element in managing these interfaces and all parties involved require a clear understanding of what constitutes temporary works and the potential risks involved. There is a tendency among designers of all disciplines to focus on the final design without understanding the construction steps required to construct their design. All engineering disciplines should consider how to eliminate or reduce risks and how their designs are to be constructed in a safe, efficient and timely manner.

Some mechanical designers may not be familiar with temporary works and may not be fully aware of their duties and responsibilities. When designing plant or equipment, the following typical examples should be considered.

- Some items can be large and heavy and fabricated off site. How will they be transported safely to site? Logistical issues need to be considered and temporary transportation stillages should be designed to secure the load.
- How will the item they have designed be manoeuvred and positioned within a structure (and removed if being maintained or replaced)? This should include the provision of lifting points and consideration of the position of the centre of gravity in relation to the lifting points to aid installation. Once the item has been positioned, is it stable or are additional temporary supports required?
- If an item is to be jacked into position, have the jacking locations been identified and designed?
- During transportation and positioning, the items may experience additional stresses and strains which need to be considered (to avoid damage).
- Large and heavy items (such as turbines) may be placed into the final position by crane and then the structure completed around them. Careful consideration should be given to how these items can be replaced (if necessary) within the confines of a completed structure. Craneage can be incorporated into the structure, providing the resulting loadings are adequately considered.
- How will access be provided and are access platforms / edge protection required?
- If the items are modular, how is stability of each section ensured throughout the erection process?
- Are temporary supporting brackets required and, if so, have they been suitably designed and checked?
- Clear communication of loading information (factored or unfactored) is key to avoid confusion for organisations involved in coordinating designs or checking designs.
- Are any residual risks suitably communicated to the site team so they can manage these risks appropriately?

Overall management coordination and integration of mechanical designs with civil / structural designs is required.

10.6. Nuclear maintenance / outage works

Nuclear power stations have regular planned outage / maintenance work when the supply of power is not as high, and the station can stop producing power while the required maintenance work is being undertaken. Some of this work is routine and will require little or no temporary works but some may be more major and require extensive temporary works and suitable long-term planning. This type of work has time constraints and there will be agreed shut-down and start-up dates for power production and supply of power back to the grid.

Major work may require the removal and replacement of some existing equipment and could have a range of temporary works. There may be opening-up and making good works required to facilitate the removal of equipment and replacement. Access scaffolding and temporary propping are often required, as is lifting machinery. These requirements need to be identified (along with any potential clashes with permanent works) as early as possible to ensure there is sufficient working space, adequate access, and appropriate equipment is available. Where possible, the temporary works solutions should be standardised to ensure efficiency can be maximised within the limited time available to carry out the work. This type of work is undertaken yearly across the country at different levels of complexity at all power stations.

10.7. Nuclear decommissioning / demolition

Decommissioning / demolition of aged plants has its own risks, in particular the possibility of dealing with contaminated waste materials and components to minimise the risk to human health and the environment. The requirement to decommission / demolish existing nuclear power stations is going to increase as the old UK fleet of power stations are progressively turned off. This process will bring its own temporary works challenges that have very significant nuclear safety considerations. This type of work is very prominent at Sellafield in Cumbria, UK, and there have been some significant examples of industry-leading work from projects delivered there, such as the Windscale Pile 1 filter gallery demolition. Working around the remnants of live nuclear material has its own challenges. Workers in these environments will work to the requirements of the Ionising Radiations Regulations, IRR17 (HMG, 2017) and will be categorised as 'classified workers', depending on their exposure limits for the work being undertaken. This comes with suitable controls and precautions that must be considered for any task being planned, and any required temporary works will have to take this into account. Potential working areas need to be carefully assessed prior to work commencing, with a clearly defined plan and control measures determined (including for the necessary temporary works). Works areas need to be clearly defined and controlled, with extensive monitoring to ensure agreed trigger levels are not exceeded. The use of applied coatings can assist with decontamination and reduce the resulting waste classification.

The consideration of possible contamination or contamination prevention is very prominent and must ensure any possible contamination of equipment is minimised and, as a result, any required waste control is minimised. Another consideration is when temporary works equipment such as props are required and operatives are working in a controlled area with suitable radiation controls in place (such as protective suits and clothing), whether this equipment can be safely positioned, used and removed. A further consideration is whether the temporary works design has been suitably considered to minimise the risk of contamination to the operatives throughout this process.

If temporary works equipment is to be erected by personnel in controlled areas, protective suits, full-face masks with suitable air filters and personal dosimeters will be required

Figure 10.4 Temporary works being erected in a controlled area (reproduced with kind permission of R O'Driscoll)

(see Figure 10.4). Operatives will have restricted working times imposed and the task becomes more difficult due to reduced visibility, risk to protective clothing being damaged or torn, and time constraints. The selection of temporary works equipment and sequencing of the work becomes even more critical.

10.8. Specific HPC project requirements

There are several specific items that can differentiate nuclear new build projects from 'more typical' construction projects. Some examples are provided.

- The complexity of these projects and the number of contractors involved (UK and international) will create a multitude of complex interfaces between trades, components with differing tolerances and contractors, requiring careful management and coordination.
- There is extensive site security, which can include robust temporary security fencing (may become permanent), entry / exit turnstiles, site facilities (welfare, parking, temporary accommodation, canteens, etc.) for several thousand operatives, as well as lay down and storage areas and temporary roads with a network of onsite transportation for goods and operatives.
- There are extensive and robust storage areas for heavy materials, machinery and fuel, which will be used for many years so require regular maintenance.
- There are onsite batching plants for concrete with associated temporary storage bins for aggregates, sand, cement, additives and so on.

- Extensive temporary site services are supplied, such as power, communication and water. Temporary storage lagoons for water have been constructed.
- The density of reinforcement in concrete is increased, including the use of larger diameter bars (32 mm and 40 mm are common). This additional density and diameter can add complications when tying formwork panels to avoid clashes with reinforcement and to avoid sensitive / critical areas. Often the panels may need to be secured directly to the reinforcement cages. The location of any temporary fixings that are left in place in the finished concrete should be recorded, along with confirmation that the material left in place has been approved as acceptable by the PWD.
- Thicker slabs supporting heavy machinery and increased density of concrete (plus increased percentage of reinforcement) add to the overall weight to be supported on falsework, back-propping and lifting appliances. Consideration must be given to back-propping of reinforced concrete slabs that are densely reinforced and have depths that exceed 350 mm where there is currently no industry guidance on sharing of propping loads to the levels below. In these instances, there must be close collaboration with the construction teams and the PWD to confirm what propping has been allowed in the design and when that propping is to be removed.
- There will be numerous heavy and complex items to be lifted into position. This will occur during all phases (civils construction phase, mechanical phase, maintenance / upgrading and eventual decommissioning / demolition). Purpose-designed and manufactured temporary lifting frames with multiple lifting points will be required. Load compensation devices are frequently used to ensure equalisation of loading at each lifting point. Many items may need to be lifted and rotated from the horizontal to vertical plane. Consideration should be given to prevent excessive deformation and the build-up of stresses in any object being lifted.
- Many reinforcement cages are likely to be large and complex to construct. Cage designers and detailers need to carefully consider assembly sequences, temporary stability of cages built in situ or prefabricated cages being lifted into position, safety of operatives building cages and the avoidance of clashes. Temporary Works Forum has produced two guidance documents (TWf, 2020, 2021).
- Normally simple requirements to drill and fix temporary works fixings into concrete require more detailed consideration and, if drilling and fixing cannot be avoided, the process to identify suitable drilling locations and the required remedial works must be suitably managed and controlled.
- Seismic loading is an important consideration in the design of permanent reinforced concrete, and temporary works requirements will reflect this.
- Some items of temporary works will be in use for many years (perhaps 10 years) in a safety-critical environment next to the sea. Designers should consider longevity and robustness in their designs.
- Locked-in unfavourable stresses and strains due to externally applied temporary works loads in reinforced concrete structures should be avoided. A simple example of this could be the construction of a multi-storey reinforced concrete structure, with large distances between floors or from floor to roof slab. Double-sided formwork with through ties and a cantilevered working platform on one side will be used to construct the wall. At higher levels, raking props for stability will not be used and the wall will have an off-centre load applied from the access platform loading. This may lock a stress

into this wall prior to the casting of the concrete slab above. It is noted that this loading may very well be insignificant, depending on the wall construction, but the PWD must ensure that if any stresses are locked into the construction of the wall, they are acceptable for the design life of the structure.

10.9. Modular nuclear construction

An issue with nuclear new build projects relates to the risk, cost and time taken from the planning and concept stage to generating power, along with waste and decommissioning issues. Modular construction and particularly the proposed 'small modular reactors' (SMRs) can reduce some of the construction issues. The concept is to prefabricate components and transport them to their final location. The concept relies on standardisation of design and repetition in manufacturing in controlled factory conditions. This can improve efficiency and quality, and reduce construction time, risk and costs. However, careful planning of sequencing and transportation logistics is required. To maximise the benefits of these modular systems, careful planning and coordination of civils, mechanical, commissioning and decommissioning elements are critical. Temporary works must be considered as early as possible in the planning and design stage and ideally simplified or eliminated (as far as practicable). The following are some typical items that should be considered.

- Identify / construct factories (some factories could be temporary) with sufficient skills and capacity to provide the required modular units. These factories will require significant craneage, temporary lifting frames and temporary stillages for moving units and for transportation.
- Have temporary lifting points been designed into the modular units for all the required load cases and the direction of the load when manoeuvring the modules? Often modules will be rotated from a horizontal transport position to vertical, and consideration should be given to any resulting stresses and strains transferred into the module itself (unfavourable stresses and strains could also be created during transportation).
- Construction tolerances need to be identified and considered in the overall design when modules are connected as there will be major differences between civils and mechanical elements.
- Temporary jacking / support may be required, with suitable access being provided for the installation of jacks and tirfors. The design should ensure there are no detrimental stresses and strains created in the elements created by jacking or temporary support.
- Careful consideration of the construction sequence is required. Once a module is located and connected in its final position, can the redundant lifting points then be utilised to provide connection to the next module to assist in positioning and fixing that module in location, reducing the need for further temporary works?
- The construction sequence should integrate, where possible, the civils and mechanical elements, along with their respective temporary works requirements.
- The construction sequence should allow for suitable positioning of lifting equipment, such as crawler or mobile cranes, to reduce the required lift radius and maximise the utilisation of the machinery. This should also reduce working platform thickness, as well as reduce the need for special accessories for lifting (i.e. super lift or similar).
- Modules should be designed to be self-supporting and have temporary stability in their final position to eliminate the need for additional temporary support.

- In the longer term, the requirement for temporary works during the maintenance and decommissioning / demolition phases needs to be identified and allowed for to maximise safety and efficiency.

Small modular reactors (SMRs) are advanced nuclear reactors which have been developed to further simplify construction and reduce risk in nuclear new build projects. Physically, they occupy less space than a conventional nuclear power station (they may be suitable at locations where conventional power stations are not), but consequently generate less power. By making the systems modular, there can be benefits as described above, and they can be constructed incrementally to match an increase in power requirements.

10.10. Industry education

The UK has significant skills and experience in the delivery of nuclear new build projects as well as decommissioning / demolition. These skills need to be continuously refined and improved. Education is essential to ensure there is a clear understanding of what temporary works are and how they should be managed at all levels, from senior management to site operatives. This has been evident at HPC, and this requirement will continue with other proposed nuclear new build projects. There is a skills shortage in the UK construction sector, and major projects such as nuclear new build can play an important role in trying to redress this. Apprenticeships are widely used to train the next generation of operatives in both civils and mechanical skills. There is also an ongoing requirement to continue educating PWDs (civils or mechanical) to recognise their duties in relation to CDM Regulations (HMG, 2015) (see chapter 2). They should follow the principles of prevention and identify safe and efficient construction sequences for their design. This principle applies to all phases of a project, from design through to construction and eventual decommissioning / demolition. They should better recognise what temporary works are typically required and refine their permanent designs to maximise efficiencies in temporary works.

To assist all designers (civil / structural / mechanical), suitable temporary works training is required and all engineers need a working understanding of BS 5975 (BSI, 2024). There are many useful additional temporary works resources which have been highlighted in this publication in the References, Useful websites and Further reading sections.

REFERENCES

BSI (2019) BS 5975:2019: Code of practice for temporary works and permissible stress design of falsework. BSI, London, UK.

BSI (2024) BS 5975-1:2024: Management procedures for the control of temporary works. Code of practice. BSI, London, UK.

O'Driscoll R (2022) Construction Engineering: A Guide to Temporary Works in Construction and BS 5975:2019. www.constructionengineering.uk (accessed 07/01/2025).

TWf (2020) TWf2020:03: Temporary Condition of Reinforcement Cages Prior to Concreting: Part 1 – Management Guidance. https://www.twforum.org.uk/viewdocument/rebar-guide-part-one-management (accessed 07/01/2025).

TWf (2021) TWf2021:01: Temporary Condition of Reinforcement Cages Prior to Concreting: Part 2 – Technical Guidance. https://www.twforum.org.uk/viewdocument/rebar-guide-part-two-technical (accessed 07/01/2025).

HMG (Her Majesty's Government) (2013) Energy Act. The Stationery Office, London, UK.
HMG (2015) Construction (Design and Management) Regulations. The Stationery Office, London, UK.
HMG (1998) Lifting Operations and Lifting Equipment Regulations. The Stationery Office, London, UK.
HMG (2017) IRR17: Ionising Radiation Regulations. The Stationery Office, London, UK.

USEFUL WEBSITES

EDF Energy – www.edfenergy.com
ONR (Office of Nuclear Regulation) – www.onr.org.uk
TWf (Temporary Works Forum) – www.twforum.org.uk

FURTHER READING

Pallett PF and Filip RK (2019) *Temporary Works: Principles of Design and Construction*, 2nd edn. ICE Publishing, London, UK.
Webster M (2023) *Report RR1198: Implementation of the Principal Designer's Role within CDM 2015*. HSE (Health and Safety Executive), The Stationery Office, London, UK.

Ray Filip
ISBN 978-1-83608-525-6
https://doi.org/10.1108/978-1-83608-524-920251012

Chapter 11
Temporary works in a marine environment

Jan de Klerk
TWD Ltd, UK

11.1. Introduction

For many projects, working over water is unavoidable; these projects could be close to land or further offshore. Examples of projects close to land (near-shore or river projects) include harbour extensions, quay construction and land reclamation. Further from land, the complexity and risk tend to increase, and examples of these projects include the construction of offshore wind turbines, offshore platforms and deep-sea cable laying. Often overlooked but relevant for both land and marine projects is the transportation at sea of large components. The temporary works required for successful execution can be significant and can affect the permanent works design.

Working over water creates significant challenges when compared to working on land. Often, stable ground is not sufficiently close to carry out the work from, so a site needs to be established on water using vessels or temporary works like bunds or platforms. Consideration also needs to be given to the additional loading that can be generated by water, which needs to be factored into the design of both permanent and temporary works, as well as the construction sequence / installation methodology. Some of the topics relevant to marine projects, like floating plant and cofferdam construction, have been discussed in *Temporary Works: Principles of Design and Construction* (Pallett and Filip, 2019, chapters 17 and 18), so this chapter will focus on the execution of marine works and the temporary works required for them. Temporary works in the marine environment are among the largest and most complex elements in any sphere of temporary works and need to be designed robustly to be able to accommodate and withstand the forces of nature that can be experienced offshore.

11.2. Challenges in a marine environment

The main challenge with working over water is the absence of stable ground to support plant and materials and alternative solutions need to be considered to set up a project site. These could include:

- floating plant such as barges or vessels
- jack-up barges
- temporary platforms constructed from soil, timber, steel and / or concrete
- using the permanent works that have been partially built in an earlier phase.

All these options have their own benefits, but a common issue is that significant resources may be necessary, even on a relatively small project site, with extensive planning and smart design solutions required. In addition, planning logistics can be challenging and might require back-up solutions, as weather conditions will make air or sea travel impossible at times.

The limited working space and relatively challenging conditions add complexity to marine projects. This results in a desire to pre-fabricate at least part of the permanent works on land where facilities are more accessible. This adds the requirement for heavy transport and heavy lifting to many marine projects. So-called 'sea-fastening designs' are required to prevent items from sliding and tipping on a moving vessel, while also distributing the load to deck. Lifting operations need to consider the additional dynamics expected offshore as items are lifted to or from floating vessels that move due to currents, waves and swells.

If the project is near a residential area, additional restrictions can be imposed relating to noise, vibration, light pollution and working hours. Additionally, further special consideration must be given to environmental issues, as many marine areas will be important wildlife habitats for both land-based and marine life, which require protection. Pollution and noise can travel far and easily in water, so mitigation measures, such as temporary bunds or screens, can have a significant effect on the planning and execution of the project.

11.2.1 Inshore and near-shore project management

The management of inshore and near-shore marine temporary works should be carried out similarly to projects on land. Designers and contractors of temporary works will have to comply with Construction (Design and Management) Regulations (CDM) (HMG, 2015) and the management recommendations of BS 5975-1:2024 (BSI, 2024a).

Temporary works items can generally be designed in accordance with BS 5975-2:2024 (BSI, 2024b). However, depending on the complexity of the marine aspects of the project, it can be beneficial to also select an offshore design code for the design, such as Det Norske Veritas (DNV) code for marine operations DNV-ST-N001 (DNV, 2023), or at least take inspiration from it, as it will give a better guide to all aspects involved in a marine project.

11.2.2 Offshore project management

The legal requirements for managing offshore projects can be complex and depend on the exact location and purpose of the project. The Health and Safety at Work etc, Act (HMG, 1974) and CDM regulations (HMG, 2015) apply in UK territorial waters but working further offshore usually means that a seagoing vessel is involved, where health and safety on board is regulated by the Merchant Shipping Act (HMG, 1995) and related regulations. These regulations might not say the exact same thing, but the general goal of the regulations will not be different from onshore projects – that is, plan and carry out the work safely at all stages.

The design and execution of temporary works further offshore typically follow different regulations and design codes from onshore projects (see section 11.2.5). These works often involve a marine warranty surveyor (see section 11.2.3) and the use of an offshore design code, which are explained further below.

11.2.3 Marine warranty surveyor (MWS)

The MWS is a recognised review party that provides independent third-party technical reviews and approvals for high-value and high-risk marine projects and their temporary phase marine operations. Well-known examples are Det Norske Veritas (DNV) and American Bureau of Shipping (ABS), but there are many more. The involvement of the MWS is usually a requirement of the insurance policy obtained for the project. The MWS can be involved from the planning and set-up phase to the execution phase, so will review both permanent and temporary works.

11.2.4 Ship classification society

Sometimes the structure of a vessel may need to be modified, such as deck strengthening or hull attachments. This may be required where the loads on the vessel are extremely large (e.g. from very heavy cargo or equipment), or if strong attachment points are required for attaching equipment to the hull. The ship classification society that originally certified the vessel will need to be involved to ensure the vessel will stay compliant with their standards. This is required to keep the vessel registered and insured. Well-known ship classification societies include Det Norske Veritas (DNV) and Bureau Veritas (BV); examples of their classification rules can be found in DNV (2024) and BV (2024). Some of these societies also offer MWS services, but this can be dependent on the nature of the project and the vessel if the party fulfils both roles on the same project.

11.2.5 Standards and codes of practice

For offshore projects the standards from DNV or ABS are often selected as design guidelines. Examples include DNV-ST-N001 (DNV, 2023) and the ABS MOU guide (ABS, 2021). They offer an extensive set of regulations for offshore operations, ranging from operational planning to structural design and material selection. Because of their origin in the marine industry, they offer a complete guideline to working offshore. In general, the factors of safety used in these codes would be similar to BS 5975-2:2024 (BSI, 2024b).

11.2.6 Personnel and training

Personnel involved in marine projects require specific competencies and skills relating to safety and survival. Specific training and certification are available to prepare people for works offshore. There is an overall lack of consistency, as the certification requirements for personnel can differ immensely between projects, mostly depending on whether a project is related to oil and gas, offshore wind or other offshore activities. This lack of consistency requires significant additional resource, time and expense to ensure personnel are appropriately trained and certified for the project they are intending to work on. More specialist examples can include specific diving skills for underwater works or for operators of specialist equipment like remote control vehicles. The captain of the vessel has a unique and safety-critical role, as they have the last word over what happens on board.

11.3. Design considerations

11.3.1 Vessels

In many marine projects, the vessel serves as the project site on which the equipment and materials are placed and transported (some of these vessels are discussed in Pallett and Filip, 2019, chapter 18). In general, vessels used for projects further offshore tend to be larger and stronger, so more cargo can be carried. The larger size also offers improved stability characteristics in heavy sea conditions. Some typical offshore work vessels include the following.

- Heavy lift jack-up vessel – often used for offshore wind turbine installations. The vessel can be jacked above the water to provide a stable platform for the integrated crane. Lifts of over 2500 tons are possible, while the vessel deck offers plenty of space for item transportation.
- Floating heavy lift crane vessel – provides larger crane and deck capacities than a typical jack-up. Dynamic positioning systems can keep the vessel at a fixed location with accuracy. However, the motion of the floating vessel can make installation more challenging.
- Semi-submersible crane vessel – offers the largest crane capacity of up 10 000 tons per crane. These vessels have an unusually deep hull which allows them to partially submerse themselves to offer improved stability and reduced motion.
- Semi-submersible lift vessel – these vessels do not have a crane but have the capability of partially submersing themselves and positioning their strong, flat deck below the item that needs to be lifted. By pumping out ballast, the vessel can increase its buoyancy to lift and transport the item. Lifts over 100 000 tons are possible.

The day rates for a typical heavy lift vessel in full operational mode (including crew, fuel, insurance etc.) can be hundreds of thousands of pounds per day, so exact and efficient project planning is essential to keep costs manageable.

The stability of the vessel is critical to prevent tipping during the works (see Pallett and Filip, 2019, chapter 18). Transporting items on a vessel, or even working from a floating vessel in wave conditions, will introduce additional loads on all on-board items due to the motion of the vessel. The lateral and angular movements of the vessel on the waves will cause accelerations in all directions. These will need to be calculated and accounted for in the design – for example, by following the calculations for ship motion response in DNV-ST-N001 (DNV, 2023).

Space is often at a premium on a typical marine project and cranes are essential to move items around. Some vessels have integrated cranes; if not, then it is common to place a crawler crane on the deck. Careful consideration needs to be given to the selection of the crane as some cranes are not suitable for lifting on a floating vessel at sea and those that are would be required to operate at significantly reduced capacities to allow for the motion. This is even more the case when the item needs to be lifted through the waterline.

Positional accuracy can be a major challenge as the movement of items during lifting on water also makes it more difficult to install items in their correct position. This issue may be resolved by ensuring sufficiently large tolerances are incorporated into the design, but more generally, temporary works items such as piling gates or other robust positioning templates would be necessary to guide items into their final position. Robustness is essential as accidental impact and direct wave loading can be significant.

Equipment supply over water can offer the opportunity to transport larger and heavier items directly to site, as road transport size limitations do not apply, but limitations in storage capacity on a typical marine project can be a challenge. A suitable quay side where project items can be stored needs to be identified. Transportation of items can be carried out by the vessel being used for the project. The vessel can travel between the quay and project site; however, as mentioned above, these vessels are expensive and should be utilised for the project work itself as much as possible, so it may be more cost-effective to utilise smaller supply vessels.

11.3.2 Loads on marine works and weather windows

When designing temporary works for marine environments, well-known load types, such as gravity, wind and live loads, need to be considered. Additionally, the marine environment introduces loads due to vessel motion and loads due to water, such as hydrostatic pressure, current and waves.

BS 5975-2:2024 (BSI, 2024), annex A has useful guidance on how to determine some types of marine loading. Most MWS codes provide further guidance on how to determine these marine loads; an example is DNV-RP-C205 (DNV, 2021). An opportunity can arise in that the magnitude of these loads can be influenced by selecting a suitable working weather and sea condition window (known as 'workability limit') for the operation. By only allowing the operation to take place in relatively favourable circumstances – for example, at wind speeds below 15 m/s and a wave height below 2 m – the loads will be significantly lower than for storm conditions. This can save a lot of weight and complexity in the design of the temporary works required. However, this does come at the cost of higher risk of delays in the operations. For standard pieces of equipment, certain conditions cannot be changed or influenced – for example, the operational limit of a crane (often between 12 m/s and 16 m/s wind speed) – but limits for bespoke designed equipment can be selected to suit the project. So-called workability studies can be performed to ensure that a certain probability of workability can be obtained in the required installation season. If adequate and accurate consideration has been given to the specific probable conditions, this will lead to an optimal balance between equipment cost, required vessel chartering time and operational risk.

When selecting a workability limit, it is important to consider the length of the required work window – that is, how long will the favourable conditions be required? Weather below a certain limit might not occur for long stretches of time, or at least the chance they occur for a long period will be limited. This means that for short duration operations of up to 1 day, it will be easier to set low weather limits than for long operations. For long operations it might be unavoidable to design the equipment for the most unfavourable conditions possible.

11.4. Functional design method

Designing equipment for installations in a marine environment requires different engineering disciplines to be combined. Interfacing between civils, marine and mechanical engineering is required. Bringing these disciplines together into one effective design can be challenging and requires a structured design approach. This can be achieved by using a functional design methodology, which achieves an effective design approach by making idea generation an analytical and systemic process. This is based on the VDI 2222 (VDI, 1997) standard for product design. When temporary works become more complex and need to fulfil multiple functions, it can be very helpful to break the design process down into the following steps.

- Clearly identify the main function and purpose of the design. For example, design a tool that can accurately guide a foundation pile into its final position when it is being installed from a vessel.
- Clearly define the exact requirements and specific goals for the design. For example:
  - it must be able to handle pile diameters between 800 mm and 1200 mm
  - obtain a pile positioning accuracy of +/−100 mm
  - design weight cannot exceed 50 ton.
  - The requirements can be divided into 'essential requirements' and 'possible requirements' to make evaluation of the potential solutions easier.

- The overall design goal should be broken down into a series of practical 'sub-problems' which should be described as simply as possible. For the above pile guide example, this could be:
  - pile diameter adjustment
  - horizontal position correction
  - interface to pile.
  - When complete, this list gives a good functional overview of the design required. For each sub-problem, multiple solutions can be found. Breaking up the overall problem into smaller 'sub-problems' allows engineers with certain expertise to contribute effectively together.
- Potential solutions to each sub-problem may not be compatible when assessed for the overall problem. A way to visualise this is to create a morphological chart (Figure 11.1). Each sub-problem has its own row, and the columns can be used to present the solutions to the sub-problem. Potential solutions can be connected to provide an overall solution to the main problem.
- The overall solution can now be verified by calculation and detailing.
- The final evaluation should be based on the initially defined requirements. They can be combined in a trade-off table where each concept can be scored for its expected performance on each requirement. Optionally, each requirement can be weighed differently. The best scoring concept should be selected for detailed design.

Figure 11.1 Schematic diagram of the functional design method (reproduced with kind permission of TWD Ltd)

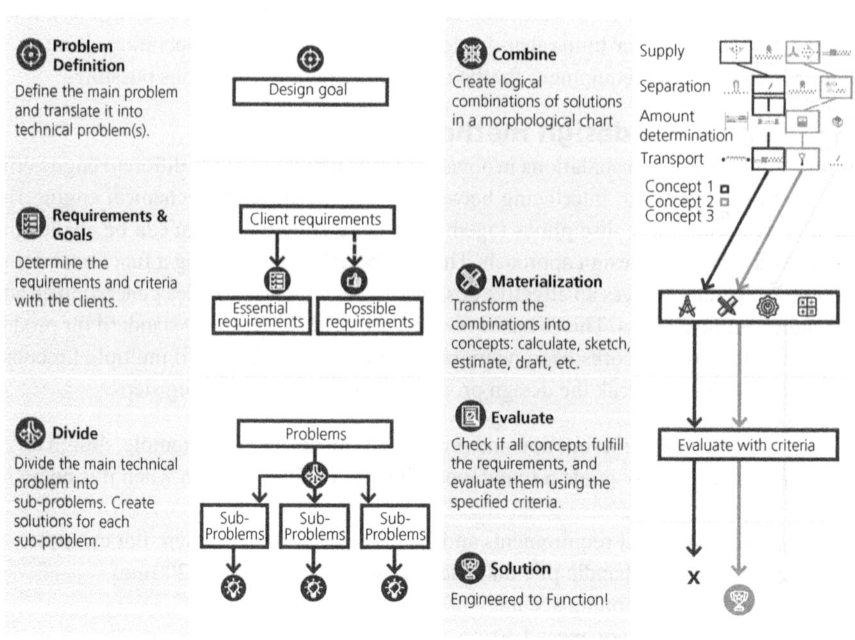

11.5. Examples

The examples in this chapter are typical marine temporary works and emphasise how varied the temporary works for marine projects can be. All these examples are temporary works in the sense that they are designed and constructed to facilitate the construction of permanent works and most of them are bespoke project equipment. However, the level of complexity and the operational lifetime would require these items to be utilised for a significant time, so corrosion, fatigue and robustness must be considered in the design.

11.5.1 Pile installation – onshore and near-shore piling gates

Piling gates are used to guide piles into their correct final position during lowering and the driving process. They can be simple frames attached to a floating or jacked up vessel. Two alternative options are shown in Figures 11.2 and 11.3: the piling gate itself can be made from floating modules or can be supported on the previously installed piles. Other design considerations can include the following.

Figure 11.2 Floating piling gate used for sheet pile installation (reproduced with kind permission of Graham Construction)

Figure 11.3 Double-layered piling gate for installing tubular piles. The gate is supported on previously installed piles and can move itself to the next pile position (reproduced with kind permission of Volker Stevin)

- Number of gate layers – a single-layered gate will result in a light and simple construction, whereas a double-layered gate provides better guidance accuracy on the rake of the pile. Double-layered gates are required for raked piles and for pile installation in challenging conditions.
- Interfacing with drill or hammer – when using a piling hammer, the target pile level might require the piling gate to open to allow the hammer to pass. Alternatively, a temporary pile extension in the form of a follower can be used. If a drill is also used (e.g. when the pile is to be embedded into bedrock and a rock socket is to be drilled out), it is important to know whether the drill must be supported by the gate, potentially resulting in large additional loads.
- Positioning – the piling gate can have adjustable pile interfaces to correct for pile position. Alternatively, the complete gate could be moved to the next pile location by crane or by moving the barge, or by built-in hydraulic actuators for self-movement. When a project consists of a large number of regularly spaced piles, an automatic moving system could be a cost-effective solution.

11.5.2 Monopile grippers

Monopile foundations are used extensively for wind turbines and platforms. They are commonly hammered into the seabed and drilling may be used to aid installation into harder soils or rocks. Guiding these piles during their installation process is done with grippers, which are larger and more advanced versions of a pile gate. Design considerations can include the following.

Figure 11.4 Monopile gripper attached to a jack-up vessel (reproduced with kind permission of Van Oord Offshore Wind)

- As the monopile usually forms the complete foundation and does not connect to other piles, the positional tolerance is not as important as the vertical tolerance. The weight of the monopile is so large that in many cases it will straighten itself out during the lift. When mounted to a jack-up vessel, a single-layered gripper is sufficient to achieve required tolerances. An example is shown in Figure 11.4. As the gripper is operated overboard, it is often designed to be retractable to improve the vessel's sailing and berthing capabilities.
- Installing monopiles from a floating vessel can be significantly more complicated. The vessel can hold itself in place using dynamic positioning systems, but small movements in all degrees of freedom will still occur. In this case, the gripper can allow for motion compensation. This involves designing extensive control systems and software that can compensate for the motion of the vessel. Comprehensive understanding of the dynamic properties of vessel, gripper, monopile and water are necessary. The example in Figure 11.5 shows the complexity involved: a motion-compensated gripper that also functions as a monopile upending system.

11.5.3 Jacket pin pile templates

Jackets are a type of offshore foundation historically used in the oil and gas sector. In the offshore wind sector, they are the foundation of choice for a rocky seabed. These large truss

Figure 11.5 Combined double-layered monopile gripper and upending system which is also motion-compensated for installation on a floating vessel (reproduced with kind permission of DEME Group)

structures are anchored using pin piles that are either hammered or drilled into the seabed. In the case where the pin piles are installed prior to the installation of the jacket, the positional tolerances are tight to ensure that the prefabricated jacket fits over the pin piles. To achieve these tolerances, piling or drilling templates are used. These seabed templates appear to be very large and heavy structures due to the required span and strength, but they are precision instruments that guide piles to their final position at the required degree of verticality. Typical design considerations can include the following.

- Soil properties – these determine whether piles can be hammered or whether drilling is required, resulting in a potential requirement for a drill interface on the template. In softer soils, large 'mud mats' will be required to prevent the template sinking into the soil. Figure 11.6 shows a pin pile drilling template.
- Installation tolerance – the pile guides in the template need to be tight enough to ensure positional accuracy and verticality of the piles. They also need to accommodate the largest pile diameter possible. Adjustable pile interfaces can be used to correct further, if required.
- Underwater operations – these make visibility challenging; cameras can help but visibility might still be poor. Templates will require sensors to confirm position and level. Thrusters are sometimes used to guide the template to its intended position when local current is high. Lifting guides are required on the template to guide the piles into the correct position during lifting.

11.5.4 Offshore lifting

Lifting heavy items offshore can be challenging due to additional motions and loads on the lifted item and the crane. Other considerations can include the following.

Figure 11.6 Pin pile drilling template as used for the St-Brieuc offshore wind farm project (reproduced with kind permission of Van Oord Offshore Wind)

- Often the lift points are hard to access and poorly visible due to being high up or underwater. Remote connect and release lift attachment points can help. These can be off-the-shelf or specifically designed for the project.
- To guide the item into place, large bumper and guiding structures can be required.

11.5.5 Marine transportation

Transportation of heavy cargo items over sea requires the design of a temporary sea-fastening structure (Figure 11.7) that secures the cargo (prevents movement, rolling and potential damage) and correctly transfers the load to the deck. Depending on the shape of the cargo and the suitability of the vessel, these designs can become quite complex.

- Position on the vessel – cargo with a centre of gravity close to the centre of the vessel will experience less acceleration from vessel motion, so the sea-fastening design can be lighter. However, contractors often want to maximise the cargo taken aboard. Figure 11.8 shows an outboard sea-fastening design for wind turbine blades, freeing up deck space for other cargo.
- Interface points – the sea-fastening is required to transfer the loads from suitably strong points on the cargo to the strong points in the deck. This becomes even more important when the loads are such that the sea-fastening will start pulling on the vessel deck. In this case, the sea-fastening structures are welded to the deck and exact alignment between welds above and below deck is required to obtain a direct load path. If this is not done correctly the deck could potentially be pulled from its internal structure.

Figure 11.7 Temporary sea-fastening structures on two barges used for the transportation of wind turbine components (reproduced with kind permission of DEME Group)

Figure 11.8 Outboard temporary sea-fastening system for wind turbine blades (reproduced with kind permission of DEME Group)

- Fixing of cargo to sea-fastening – it is not always possible to weld the cargo to the sea-fastening structure, perhaps due to structural reasons or because quick attachment and release is required. For lighter cargo, lashing systems are possible, but for heavier items it may be necessary to design adjustable stopper and clamping systems. Friction can often not be relied on so standards might advise to ignore it for lighter cargos or when sea-fastening calculations are not accurate enough (see DNV-ST-N001, DNV, 2023).
- Reusability – a sea-fastening structure might have to be used for multiple trips with similar cargo. If the cargo items are not identical, the sea-fastening will need to be designed with some adjustability. It should be kept in mind that welding multiple times at the same spot is not beneficial for material quality, so other attachment systems will be required.

REFERENCES

ABS (American Bureau of Shipping) (2021) *ABS rules for building and classifying mobile offshore units (MOU rules)*. ABS, Houston, TX, USA.

BSI (2024a) BS 5975-1:2024: Management procedures for the control of temporary works. Code of practice. BSI, London, UK.

BSI (2024b) BS 5975-2:2024: Temporary Works – Falsework: design and implementation. Code of practice. BSI, London, UK.

BV (Bureau Veritas) (2024) NR467: Rules for the classification of steel ships. BV, Birmingham, UK.

DNV (Det Norske Veritas) (2021) DNV-RP-C205: Recommended practice on environmental conditions and environmental loads. DNV, Høvik, Norway.

DNV (2023) DNV-ST-N001: Marine operations and marine warranty, 2023-12 edn. DNV, Høvik, Norway.

DNV (2024) Class rules for ships and offshore units. DNV, Høvik, Norway.

HMG (Her Majesty's Government) (1974) Health and Safety at Work etc. Act. The Stationery Office, London, UK.

HMG (1995) Merchant Shipping Act. The Stationery Office, London, UK.

HMG (2015) Construction (Design and Management) Regulations. The Stationery Office, London, UK.

Pallett PF and Filip RK (2019) *Temporary Works: Principles of Design and Construction*, 2nd edn. ICE Publishing, London, UK.

VDI (Verein Deutscher Ingenieure) (1997) *VDI Guideline 2222, Blatt 1: Methodical Development of Solution Principles*. VDI, Düsseldorf, Germany.

USEFUL WEBSITES

ABL Group – www.abl-group.com
ArcelorMittal – www.projects.arcelormittal.com
BSI (British Standards Institution) – www.bsigroup.com
BV (Marine Offshore Bureau Veritas) – www.marine-offshore-bureauveritas.com
DEME Group – www.deme-group.com
DNV (Det Norske Veritas) – www.dnv.com
OCIMF (Oil Companies International Marine Forum) – www.ocimf.org
Trelleborg – www.trelleborg.com
TWD Ltd – www.twd.nl
TWf (Temporary Works Forum) – www.twforum.org.uk
VDI (Verein Deutscher Ingenieure) – www.vdi.de

FURTHER READING

ArcelorMittal (2022) *Piling Handbook*, 9th edn. Arcelor Mittal. https://sheetpiling.arcelormittal.com/download-center/piling-handbook (accessed 07/01/2025).

BSI (2016) PAS 8812:2016: Temporary works. Application of European standards in design, group 5 – marine temporary works. BSI, London, UK.

BSI (2021) BS 56349:2021: Design of marine structures. BSI, London, UK.

Ligteringen H and Velsinki H (2022) *Ports and Terminals*. Delft University of Technology, Faculty of Civil Engineering and Geosciences, Delft, the Netherlands.

OCIMF (Oil Companies International Marine Forum) (2018) *MEG4 – Mooring Equipment Guidelines*, 4th edn. OCIMF, London, UK.

Thorsen CA (2014) *Port Designers Handbook*, 4th edn. ICE Publishing, London, UK.

Trelleborg (2024) *Marine and Infrastructure – Design Manuals and Guides*. Trelleborg, Sweden.

TWfNZ (Temporary Works Forum New Zealand) (2022) TWfNZ TGN02:22: Technical Guidance Note – Temporary Works in a Marine Environment: New Zealand Best Practice Guidance. https://d2rjvl4n5h2b61.cloudfront.net/media/documents/TWf_NZ_-_TGN_02.22_Temporary_Works_in_a_Marine_Environment_-_FINAL_29-06-22.pdf (accessed 10/01/2025).

Ray Filip
ISBN 978-1-83608-525-6
https://doi.org/10.1108/978-1-83608-524-920251013

Chapter 12
Rail

Jeff Mahony
SCS, HS2 Project, UK

Steve Williams
Network Rail, UK

The authors would like to thank Jayna Gorsia (graduate engineer, City University, London), Mark Walker (Costain Group Rail Technical Authority), and Calum Barrett (Transport For London (TfL) Head of Construction Engineering and Assurance) for their contribution to the writing of this chapter.

12.1. Introduction

Using temporary works on or adjacent to a railway can be complex and high risk, with multiple interfaces and multiple adjacent assets / landowners. Asset protection and causing minimal disruption to the service are paramount and the consequences of failure could be catastrophic. The primary aim of this chapter is to focus on the differences and variances that occur when managing temporary works in the rail environment. It will focus on rail processes and procedures that the temporary works professional will need to be aware of. It will touch on these processes and procedures in summary and will direct the reader to the correct documents for further reference and knowledge gain. It will also focus on the different interfaces, assurance and rail-specific roles the temporary works professional will typically have to engage with on a typical rail project. It will highlight the importance of protecting railway assets / structures, the extended design, construction methods and monitoring techniques typically adopted and implemented in the rail sector, which will include good practice. Rail works are as defined in the Railways and Other Guided Transport Systems (Safety) Regulations (ROGS) (HMG, 2006). In relation to ROGS, infrastructure owners will have their own safety authorisation and certification process.

This chapter will also provide a summary of lessons learned from temporary works incidents in the rail industry in recent years. These provide further appreciation and focus to the rail industry's processes and procedures, required for the effective management of temporary works. The railway environment can be one of increased risk, especially relating to temporary works operations.

12.2. Roles, responsibilities and competency assessments

On rail projects there are a number of roles which you may not find in other aspects of the construction industry. Different infrastructure managers, such as Network Rail Ltd (NR), London Underground Ltd (LUL) and High Speed 1 Ltd (HS1), have different roles but, in order to simplify this here, we will concentrate on roles from NR engineering assurance standards as would be applied to Network Rail Managed Infrastructure (NRMI). All NR projects (e.g. where NR is the client or for works affecting or potentially affecting NRMI in liaison with NR Asset Protection and Optimisation (ASPRO)) are deemed Common Safety Method for Risk Evaluation and Analysis (CSM-RA) significant. This is in addition to or parallel with Construction (Design and Management) Regulations (CDM) (HMG, 2015) compliance and NR/L2/RSE/100/02 (Network Rail, 2015). Within NR, these roles and their responsibilities are outlined within standards NR/L2/RSE/02009 (Network Rail, 2021c) and NR/L2/CIV/003 (Network Rail, 2020a), where specific tables of roles and corresponding responsibility, accountability, consultation and information (RACI) matrices provide further information on engineering roles and responsibilities.

Some of the key engineering roles which will typically be involved in temporary works are as follows.

- Asset manager (AM) – appointed by the infrastructure manager to manage all assets within the area and for a particular discipline. For major temporary works (TW) schemes, an approval in principle (AiP) will be required for endorsement by the AM, using the form A engineering assurance process as required by NR/L2/CIV/003 (Network Rail, 2020a). Multiple TW submissions as part of a larger scheme of works, subject to acceptance, can be submitted to the AM under a single form A. For lower-risk TW designs, the AM might defer to the project engineer or designated project engineer for approval and management of the same.
- Designated project engineer (DPE) – a mandatory appointment by NR as their engineering representative who will be assisted by the design and construction team, where appointed, through construction assurance activities. They will support the project manager in accepting the constructed asset quality, in line with the deliverable's specification. They must maintain a technical and commercial independence from the contracted organisation and demonstrate the ability to meet the non-technical skills and technical competence requirements required for the role in NR/L2/RSE/02009 (Network Rail, 2021c) (for NR projects). One DPE will be appointed on every project, and they will not undertake the role of contractor engineering manager (CEM) or contractor's responsible engineer (CRE) on any other project. The DPE, as the principal engineering representative, leads on all engineering issues associated with the execution of the principal contractor's (PC) role. The DPE will coordinate the engineering activities for all construction organisations during the construction phase with the full cooperation of all CEMs and CREs in undertaking these duties. They will document any additional process or requirements for discharging these duties, included in all contract requirements technical (CR-T[Note 1]) (or equivalent) documentation, for each package of work. They will put in place a suitable means of documenting the essential requirements of the national technical specification notices (NTSN[Note 2]) and provide the technical assurance process that provides the evidence that these requirements have been met. They are accountable for the technical integration of the works and for identifying the principal options along with the bespoke arrangements to be used for the technical integration of the works. They will document the technical integration requirements in the CR-T (or equivalent) for all packages of work and include the arrangement details in the project engineering and construction management plan. All variations and the

management of the same are typically carried out by the DPE in collaboration with the CEM / CRE. The DPE is the key role in accepting all engineering deliverables required to be submitted by the CEM and is typically the role which manages and executes interdisciplinary checks / reviews (IDC / IDR), which are a fundamentally important process in managing rail-related design.The DPE will undertake the acceptance of 'as-built' records and ensure that they accurately reflect the status of the infrastructure prior to being taken into operational use by the infrastructure manager. A health and safety file or close-out for any project cannot take place until the DPE has confirmed that all engineering deliverables have been submitted and accepted. The DPE will support the project manager (PM) in undertaking the 'entry into service' process in accordance with NR/L2/INI/CP0075 (Network Rail, 2019a).

Note 1: a form of agreement and specification for works, appended to a contract, which creates and defines obligations between a project and a design and / or construction organisation, in the form of technical requirements.

Note 2: from the outset, the project engineer will have determined which specifications are relevant and to what extent they apply.

- Contractor's engineering manager (CEM) – senior engineering person who is typically appointed within each design and / or construction organisation contracted on rail projects with overall accountability for all that organisation's engineering activities. The CEM will also likely have their scope of responsibility extended to include those subcontracted organisations managed by the principal contractor. The CEM for a single organisation will implement the process that integrates all designs (including temporary works design) for a project. The CEM, depending on the scale of the project, may not be able to dedicate sufficient time to TWs and may rely on the appointment of a suitably qualified CRE (with formal TWC appointment) to maintain responsibility for TWs, who would in turn report to the CEM on a regular basis. The DPE and discipline project engineer (DiPE) will review and, where suitable, accept the nomination of the CEM for their project.
- Contractor's responsible engineer (CRE) – the CEM for a contracted organisation, internal or external to the infrastructure manager, will, with the support of discipline specialists where required internally, identify the discipline (civil, mechanical and electrical, overhead line equipment (OLE), permanent way (PWay) and, depending on scale of the project, a CRE specifically covering TWs), CRE design and / or construction resources and other non-appointed supporting discipline engineering resources required for the delivery of a project with respect to the contracted organisation's scope of works. The CEM / CRE will demonstrate the ability to meet the behavioural and technical competence requirements required for the discipline project engineering role as per NR/L2/RSE/02009 (Network Rail, 2021c) (for NR projects). The appointed CEM may also act as the CRE for a specific engineering discipline, subject to the individual demonstrating their competence (under NR/LUL/TfL competency management systems) and availability to carry out all the required duties. There will be at least one CRE for each engineering discipline that will undertake works on a contract. Where a CRE has been appointed for an engineering discipline, they will manage the interface between design, installation and testing activities. Where separate CREs (design) and CREs (construction) are appointed for the same discipline on a project, the CEM will define and document the division of responsibilities between the two interfacing CREs. The CEM and all CREs retain full accountability for their design,

including its capability to interface correctly with other designs, and submit deliverables to the infrastructure manager for acceptance. The DPE and DiPE will review and, where suitable, accept the nomination by the CEM of the CRE(s) for their project.

- Project engineer (PE) – responsible for undertaking PE's duties for a specific discipline as required by NR/L2/RSE/02009 (Network Rail, 2021c). The PE will support the DPE in the PE's field of expertise and support the engineering assurance of deliverables to the appropriate engineering assurance standard.

Key non-engineering roles for projects on NRMI include sponsors and project managers, which are described below.

- Sponsor – the lead and guiding mind throughout the project / programme lifecycle and the link between client, funder, system operator, delivery agent and key stakeholders. They make sure things happen and get things done. They are vital in making the case for rail as a safe, efficient and reliable transport option for passengers and freight users alike. They are ultimately accountable for the success of a project, including enablement or realisation of benefits in line with the client's needs.
- Project manager – manages the project on a day-to-day basis, coordinating all project activities to deliver the required outcome to time, cost, quality and budget. The day-to-day running of a project is typically in accordance with NR standard NR/L2/P3M/201: *Project Acceleration in a Controlled Environment (PACE)* (Network Rail, 2021b), which comprises phases of the project lifecycle illustrated in Table 12.1.

Table 12.1 Network Rail guidance on temporary works – the PACE phases (reproduced with kind permission of Network Rail) (continued on next page)

Phase		Aim / Outcomes
A	Project Initiation	The purpose of this phase is to: ■ appoint the project's sponsor; ■ appoint the team required to plan and enter the next phase; and ■ prepare the Project Management Plan.
1	Strategic Development & Project Selection	The purpose of this phase is to: ■ determine and baseline the client requirements and value drivers for the project; ■ identify constraints that will impact the feasibility of the project; and ■ determine an MVP single option (the project) that meets the client requirements within the identified constraints.

Table 12.1 Continued

Phase		Aim / Outcomes
2	Project Design	The purpose of this phase is to: ■ undertake development of the MVP single option to agree Approval in Principle and standards to which the project will be constructed; and ■ produce an approved ready for construction design which meets the client's requirements.
3	Project Delivery	The purpose of this phase is to: ■ safely and efficiently deliver the project to the design: ■ undertake final verification and validation of requirements, testing, and commissioning; and ■ asset enters service.
4	Project Close	The purpose of this phase is to: ■ transfer of asset from the project team to the Infrastructure Manager; ■ project manager closes project systems and demobilises; and ■ sponsor demonstrates delivery of the MVP option to client's requirements based on their value driver and formally closes the project and related support systems.

Note: this points to the early consideration of constructability and temporary works from early *Project Acceleration in a Controlled Environment* stages.

12.3. Rail process and procedures including engagement and interface

The purpose of this section is to provide a summary of the relationship between the principal contractor (PC), contractors and the infrastructure manager and how, on a typical rail project, each party is to enable the works to be constructed in accordance with the project scope and programme, while having the appropriate consents in place, without causing any detrimental effects to the operational railway. This of course includes temporary works as an integral aspect of any construction works. The infrastructure manager will manage the access to the network for external parties (including PCs) carrying out works that have an interface with the rail infrastructure by arranging all forms of access arrangements, consents and licenses for work, including safety-critical work. The costs for engineering access to a section of track as requested will be recoverable from the external party. What is known as the 'rail sponsor' is typically the person / role within the infrastructure management who will be the primary point of contact between any party wishing to specify and / or carry out changes that may affect the rail infrastructure and who provides the project remit to the appointed PC's rail interface manager. All externally delivered projects (whether third party or outside party) that are being supported by any rail authority function are considered under the asset protection and optimisation (ASPRO) service protocol and process outlined by the rail authority (see section 12.5). Across most work packages within a rail-associated project, there are interfaces with the infrastructure manager of varying types, such as:

- permanent and temporary works design interfaces with rail infrastructure
- construction interfaces where the PC works alongside the operational railway
- construction interfaces where the PC is required to maintain access to the rail infrastructure
- construction interfaces where the PC works on the operational railway in possession both within the PC's own worksite and where the PC works alongside other rail contractors.

The infrastructure manager will require the PC to maintain an interface register (which should include a management mechanism), which will include input from the temporary works team. The CEM, supported by the CRE(s), principal contractor's temporary works coordinator (PC's TWC) and other temporary works coordinators (TWCs), will coordinate all interfaces between the permanent works design and the temporary works design – see Health & Safety by Design guidance (H&SbD, ORR, n.d.). This requires close collaboration and close coordination between the parties. Indeed, the coordination and engagement of the permanent works designs with temporary works designs often lead to efficient construction, minimising costs and carbon footprint. To manage the various interfaces, a variety of mechanisms can be used, as shown in Table 12.2.

Table 12.2 Interface management mechanisms

Design interface – infrastructure manager railway infrastructure manager relevant works	Managed through a project-specific design assurance process
Design interface – railway infrastructure manager infrastructure manager property works	Managed through the NR/L2/RSE/02009/02 (Network Rail, 2021e) engineering management for projects process
Construction interface – working in possession* from the PC's own worksite	Interface managed through works package plans and delivering work within possession (DWWP) process to liaise with other works in same area
Construction interface – working on railway within possession	Interface managed through works package plans and DWWP process to liaise with other contractors in same area
Construction interface – shared access or use of railway infrastructure manager infrastructure manager land	Managed by the production of an interface control document and agreements in the land and property agreements
Railway infrastructure manager infrastructure manager land and property	Managed through the PC's land and property process. The PC's land and property team will submit asset protection and optimisation (ASPRO) queries to the PC's rail interface team for response to assist in addressing any requirements, agreements and licenses

*Note: possession is when a section of track is required for maintenance, or a construction operation where trains cannot run. It is handed over by the operators to the engineers, who 'take possession'. Special protective measures are used to prevent access by unauthorised trains. When the track is returned to the operators, the engineers 'give up possession' and normal running returns.

12.4. Rail engineering assurance

Engineering assurance may be defined as the process by which the technical integrity of a product, process or system is assured by the contractor as it is developed and subsequently monitored and maintained. This is a method of consistently verifying that building and civil engineering infrastructure designs:

- meet client, project and stakeholder requirements
- are suitable in terms of safety, security, constructability, functionality, appearance and other architectural aspects, planning conditions and requirements, heritage requirements, spatial capacity, human factors / ergonomics, wayfinding and navigation, accessibility, maintainability, sustainability (including sustaining the condition or operational serviceability of an asset over time), environmental sustainability (including decarbonisation and social value), specification, cost, decommissioning and demolition.

Quality design, checking, execution and assurance, by competent individuals, are critical to controlling business and safety risks and opportunities throughout the asset lifecycle. To provide the requisite documentation and technical assurance to the infrastructure manager, the various interfaces and assurances with the conventional railway must be identified and categorised. The works are categorised according to the definitions in the relevant works and rail property works. Relevant works are defined as works carried out by the PC, which are situated on, across, under or within the NR property, or that may adversely affect infrastructure.

The project team will seek input from the PE / DPE to establish agreement as to the route to assurance for the engineering deliverables and the level of design check category (DCC), in accordance with BS 5975-1:2024 (BSI, 2024), prior to AiP. NR standard NR/L3/CIV/0063 (Network Rail, 2021d) details the requirements regarding temporary works for piling, drilling, craneage, other mechanical equipment and similar operations on or adjacent to the railway, where such plant might collapse closer than 4 m from the boundary of NR-managed infrastructure and critical boundaries within LUL. Engineering assurance supports the effective delivery of 'right first time' products by providing evidence that the products generated by technical activities are safe, satisfy requirements, achieve the design intent and comply with regulations, standards and specifications.

Engineering assurance in the context of the PC and their supply chain is the planning and collection of evidence, at suitable periods within a technical lifecycle, to provide progressive assurance that the design intent and requirements are satisfied. By following a progressive assurance approach, the PC can assure themselves and their stakeholders that they are following the agreed procedures, undertaking the relevant checking and approval, have confirmed compliance with the project requirements, interfaces, standards and constraints and have sought acceptance of any departures in a timely manner. The CEM, working with the DPE, typically leads on the coordination, submission, compliance and management of all assurance activities and the resolution of assurance-related issues. They are responsible for preparing, implementing and maintaining along with keeping up to date the PC's technical assurance plan. This role can be aligned with the quality and assurance manager role if such a role has been appointed to the project.

In relation to temporary works on NRMI, all such works will be assured under NR standard NR/L2/RSE/02009 (Network Rail, 2021c) and standard NR/L2/CIV/003 (Network Rail, 2020a).

Note: other infrastructure managers will be assured under their own infrastructure manager assurance procedures, such as TFL's TW standard S1062 (TfL, 2015), assurance standard S1538 (TfL, 2022) and outside parties standard S1023 (TfL, 2017). TFL's challenges align with NR and there is a huge benefit from early engagement. Similar to NR, TfL seeks to mitigate design and construction challenges with their PCs and supply chain through early engagement. TfL also utilises competency-assessed individuals and ensures that risks are mitigated as low as reasonably practicable (ALARP) as per the requirements of ROGS (HMG, 2006).

Consultation will include discussions on the designs / options being considered as well as subsequent design developments. The DPE will engage with the PE to agree the assurance arrangements and the design check category(s) for detailed design stage with the CEM / CRE prior to issue of forms A, B and C on NR projects. The technical assurance of the detailed design of works to building and civil engineering infrastructure will be documented and accepted within NR/L2/CIV/003/FormB (Network Rail, 2020a), *Certificate of Design and Check for Permanent Works*. This form will be submitted by the CRE and confirm that the design has been:

- carried out in accordance with NR/L2/CIV/003/Form A (Network Rail, 2020a)
- checked in accordance with the agreed design check category (DCC).

In addition, the CRE will confirm the matters that have been reviewed in the design by completing the design checklist in NR/L2/CIV/003 (Network Rail, 2020a). There is no need to complete NR/L2/CIV/003/Form B (Network Rail, 2020a) for a permanent / temporary works design submission if the asset manager and / or PE has accepted in writing the selection of DCC 0. In such cases, the simplified NR/L2/CIV/003/Form H (Network Rail, 2020a) will be used. A design check category of 0 will not be adopted for designs that vary from the requirements of the infrastructure manager's standards or railway group standards. Engineering assurance of the temporary works detailed design will be documented and accepted within NR/L2/CIV/003/Form C (Network Rail, 2020a). This form will be submitted by the CRE and confirm that the design has been:

- carried out in accordance with part 1 of NR/L2/CIV/003/Form C (Network Rail, 2020a)
- checked in accordance with the agreed DCC
- signed by the CRE of the organisation responsible for the design
- submitted to NR for approval.

In addition, the CRE will confirm the matters that have been reviewed in the design by completing the design checklist in NR/L2/CIV/003/Form C (Network Rail, 2020a). Any significant site changes to temporary works which might affect rail authority-managed infrastructure will be agreed with the designer prior to execution, with redesign, checking and assurance as appropriate. From an NR perspective, the assurance standards include the following:

- NR/L2/RSE/02009 (Network Rail, 2021c)
- NR/L2/SIG/30035 (Network Rail, 2019c)

- NR/L2/TRK/2500 (Network Rail, 2020d)
- NR/L2/ELP/27311 (Network Rail, 2020c)
- NR/L2/TEL/30022 (Network Rail, 2011b)
- NR/L2/ERG/24020 (Network Rail, 2011a)
- System integration deliverables for System Safety, Interoperability, System security, Performance, Reliability, Operational readiness and Maintenance readiness.

Projects will allow for adequate supervision and construction assurance. Constructability reviews will be carried out during the design and will include specialist subcontractors, principal contractor, designers and principal designer to provide guidance on viability, capacity and size of units needed for the proposed temporary works schemes. The TWC should be a key role included in these reviews. If a TWC is not available, then an engineering manager with TW experience and competence should be present to cover temporary works as a discipline. The TW permit is the foremost construction assurance document / process which manages the safe delivery of temporary works on the ground. It can be supported by a formal ITP (inspection and test plan) as a supporting document, which should be appended. NR is also a railway duty holder and is required to use the common safety method (CSM) for risk assessment. A project hazard record, as required under CSM risk assessment and evaluation requirements, could be used to identify hazards that need to be controlled by suitable and sufficient risk assessment.

12.5. Guidance on protection of outside and third-party assets

There are many developments which take place on or near the railway by external parties to NR. Asset protection and optimisation teams (ASPRO) provide expert railway assistance and support to those who are planning activities on or near the railway. They help with safety, construction, planning, legal matters and utilities. All externally delivered projects (whether third-party or outside party*) that are being supported by any infrastructure manager function are considered under the ASPRO service protocol and process outlined by the infrastructure manager.

*Note: an outside party is an individual, member of the public or organisation other than NR which promotes, funds, designs, constructs, owns and maintains the works but will not physically change the operation of railway infrastructure and will interface with the railway infrastructure; a third party is an organisation other than NR intending to undertake works that physically change or impact the operation, assets, design or other functionality of NR's infrastructure. Third parties may include: train and freight operation companies and station operators, local passenger transport executives, the Department for Transport, design build finance operate, private finance initiative or commercial developers and so on.

Responsibilities of the ASPRO teams include the following.

- Engage with all contractors, attend meetings, review risk assessments and method statements (RAMS), review design submissions and attend site.
- Link to other rail teams, IP, maintenance, route asset managers, property and so on.
- Manage access to the conventional railway (possessions), both disruptive and non-disruptive, on behalf of the PC.

- Negotiate disruptive track access in line with infrastructure manager requirements to ensure less disruption to the fare-paying passenger; this will also reduce the amount of compensation paid by the PC for track access.
- Negotiate all disruptive access and changes to the rules of the route (ROR) with the train operators and attend the necessary meetings for all possession planning meetings across the routes.

All route ASPRO teams will have a mechanism in place to identify, engage with and manage both internal and external stakeholders. This will include a clearly documented set of processes and controls relating to stakeholder identification and interface for projects and referrals. For further information refer to the links at the end of this chapter.

12.6. Health and safety by design – best practice in rail design and construction

The Office of Rail and Road (ORR) strategy for Health and Safety by Design (H&SbD) (ORR, n.d.) states

> H&SbD is the elimination or control of H&S risks in infrastructure, products or processes by early consideration and addressing those potential risks at the design stage.

Too often the planning of new works, or the modification of existing assets, fail to take simple steps to avoid future hazards to the health and safety of workers and the public, contrary to CDM 2015 (HMG, 2015). Better planning, more consistent application of standards, integration of temporary works with permanent works, and incorporating advice from lessons learnt and good practice will help to improve safety, reduce error and improve efficiency. An additional benefit of improved efficiency and reduction in error is to intrinsically reduce the carbon footprint.

NR and its supply chain have worked collaboratively over many years to develop H&SbD guidance. H&SbD principles have been developed to enable safe design, including:

- having plans and permits in place
- ensuring competency
- identifying hazards and eliminating or mitigating them, using the CSM for risk evaluation and assessment
- seeking to improve H&SbD performance by sharing lessons learnt, good practice, education, new technology, innovation and any other appropriate initiative, for whole industry benefit, fostering a culture of continuous learning, without apportioning blame
- assessing the safety implications of design changes and implementing mitigation when design change imports hazards; updating hazard logs
- fostering open and honest dialogue and collaboration with clients, contractors, designers and other stakeholders in order to ensure that designs are safe; listening to stakeholders
- fully understanding the site, its vicinity to the operational railway, and assets which may be affected by the designs, ensuring sufficient knowledge of the site and its vicinity
- upholding the values of relevant professional institutions and companies; doing what is considered to be the right thing; demonstrating leadership in H&SbD
- refusing to carry out design and / or execution which is believed to be unsafe and escalating any situation which is believed to be the case.

The principles are now embedded in H&SbD activities and the Buildings & Civils Working Group (H&SbD B&C WG) has enabled the production of non-mandatory guidance that helps the wider railway community. Examples include the following.

- H&SbD B&C WG *Guidance Note: Early Focus on Constructability and Temporary Works* (Network Rail, 2023)

A disproportionate number of incidents / problems within railway construction projects relate to constructability and lack of early consideration of temporary works. Earlier focus in these areas leads to reduced safety risk and improved execution of design and implementation, together with a reduction in rework. This is key guidance as it is a compilation of over 15 years of industry good practice, incorporating many lessons learnt and examples of good practice. Issue 4 was published in November 2023.

- H&SbD B&C WG *Guidance for the Use of Warning Triangles* (Network Rail, 2022)

This guidance has been through a comprehensive review by the B&C WG and provides non-mandatory guidance on the provision of risk warning triangles on design drawings to complement good industry practice and provide consistency within the rail environment. The provision of suitable warning triangles helps to focus the reader on unusual aspects of the design which may demand closer attention.

- H&SbD B&C WG lifesaving rules (LSR) aide memoire (Safety Central, n.d.)

This guidance enables temporary works and permanent works solutions to embed the Network Rail lifesaving rules within designs.

All the above guides are freely available to download on the H&SbD page within the NR Safety Central (https://safety.networkrail.co.uk/safety/prevention-through-engineering-and-design/safe-by-design-groups/building-and-civils/).

12.7. Temporary works and working on the railway

The railway is a complex system providing an environment where temporary works require special consideration. The railway has a number of sub-systems including rolling stock, permanent way, signalling systems, AC overhead line power systems, DC power systems and buildings / civils structures, including stations, platforms, canopies, retaining walls, cuttings, embankments, tunnels, footbridges, bridges, drainage systems and so on. As such, temporary works requirements need to account for, and not adversely affect, all these systems. Key aspects of design and execution of temporary works in a railway system environment include the following.

12.7.1 Risk from adjacent sites

It is important to reduce risks to temporary works from adjacent sites which could adversely affect railway infrastructure and trains. Standards are in place to mitigate the risks of temporary works adversely affecting the railway, through use or failure, and need to be adhered to. For instance, NR standard NR/L2/CIV/0063 (Network Rail, 2021d) and ASPRO standards give requirements for mitigating risks, should use or failure of temporary works be within 4 m of the boundary of NRMI.

12.7.2 Maintaining rail operations

The major challenge of closing the railway for any scheduled work requires extensive planning and coordination between many interested parties. Shutting down of the railway to carry out work is costly; unscheduled over-running of work can be very costly. Work carried out in such circumstances normally has strict time constraints and detailed half-hour increment planning is normally required to ensure that 'hand back' of the railway infrastructure can be safely achieved on time.

12.7.3 Understanding the site

Railway sites can have significant access restrictions and many safety-critical buried services. It is essential that all details of existing assets are provided as part of any design. The CDM pre-construction information should be adequate and complete; see *Managing Health and Safety in Construction: Guidance on Construction (Design and Management) Regulations* (HSE, 2015) and NR/L2/OHS/0047 (Network Rail, 2019b). Asset managers and maintainers, who often have detailed records and knowledge of the infrastructure, should be contacted at an early stage once this is received. No design should take place without an appropriate ground investigation and site inspection since without this it cannot be assumed that the design problem and scheme constraints have been fully understood. For the most part, the UK's railways contain aging infrastructure which often contain hidden obstructions, such as hidden utilities that may not be obvious. In planning ground investigations, reference should be made to the requirements of NR/L2/CIV/071; design of geotechnical structures (Network rail 2011c). All surveys and investigations should be targeted to address risks to be resolved or refined. Surveys, particularly when intrusive, can be disproportionately expensive and, without proper gap analysis, can add little value.

12.7.4 Constructability reviews

Designers should undertake constructability reviews at various stages of a project. The chosen stages may vary to suit complexity, design development or changes to access or site conditions. Attendance at constructability reviews should not be limited to the design team, and the involvement of contractors and / or other independent construction professionals or peers is encouraged. Constructability reviews will highlight issues which may affect safety, operations, temporary works requirements and so on and will inform designer risk assessments. They should also consider the likely impact on railway neighbours and may include possession availability and design change. With a constructability review in place, the IDC / IDR will be able to consider both permanent works and temporary works interfaces with other disciplines. An example of (mandatory) IDC / IDR activity is the provision of a complex scaffold structure to gain access to the roof above a concourse of a major station for roof repair / refurbishment: while the scaffolding system may be reasonably straightforward, the team will need to consider interdisciplinary matters such as, but not limited to:

- fire assessment to confirm station fire strategy will not be compromised – or what the temporary mitigations are. There are additional requirements for stations that are subsurface which are covered by LUL standard S1088, section 12 (TfL, 2020)
- security team input to location of hoardings, doors, inspection windows and other means of access to critical safety systems
- protection of heritage features
- confirmation that hoarding lines do not significantly affect pedestrian flows and that mitigation measures are in place

- temporary lighting design (main and emergency) may be significantly affected by the scaffold and may require building services changes
- impact on fire detection and smoke extraction systems
- impact on station information security systems (SISS) equipment
- a typical scaffold design often stops at support level and the foundations require further assesment and design submission to ensure there is no detrimental effect on suspended platform floor structures and so on.

12.7.5 Scaffolding

NR does not have any specific standards for scaffolding design but relies on National Access and Scaffolding Confederation (NASC) and British Standards. Within the railway environment there are many considerations that influence the design and construction of scaffolding access, protection and shoring. Equally, selection of the most appropriate scaffolding solution may also be heavily influenced by the environment within the railway infrastructure. Operational equipment, signal sighting and so on cannot remain hidden or disrupted due to scaffold erection around stations and other rail assets. It is key that a competent and relevant CRE with required railway experience reviews all such designs to check that all interfaces are included and captured. When working close to overhead line equipment (OLE) and other exposed live electrical equipment, the use of glass reinforced plastic (GRP) or other non-conducting materials is necessary. Design of bonding will be required for metallic items in some situations adjacent to OLE areas. Interface and support from an electrification and power (E&P) project engineer should always be sought and form part of an approved designed solution.

12.7.6 Working platforms and hoardings

The requirements for piling, drilling, cranes and other mechanical equipment along with similar operations on or adjacent to the railway are set out in NR/L3/CIV/0063 (Network Rail, 2021d). Particular attention is drawn to section 8 which sets out the requirements for management of the working platform. Additional guidance can also be found through the Temporary Works Forum working platform guidance document (TWf, 2019). A properly designed and maintained piling / crane platform is a crucial element of temporary works and is key to managing risks associated with these works.

Hoardings and protection to / from active work are required at all interfaces between construction and non-construction / project personnel. Hoardings have the dual purpose of providing a barrier that protects construction operatives from potential danger, such as live OLE, third rail and moving rolling stock, as well as protecting non-construction / project personnel during construction activities. Specifications for hoarding and protection barriers can be found in British and NR standards. Typical key issues for design are:

- fire and flame spread properties
- impact, wind and crowd loading
- dynamic train pressures
- loading from mobile plant / station platform equipment
- protection against materials being thrown over a barrier / hoarding
- ensuring that electrical and rail vehicle gauge clearance are not compromised
- ensuring that signals are not obscured

- ensuring that sight lines for train despatch staff are not impeded at all locations where despatch staff are located
- ensuring that operational equipment remains accessible and access for maintenance is not prevented
- ensuring that pedestrian flows are safe – it may be necessary to carry out pedestrian flow modelling to validate the effect of building a hoarding, particularly when working in and around railway stations. When hoardings affect circulation of people, consideration should be given to trialling the new layout in advance under controlled conditions in order to fully understand the impacts. This may be done by using tape or easily removable barriers and observation, particularly at peak times, and is a good way of confirming pedestrian flow predictions
- checking that any equipment to be fixed to hoardings post-erection does not affect clearances and / or pedestrian flow
- when working on stations, hoarding proposals require the agreement of train operating companies (TOCs) and / or the NR major station management team
- from an LUL perspective, there is also an interface on LUL underground stations with train doors and not obstructing access for passengers – clear signage and minimum platform widths and so on. This might result in a changed operation of trains for the affected station. If the hoarding is protecting the station from the external environment, then water ingress proves to be a challenge for all concerned.

12.7.7 Protection decks

In areas where construction activities are to be carried out over a live railway or station while maintaining 'business as usual' operations, a protective / working barrier / deck may need to be installed. The functionality of this barrier can change throughout the duration of any works. Early understanding of the full functionality of the proposed protection / working deck is essential if an effective and safe system of work is to be established for all aspects of the work to be undertaken.

12.7.8 Earthing and bonding of metallic structures, gauging and electrical clearances

Earthing and bonding requirements for temporary (and permanent) metallic structures are frequently overlooked by designers and many NR standards must be complied with e.g. NR/L2/ELP/21085 (Network Rail, 2020b) for AC lines and NR/L3/MTC/EP0152 (Network Rail, 2011d) for DC lines. Multi-disciplinary teams must be in place to ensure adequate earthing and bonding. Gauging is the activity of demonstrating compatibility between train and train, and between train and infrastructure, to ensure that sufficient space exists around a moving train (referred to as clearance), ensuring safe operation. The contractor must be satisfied and satisfy the employer's representative that:

- all temporary works are correctly designed for the expected loading conditions
- all gauging requirements are fully satisfied whenever track is open to traffic with TW in place. This includes the envelope above any train and especially the electrical clearance to the pantograph horns in 25 kV areas. There are many NR and Rail Safety and Standards Board (RSSB) Group standards which need to be referred to in order to ensure temporary works do not compromise gauging and electrical clearances.

12.7.9 Temporary works affecting assets (e.g. track support zone)

The risk of asset instability due to temporary excavations during the works requires careful consideration and should be explicitly addressed during the design and design risk assessment process and then onto the construction phase through the health and safety plan, work package plans and task briefings for all applicable projects. Temporary works either within the track support zone (TSZ) (see definition within NR/L2/CIV/177, Network Rail, 2021a) or affecting the stability of the TSZ have the potential to cause changes to track geometry leading to train derailments. Key aspects of this standard include the need to understand the existing track geometry and position in relation to allowable tolerance. The contractor is to carry out a risk assessment and agree a track monitoring plan (including trigger levels and interventions) with infrastructure maintenance engineers in advance of the works, where the works might affect track support, restraint or geometry. Where the works could potentially lead to buckling of the rail, exacerbated by higher temperatures, the critical rail temperature should also feature in calculations, monitoring and management, in accordance with NR/L2/TRK/7012 – (Network Rail,2020e).

12.8. Rail-related plant and equipment: planning and considerations for temporary works

There are many types of railway-specific plant and equipment which are used on the railway to undertake works activity – for example, 'Kirow' cranes and road rail vehicles (RRVs) which are adapted to run on the road or railway. Over one hundred current NR standards exist and guidance is given in NR's NR/GN/RMVP/0200 (Network Rail, 2018) and associated briefings, such as:

- module P301 for road rail access points (RRAP)
- module P503 for lifting operations
- Rail Safety and Standards Board (RSSB)-owned mechanical and electrical engineering (M&EE) codes of practice, such as COP0032 for adjacent line open (ALO) works (RSSB, 2020) and COP0044 (RSSB, 2024) for re-railing Kirow cranes.

It is essential that the requirements of these standards are followed so that plant and equipment are carefully considered by competent persons for TW operations and the temporary condition of permanent works in order to ensure the safety of the railway.

REFERENCES*

BSI (2024) BS 5975-1:2024: Management procedures for the control of temporary works. Code of practice. BSI, London, UK.

HMG (Her Majesty's Government) (2006) The Railways and Other Guided Transport Systems Regulations. The Stationery Office, London, UK.

HMG (2015) Construction (Design and Management) Regulations. The Stationery Office, London, UK.

HSE (Health and Safety Executive) (2015) *L153: Managing Health and Safety in Construction: Guidance on Construction (Design and Management) Regulations*. HSE, London, UK.

*The various standards cited in this chapter are under continuous improvement, hence may be subject to change. Reference must be made to current standards at all times.

TWf (Temporary Works Forum) (2019) Working platforms – Design of granular working platforms for construction plant: A guide to good practice. https://www.twforum.org.uk/viewdocument/working-platforms-design-of-granu (accessed 10/01/2025).

SPECIFIC RAIL REFERENCES

European Union Agency for Railways (n.d.) *Common Safety Method (CSM)*. European Union Agency for Railways, Valenciennes, France.

H&SbD B&C WG (Network Rail Health & Safety by Design; Building and Civils Working Group) (n.d.) https://safety.networkrail.co.uk/safety/prevention-through-engineering-and-design/safe-by-design-groups/building-and-civils/ (accessed 10/01/2025).

Network Rail (2011a) NR/L2/ERG/24020 – *Engineering Assurance Requirements for Ergonomics Within Design and Development Projects*. Network Rail, London, UK.

Network Rail (2011b) NR/L2/TEL/30022 – *Engineering Assurance Arrangements for the Design and Implementation of Electrical Power*. Network Rail, London, UK.

Network Rail (2011c) NR/L2/CIV/071 *Design of Geotechnical structures*. Network Rail, London, UK.

Network Rail (2011d) NR/L3/MTC/EP0152 – *Working on or Adjacent to Conductor Rail*. Network Rail, London, UK.

Network Rail (2015) NR/L2/RSE/100/02 – *Application of the Common Safety Method for Risk Evaluation and Assessment*. Network Rail, London, UK.

Network Rail (2018) NR/GN/RMVP/0200 – *Infrastructure Plant Manual Guidance*. Network Rail, London, UK.

Network Rail (2019a) NR/L2/INI/CP0075 – *Entry into Operational Service*. Network Rail, London, UK.

Network Rail (2019b) NR/L2/OHS/0047 – *Managing Health and Safety in Construction (Application of the Construction (Design and Management) Regulations to Network Rail)*. Network Rail, London, UK.

Network Rail (2019c) NR/L2/SIG/30035 – *Signalling and Level Crossing Approval Process*. Network Rail, London, UK.

Network Rail (2020a) NR/L2/CIV/003 – *Engineering and Architectural Assurance of Building and Civil Engineering Works*. Network Rail, London, UK.

Network Rail (2020b) NR/L2/ELP/21085 – *Earthing and Bonding on AC Electrified Railways*. Network Rail, London, UK.

Network Rail (2020c) NR/L2/ELP/27311 – *Engineering Assurance Arrangements for the Communications Engineering Schemes and Services*. Network Rail, London, UK.

Network Rail (2020d) NR/L2/TRK/2500 – *Engineering Assurance Arrangements for the Design and Construction of Track*. Network Rail, London, UK.

Network Rail (2020e) NR/L3/TRK/7012 – *Critical Rail Temperature Management for Projects*. Network Rail, London, UK.

Network Rail (2021a) NR/L2/CIV/177 – *Monitoring Track Over or Adjacent to Construction Works*. Network Rail, London, UK.

Network Rail (2021b) NR/L2/P3M/201 – *Project Acceleration in a Controlled Environment*. Network Rail, London, UK.

Network Rail (2021c) NR/L2/RSE/02009 – *Engineering Management of Projects*. Network Rail, London, UK.

Network Rail (2021d) NR/L3/CIV/0063 – *Piling, Drilling, Crane, MEWP and SPMT Operations Adjacent to the Railway*. Network Rail, London, UK.

Network Rail (2021e) NR/L2/INI/02009/02 – *Engineering Management for Projects: The Management and Review of Engineering Deliverables*. Network Rail, London, UK.

Network Rail (2022) *H&SbD B&C WG Guidance for the use of Warning Triangles to Highlight Points of Unusual and Unmitigated Hazards on Engineering Drawings*, issue 3. Network Rail, London, UK. https://safetycentral.wpenginepowered.com/wp-content/uploads/2023/11/HSbD-BC-WG-Guidance-on-Use-of-Warning-Triangles-on-Engineering-Drawings-Issue-3-5.8.22.pdf (accessed 10/01/2025).

Network Rail (2023) *H&SbD B&C WG Guidance Note: Early Focus on Constructability and Temporary Works*, issue 4. Network Rail, London, UK. https://safetycentral.wpenginepowered.com/wp-content/uploads/2023/11/HSbD-Guidance.-Early-Focus-on-Constructability-and-Temporary-Works.-Issue-4-16.11.23.pdf (accessed 10/01/2025)

ORR (Office of Rail and Road) (n.d.) *Health and Safety by Design (H&SbD)*. ORR, London, UK.

RSSB (Rail Safety Standards Board) (2020) COP0032 – *Code of Practice for Any Line Open Working*. Rail Safety Standards Board, London, UK.

RSSB (2024) COP0044 – *Code of Practice for Re-Rigging Kirow Cranes Using Outriggers*. Rail Safety Standards Board, London, UK.

Safety Central (n.d.) Building and Civils Working Group. https://safety.networkrail.co.uk/safety/prevention-through-engineering-and-design/safe-by-design-groups/building-and-civils/ (accessed 24/01/2025).

TfL (Transport for London) (2015) S1062 – *Standard for Temporary Works*. Transport for London, London, UK.

TfL (2017) S1023 – *Standard for Infrastructure Protection*. Transport for London, London, UK.

TfL (2020) S1088 – *Managing changes to stations and shafts fire precaution*. Transport for London, London, UK.

TfL (2022) S1538 – *Standard for Assurance*. Transport for London, London, UK.

Note: Network Rail standards are available to all through the NR website at the following link: https://www.networkrail.co.uk/industry-and-commercial/third-party-investors/network-rail-is-open-for-business/reviewing-our-standards/

USEFUL WEBSITES

BSI (British Standards Institution) – www.bsigroup.com

H&SbD B&C WG (Network Rail Health & Safety by Design; Building and Civils Working Group) – https://safety.networkrail.co.uk/safety/prevention-through-engineering-and-design/safe-by-design-groups/building-and-civils/

HSE (Health and Safety Executive) – www.hse.gov.uk

Network Rail ASPRO – https://www.networkrail.co.uk/running-the-railway/looking-after-the-railway/asset-protection-and-optimisation/

NR (Network Rail) – www.networkrail.co.uk

TfL (Transport For London) – www.tfl.gov.uk

TWf (Temporary Works Forum) – www.twforum.org.uk

FURTHER READING

BSI (2017) PAS 8811: Temporary works. Major infrastructure client procedures. Code of practice. BSI, London, UK.

Ray Filip
ISBN 978-1-83608-525-6
https://doi.org/10.1108/978-1-83608-524-920251014

Chapter 13
Environment, sustainability and low carbon

Ray Filip
RKF Consult Ltd, UK

Sam Hurst
Freelancer, UK

13.1. Introduction

Until recently environmental issues, sustainability and emissions were rarely formally considered in the construction industry and arguably even less in temporary works. The construction industry is a major producer of greenhouse gas emissions and has come under increasing pressure to reduce its negative environmental impact. It now operates at the interface of two critical imperatives: firstly, compliance with environmental regulations in the present day and secondly, it is a moral obligation for all of us (as a species we are all in this together) to safeguard the planet for future generations. In 2021, the UK government introduced a strategic plan with policies and pledges aimed at achieving the target of 'net zero' emissions by 2050. This ambitious goal seeks to balance greenhouse gas emissions with their removal from the atmosphere, ultimately striving to curb global warming. However, without commitment and innovation, 'net zero' may be unachievable, as shown by the 4% reduction in material use between 2018 and 2022, far below the 17% required (UKGBC, 2021). True innovation should genuinely act to benefit the wider industry and aid commercialisation while also offering a more sustainable solution. Every engineer must practice ethically and contribute to sustainability in line with the Engineering Council's Code of Conduct.

The aim of this chapter is to help those involved in temporary works to become aware of the issues and their responsibilities in this complex field where our understanding is constantly developing. Consideration of the environmental impact of temporary works is a new concept and those involved need a better understanding of some of the legislation and guidance available.

There are seven principles of sustainable construction: sustainable / green design, durability to extend lifespan, energy efficiency to minimise usage throughout the whole lifespan, waste reduction, indoor air quality, water conservation and using sustainable building materials. Most of these can be applied to temporary works.

Long-term sustainability cannot be considered within a silo and must be considered holistically, with every decision within the engineering design process used to solve technical problems, increase efficiency and productivity and improve systems. There are challenges and opportunities presented by sustainable construction practices, including resource management, carbon footprint reduction and biodiversity preservation. There should be a focus on sustainability and helping ethical suppliers, while calling out those who take unfair advantage to cause confusion for commercial gain (commonly known as 'greenwash'). This chapter considers these two interconnected concepts, providing a signposting guide for navigating the complex landscape of environmental legislation, regulations and initiatives in construction. Although many items of legislation and regulation can relate directly to permanent works, those involved in temporary works need to be aware and mindful of them.

The Environment Act (HMG, 2021) requires the Secretary of State to set at least one long-term target for each of four priority areas: air quality, biodiversity, water and waste. It sets requirements in relation to biodiversity net gain and deforestation reduction. The most comprehensive guidance is the *Construction Environmental Manual* (Construction Industry Publications, 2021) which provides further information on environmental legislation, policy, best practice, compliance and environmental risk mitigation from design to construction. Some relevant legislation, regulations and standards are listed below (Note: this is not an exhaustive list, and additional legislation and regulations may apply, depending on the specific project and location).

13.1.1 Design

- Building Act (HMG, 1984): sets standards for building safety, accessibility and sustainability for England and Wales. Building regulations are made under this Act.
- Town and Country Planning Act (HMG, 1990a): requires planning permission for development in England and Wales, including consideration of design impact on visual amenity and local character.
- Listed Buildings and Conservation Areas Act (HMG, 1990b): protects listed buildings and their settings, requiring consent for alterations or demolition.
- National Planning Policy Framework (NPPF) (HMG, 2012): provides national guidance on design principles for sustainable development.

13.1.2 Ecology

- Wildlife and Countryside Act (HMG, 1981): protects flora, fauna, wildlife and habitats in Great Britain, including designated sites (sites of special scientific interest – SSSIs) and species.
- Conservation of Habitats and Species Regulations (HMG, 2017): implements the European Union Habitats Directive, protecting European protected species and their habitats.
- National Planning Policy Framework (NPPF) (HMG, 2012): promotes biodiversity net gain and avoidance of harm to protected species and habitats; also promotes energy efficiency and renewable energy generation in development proposals.

13.1.3 Contamination, waste and nuisance

- Control of Pollution Act (HMG, 1974): makes provisions for waste disposal, water pollution, noise, atmospheric pollution and public health.

- Environmental Protection Act (HMG, 1990c): establishes a framework for identifying and remediating contaminated land. Part 2A imposes liability on polluters for remediating contaminated land. Creates offences of causing statutory nuisance, including noise, dust and odours. Also creates duty of care for waste, requiring proper management and disposal.
- Noise and Statutory Nuisance Act (HMG, 1993): includes nuisance from vehicles, machinery and equipment.
- Control of Substances Hazardous to Health Regulations (HMG, 2002): state that hazardous materials need to be adequately controlled to prevent exposing persons to health hazards.
- Contaminated Land: Statutory Guidance (HMG, 2014): provides guidance on the application of the legislation.
- Common law nuisance: provides additional protection against unreasonable interference with the enjoyment of land or property.
- Planning guidance: considers potential nuisance impacts during the planning process.
- Waste (Circular Economy) (Amendment) Regulations (HMG, 2010a): aim to prevent waste, which will reduce environmental pollution with an emphasis on efficiency and reuse.
- Waste (England and Wales) Regulations (HMG, 2011a): implement the Waste Framework Directive, regulating waste collection, treatment and disposal.
- Construction Site Waste Regulations (HMG, 2016a): encourage the reuse and recycling of construction site waste.

13.1.4 Materials

- Construction Products Regulations (HMG, 2011b): set requirements for the safety, performance and sustainability of construction products.
- 2008/98/EC Waste Framework Directive (HMG, 2008a): promotes waste prevention, reuse and recycling, with implications for material choices in construction.
- BS 8895-1 (BSI, 2013): this standard offers a simple approach for designers to take account of waste and resource efficiency as detailed above – that is, the process of identify, investigate and implement.
- Modern Methods of Construction Strategy (MMC) (HMG, 2022): encourages the use of innovative and sustainable construction materials and methods.

13.1.5 Climate change

- Climate Change Act (HMG, 2008b): sets legally binding targets for certain greenhouse gas emissions reductions (net zero 2050) and confers powers to establish trading schemes to limit such emissions.

13.1.6 Water

- Water Resources Act (HMG, 1991): regulates water abstraction, abstraction licensing and river flow.
- Flood and Water Management Act (HMG, 2010b): promotes flood risk management and sustainable water management.
- Environmental Permitting (England and Wales) Regulations (HMG, 2016b): regulate certain discharges to water and groundwater courses.

13.1.7 Other resources

- UKGBC (UK Green Building Council) has produced circular economy guidance (see Useful websites at the end of this chapter) that provides comprehensive practical guidance to support construction clients who want to ask for circular economy principles in their project briefs for non-domestic built assets.
- BS 8001:2017 (BSI, 2017): this framework for implementing circular economy principles has been introduced to assist organisations in their transition from a linear mode of operation to a sustainable business model where resources are more efficiently managed to enhance financial, environmental and social benefits.

13.2. Design considerations

The overall philosophy for designers should be to use less 'stuff' over the life of the asset, such as concrete binder, new steel, non-local stone, timber and so on. Making the 'stuff' less carbon-intense (e.g. using high-GGBS cement, recycled steel, recycled aggregate, sustainably sourced timber) is unlikely to reduce global emissions due to resource scarcity and as such is the last resort after all effort to use less 'stuff' has been exhausted. The Institution of Structural Engineers (Halliwell, 2024) clarifies this, which we have summarised below.

Use less stuff:

- Build nothing – is construction required? Challenge the brief and educate the client.
- Build less – repurpose / refurbish / reuse. Using proprietary equipment is likely to be the main aim for temporary works designers. Also consider trying to reuse existing structural components.
- Build clever – appropriate design and structural configurations.
- Build efficiently – highly utilised structures; carbon-efficient materials; offset carbon.
- Minimise waste – prefabricate and improve construction practice.

Chapter 2 highlights that there is no distinction in the Construction (Design and Management) Regulations (HMG, 2015) between permanent works designers (PWD) and temporary works designers (TWD) and both should carefully consider the environmental impact of the decisions they make in their designs. Prioritising low carbon solutions in design and considering resource limitations and biodiversity are vital for reducing the construction industry's environmental impact, while also offering potential cost savings and promoting resource conservation. This requires greater collaboration between the various stakeholders and stronger regulation at the planning stage from local authorities and finance institutions.

By understanding both the regulatory requirements and the broader ethical responsibility for environmental stewardship, construction professionals can make informed choices that ensure the industry builds a thriving future, in harmony with the natural world. Examples of practical application of these design approaches include the following.

- Design for longevity and refurbish or repair existing structures rather than demolish and construct new structures (an example could be the use of self-healing concrete).
- Apply whole-life costings.

- Design to enable prefabricated hired equipment and components. There can be significant efficiencies to be made with factory-based manufacture and this could lead to efficiencies on site.
- Design to standard sizes to minimise off-cuts and material wastage. Many construction products (e.g. plywood and plasterboard) are manufactured to 2440 mm × 1220 mm (8 foot × 4 foot); utilising multiples of these sizes will reduce wastage. Careful transportation and site distribution will also limit damage and waste.
- Design to allow the use of recycled or reused materials.
- Review construction techniques to minimise environmental impacts.
- Design to balance cut / fill to limit the effort required to remove and replace materials.
- Consider durability and low-maintenance options for the final product.
- Design for efficient procurement and specification, being mindful of resource efficiency.
- Design for application of lean waste techniques in construction.
- Specify materials that are both cleanable and durable.
- Coordinate services runs to minimise trenches in working platforms.

During the design phase, three basic steps should be applied.

- Identify design options to reduce waste / carbon over the lifecycle of a project.
- Prioritise design options to determine viability of waste / carbon reduction.
- Implement design solutions, including sharing of continuous improvement to others in the industry.

13.3. Material considerations

Although there are multiple environmental impacts (global warming potential, ozone depletion potential, acidification potential, eutrophication potential, smog formation potential and primary energy use) of materials and products, carbon is used as the proxy metric for civil engineering comparisons.

Most temporary works solutions utilise proprietary equipment (edge protection systems, scaffolding, trench boxes, falsework etc.) which are often made from steel or aluminium and are reused many times on projects, so it could be argued they are by default 'sustainable'. However, consideration should be given to the materials being used and how the equipment is manufactured. Increasingly, structural engineers are reusing existing components rather than replacing them. As discussed at the 59th quarterly meeting of Temporary Works Forum (TWf), recent projects have shown that entire existing floor plates can be jacked (using temporary works) to new locations (to allow an additional floor to be added). This solution reused the existing structural components rather than demolishing and rebuilding with new components.

Proprietary equipment and standard forms that allow this should be prioritised. Timber and plywood are often used in formwork and site hoardings; they should be sourced from local sustainable sources (plywood transportation emissions are typically 40%), with consideration given to longevity (number of pours for formwork) or reuse. An alternative to plywood is Alkus formwork sheeting which could offer up to 1500 uses, which is substantially more than traditional plywood sheets; however, it can require careful handling and, as such, when abused would not offer the required benefits. Other alternatives are steel plate or recycled plastic sheeting if adequate consideration is given to the pressure generated by the wet concrete on the formwork. TWDs should become familiar with the carbon content and properties of materials that offer environmental benefits and utilise them wherever possible (see https://circularecology.com/embodied-carbon-footprint-database.html). The Wood Recyclers Association (see Useful websites section at the end of this chapter) has useful information on waste wood, especially classification and if deemed hazardous from demolition projects. Recycling has to be seen as the preferable (and in many cases cheaper) option than landfill.

Concrete is a major source of greenhouse gases. The Low Carbon Concrete Routemap (ICE, 2022) introduced the Low Carbon Concrete Group benchmark (LCCG benchmark) which summarises different types of concrete and looks at best practice for making and using concrete. The most sustainable concrete has a lower binder content, created by removing durability requirements, such as freeze–thaw and chloride ingress, for temporary works purposes by utilising a risk-based approach. Secondary principles, such as high cement replacement, low carbon, sustainably sourced constituent materials and products such as polymer concrete (a composite of polymer resin matrix with aggregates which does not contain hydrated cement) are now regularly used in permanent works and increasingly in temporary works for items such as temporary foundations or thrust blocks. All designers should become familiar with the properties of these concretes, particularly the slower rate of strength gain and the supplier's ability to mitigate these risks with curing / maturing technology. This can cause issues with striking formwork / falsework times, plus the capacity of partially cured concrete slabs to support temporary imposed loads (material storage or loads from construction machinery), with the result that back-propping will become more likely. Lower concrete strength may become a significant issue when using 'top-restrained falsework' (Pallett and Filip, 2019, chapter 22), which relies on items of permanent works such as columns for lateral stability. *Specifying Sustainable Concrete* (The Concrete Centre, 2020) is a useful guide in optimising sustainable concrete.

Concrete kentledge blocks are often used for the stability of hoardings, façade retention or formwork. As an alternative to casting in situ blocks, which are often demolished after use, consideration should be given to reusable precast blocks. The weight of each block should be identified and they should be cast with lifting points. The blocks are generally required to prevent overturning and sliding. Sliding resistance is often the governing case and the size of the block can be reduced if friction beneath the block can be increased – for example, by using rubber mats. A similar principal is to precast guide walls for secant and contiguous piling operations.

There can be environmental problems with sand and gravel extraction (often dredging from rivers / deltas / coastlines / estuaries) for construction – particularly concrete products. Sand and gravel are often removed faster than the ecosystem can cope with and are not sufficiently replenished, causing a detrimental effect on the local ecosystem. Recycled and sustainably sourced materials should be investigated as alternatives.

There are alternatives to steel reinforcing bars, such as fibreglass reinforcement. It is lighter, has a higher tensile strength and has better longevity (does not rust so less concrete cover is required so the overall section size may be smaller) but it needs to be fabricated prior to installation so can take longer to fix. Fibre-reinforced concretes have evenly dispersed fibres which do not rust and hence also require less cover. They have a low weight-to-strength ratio but can be costly and challenging to pour / finish but may be suitable for some temporary works applications. Both man-made fibres (such as glass, carbon and polymer) and natural fibres (animal, mineral or cellulose fibres) can be used.

Steel is extensively used in temporary works in temporary sheet piling, steel framing for excavations, steel spreader mats beneath cranes or steel gantries to support site cabins, but its usage should be optimised. Steel is the world's most recycled material due to its strength, longevity and durability, with a significant industry developed around its recycling and reuse. The production of steel requires huge amounts of natural resources, with the manufacture and repurposing of steel being a hugely energy intensive process, traditionally in coal-fired plants. This process can be done using lower-carbon energy sources, such as hydrogen, and using electric arc furnaces. Steel sections and sheet piles can be manufactured from recycled steel.

Recycling, repurposing and reducing waste of construction materials for temporary works purposes can significantly reduce environmental impact. Concrete and bricks can be crushed on site as part of a demolition process and then reused for haul roads, hardstandings or working platforms for tracked machinery. It is worth noting that the emissions from diesel of transporting aggregate over 50 km is the same as extracting the virgin material. Combined with thinner platforms, the use of local resources should be prioritised over recycled products. Materials such as quarry waste (may have a high percentage of fines) and old railway ballast (if cleaned) can be suitable for haul roads and hardstanding. Once they have fulfilled their temporary purpose, the materials could further be repurposed beneath permanent block paving or tarmac. This will also have an additional benefit of reducing the lorry movements necessary to remove the rubble from site and import an alternative material. Temporary Works Forum's guide on working platforms (TWf, 2019) has further useful information on materials.

Recycled plastic can be a lightweight and robust material which has zero moisture absorption and good chemical resistance. Plastic beams can be used for fencing and formwork, while plastic sheeting can be used for reusable site hoardings. Plastic sheet piling does not rot and is lightweight, so can be used in an aquatic environment or to support the ground in trenches or shallow excavations (along with plastic framing systems). Plastic trench boxes and hybrid king post systems (plastic sheeting with timber piles) can also be used for temporary ground support or to protect those working in the ground. TWDs selecting plastic sheet piling or similar products should consider 'driveability' (how easy it is to install the element into the ground), especially into stiff or dense soils, and be aware of the limitations of use. Suppliers and specialist contractors have developed techniques to aid installation, so their advice should be sought.

While recycled materials can offer environmental benefits, their use in temporary works can present challenges, such as ensuring consistent quality, potential structural limitations and the need for specialised handling or installation procedures. For example, 6F material contaminated with asbestos can be catastrophic and requires a full excavate and rebuild.

Good record keeping and accurate 'as built' information is essential to fully utilise the benefits for the future reuse of onsite items. The above-ground part of a structure may have to be demolished but often little thought is given to what is below ground. The foundations or boundary retaining walls may be reused, especially bored piles, as these are difficult and energy-intensive to remove. If the foundations or retaining walls cannot be reused for the new structure, they might be suitable for temporary works. Strip footing could be utilised beneath cabins, scaffolding or propping and foundations piles could be utilised for tower crane bases or beneath large mobile cranes. Reusing existing foundations is not a new concept – buildings destroyed by the great fire of London in 1666 were rebuilt on their former foundations. The Institution of Structural Engineers (Tayler, 2020), Construction Industry Research and Information Association (CIRIA) C653 (Chapman *et al.*, 2007) and Building Research Establishment (Butcher *et al.*, 2006) have further information.

While exploring future technologies like 3D printing is valuable, focusing on maximising the potential of current standards and readily available solutions is perhaps the most practical approach for immediate implementation.

13.4. Environmental considerations

Pollution can be a major issue on many construction projects and temporary works can play their part in mitigating these issues. Some examples are given below.

Construction sites are not permitted to discharge unrestricted amounts of dirty or contaminated water or allow such water to spill onto roads / pavements or discharge into drains, canals or watercourses. There are two issues: firstly, how clean is the water and secondly, where will it be discharged? Surface run-off from sites may contain chemicals, effluent and hydrocarbons, which can pollute watercourses and damage the natural environment. The Environment Agency publish information on when water discharge activity permits are required. Delivery areas, fuel and chemical storage areas, areas where machinery or generators are working, and waste storage areas should be carefully controlled to prevent pollution. Machinery should be well maintained to prevent leaks and 'spill kits / oil retention pillows' should be available. Temporary drainage, bunds and ditches can be provided around fuel and chemical storage areas and to control spillage and any surface run-off. Any contaminated water collected is treated before being discharged. If hardstandings, haul roads and working platforms can be keep clear of contamination they will have a greater resale value (material does not require cleaning if it must be removed from site).

When excavating below the water table or during periods of heavy rain, pumping (dewatering) measures are likely to be required on site. Measures are required to collect and treat the water. The water is likely to contain mud, aggregate particles and possibly chemicals. The Environment Agency have updated their guidance *Temporary Dewatering from Excavations to Surface Water* (EA, 2023). Sediment can be filtered to an approved standard using temporary 'silt-busters', with variations also available for filtering oil and sediment, prior to discharging. When discharging clean water into the water system, it is important not to allow damage to the banks or beds by erosion. Dewatering filter bags or gravel beds can be used to slow down the discharge and reduce scour. Vacuum tankers are an expensive option for the removal of contaminated water where there is no alternative option.

Air pollution (e.g. dust and fumes), noise and vibration can be a health risk and nuisance to people living near a construction / demolition project, as well as to wildlife. Monitoring is often a contractual requirement to ensure compliance with statutory obligations. Temporary acoustic barriers (could be a temporary bund of excavated material) and dust screens can be installed. Around demolition projects, a designed, sheeted scaffold can be used to control noise and dust and act as a safety measure for flying debris. 'Quiet' machinery and 'quiet' construction techniques such as 'silent piling' (piles can be hydraulically jacked into the ground rather than being vibrated or driven with impact hammers) should be used where conditions allow. Diesel engines in construction machinery are a source of noise and air pollution. Inappropriate machinery selection (i.e. operating machinery in conditions too challenging for the specific machine) can increase wear, fuel consumption and pollution. Poor condition and an infrequent maintenance regime can also contribute significantly and are likely to reduce the lifespan of the machine. Newer generation engines are quieter and operate more efficiently, creating less pollution. Emissions are regulated by European Union Regulation 2016/1628 (EU, 2016). Alternative fuels from rapeseed oil, biodiesel and hydrogen are available but some sustainability claims should be treated with caution. A further advance is the development of battery-powered electric vehicles and equipment (ranging from cranes to vibrating plate compactors and hand tools). Consideration must be given to the battery life (plus the manufacture of the batteries) and how the batteries are to be recharged (i.e. from fossil fuel sources).

The 'Considerate Contractor Scheme' is a voluntary scheme for contractors and other organisations to raise standards and respect to communities, their own workers and the environment. The scheme promotes sustainable solutions, minimising waste, minimising pollution (noise, air, light, vibration) and protecting ecology, watercourses and wildlife. A common example is the diversion (avoiding residential streets) or reduction of vehicles onto and from a construction site or limiting times when vehicles can arrive / leave sites (try to avoid coinciding with school runs). Where possible, alternatives to road vehicles, such as rail or canals, can be an appropriate solution.

Construction sites are also increasingly adopting sustainable practices, incorporating eco-friendly cabins and welfare units, implementing paperless management systems and employing rainwater harvesting and other water reuse strategies to minimise their environmental impact.

Building Research Establishment (BRE, 2015) highlights the positive environmental impact of refurbishing buildings and structures, rather than demolition, disposal and rebuilding. Advice should be sought as early as possible in the planning / design process from temporary works specialists to ensure consideration is given to buildability / practicality and the effective use of items such as propping, shoring, façade retention and scaffolding.

13.5. Example of reducing materials

Reducing waste and reducing carbon are very similar, although waste (kg) is weighted to include carbon by using a carbon conversion factor (kg CO_2e/kg). These factors can be found in *Embodied Carbon in Temporary Works* (TWf, 2023).

An example of a working platform is shown below – to be used for 1 year and trimmed once during that period.

1. Plant: excavate for platform $1800\,m^3 = 1800 \times 0.0023 = 4\,tCO_2e$
2. Material: import stone $2400\,t = 2400 \times 0.017 = 41\,tCO_2e$
3. Plant: deposit and compact stone $1200\,m^3 = 1200 \times (0.00225 + 0.0009) = 4\,tCO_2e$

4. Plant: trim after 6 months $600\,m^3 = 600 \times 0.0023 = 1.5\,tCO_2e$
5. Material: import stone to replace trim $600\,t = 1200 \times 0.017 = 20\,tCO_2e$
6. Plant: deposit and compact trimmed stone $600\,m^3 = 600 \times (0.00225 + 0.0009) = 2\,tCO_2e$
7. Plant: excavate at end of project $1200\,m^3 = 1200 \times 0.0023 = 3\,tCO_2e$
8. Material: export stone off site $= 2400\,t = 2400 \times 0.005 = 12\,tCO_2e$
9. Plant: deposit muck back in $1800\,m^3 = 1800 \times (0.00225 + 0.0009) = 6\,tCO_2e$.

It is clear from the above example that efforts would be best spent on reducing the stone imported to site, either by reducing the quantity (reduced thickness) or intensity (locally sourced material). Reducing the thickness by 30% is likely to give a 30% carbon saving. This is a typical client target and shows that 'using less stuff' is the main target; however, other core principles should also be considered (in order).

1. Design for materials optimisation: the principle is that less material is used in the design – for example, application of the lean approach and less waste produced throughout the construction process. Considerations should include simplifying form and layout, designing to standard material sizes and encouraging the reuse of off-cuts.
2. Design for reuse and recovery: two aspects are relevant: firstly, the reuse of materials already on site, such as soils or demolition materials – this is key to resource efficiency – and secondly, the specification of materials with a high recycled content or the use of materials from other sites.
3. Design for deconstruction and flexibility: sometimes stated as 'design for dismantling not demolition', the final element is for designers to consider how materials will be recovered during operation and maintenance of the asset as well as when it comes to the end of its life to avoid waste as much as possible. Can the structure be designed with adaptability for use incorporated? Are there reusable / recyclable components? Can the building be easily deconstructed?
4. Design for waste-efficient procurement: as part of the design process, specialist suppliers should be involved to consider work sequencing, design and construction to assess whether waste can be designed out – for instance, if a different construction method can be employed, it may produce less waste.

13.6. Carbon calculation

The built environment is a huge producer of carbon and a recently revised version of PAS 2080 (BSI, 2023) sets out strategies to reduce carbon emissions and reach net zero. PAS 2080 is a global standard for managing infrastructure carbon with the aim to reduce carbon and reduce costs through intelligent design, construction and use. Carbon equivalent (CO_2e), calculated in kilogrammes or tonnes, is a useful measurement and provides a broad approach to the potential environmental damage. There are many carbon calculators available online. The Royal Institute of Chartered Surveyors (RICS) has produced a *Whole Life Carbon Assessment* (WLCA) (RICS, 2024) standard for consistent and accurate carbon measurement.

The carbon footprint of permanent works is generally considered, but temporary works items and activities have been ignored until recently. Clients and designers are starting to ask questions relating to embodied carbon of proprietary equipment used in temporary works schemes. The intention is for TWf to produce a series of guides specifically relating to these temporary works items and activities. They will provide example calculations (based on WLCA) to compare items, such as a supported trench with a battered trench, with consideration being given to:

- assets – machinery, temporary works and so on
- materials – permanent works pipe
- transport – off site and onto site
- energy – fuel for machinery, electricity and so on
- waste – sent to landfill.

These types of calculations give an approximation of carbon for that type of operation (based on the assumptions made) and allow for the identification of 'carbon hotspots' where temporary works designers and sites need to concentrate their efforts. Typical calculations would indicate that the largest carbon contributions come from transport (25% to 30%) and energy (30% to 40%). TWf has further information in *How to Reduce Carbon in Excavations* (TWf, 2024).

Building Research Establishment Environmental Assessment Methodology (BREEAM, 2025) is a validation and certification system for a more sustainable built environment. It considers different categories and different types of construction and provides a rating. Temporary works traditionally did not easily feed into BREEAM assessments and tended to be ignored. This is changing, with some clients starting to recognise the embodied carbon in temporary works and are expecting assessments to be made. It is hoped that the information required to make these assessments is included in design briefs and that these assessments are made regularly in order to continue to raise awareness. Perhaps a further development could be to include temporary works activities in carbon accounting and billing as standard.

REFERENCES

BRE (Building Research Establishment) (2015) *Briefing Paper: Sustainable Refurbishment – How to Better Understand Measures and Reduce the Embodied Impacts.* BRE, London, UK.

BREEAM (Building Research Establishment Environmental Assessment Methodology) (2025) https://breeam.com/about/how-breeam-works (accessed 10/01/2025)

BSI (2013) BS 8895-1: 2013: Code of practice for strategic definition and preparation and brief. Designing for material efficiency in building projects. BSI, London, UK.

BSI (2017) BS 8001:2017: Framework for implementing the principles of circular economy in organisations. BSI, London, UK.

BSI (2023) PAS2080: Carbon management in infrastructure. BSI, London, UK.

Butcher AP, Powell JJM and Skinner HD (2006) *Reuse of Foundations for Urban Sites, A Best Practice Handbook.* BRE, London, UK.

Chapman T, Anderson S and Windle J (2007) *C653: Reuse of Foundations.* CIRIA (Construction Industry Research and Information Association), London, UK.

Construction Industry Publications (2021) *Construction Environmental Manual.* Construction Industry Publications, Reading, UK.

EA (The Environment Agency) (2023) *Regulatory Position Statement 261: Temporary Dewatering from Excavations to Surface Water.* EA, London, UK.

EU (European Union) (2016) Regulation 2016/1628: Requirements Relating to Gaseous and Particulate Pollutant Emissions etc. EU, Brussels, Belgium.

Halliwell E (2024) Putting the net-zero hierarchy into practice: build nothing. *The Structural Engineer* **102(1)**: 10–11.
HMG (Her Majesty's Government) (1974) Control of Pollution Act. The Stationery Office, London, UK.
HMG (1981) Wildlife and Countryside Act. The Stationery Office, London, UK.
HMG (1984) Building Act. The Stationery Office, London, UK.
HMG (1990a) Town and Country Planning Act. The Stationery Office, London, UK.
HMG (1990b) Listed Buildings and Conservation Areas Act. The Stationery Office, London, UK.
HMG (1990c) Environmental Protection Act. The Stationery Office, London, UK.
HMG (1991) Water Resources Act. The Stationery Office, London, UK.
HMG (1993) Noise and Statutory Nuisance Act. The Stationery Office, London, UK.
HMG (2002) Control of Substances Hazardous to Health Regulations. The Stationery Office, London, UK.
HMG (2008a) 2008/98/EC Waste Framework Directive. The Stationery Office, London, UK.
HMG (2008b) Climate Charge Act. The Stationery Office, London, UK.
HMG (2010a) Waste (Circular Economy) (Amendment) Regulations. The Stationery Office, London, UK.
HMG (2010b) Flood and Water Management Act. The Stationery Office, London, UK.
HMG (2011a) Waste (England and Wales) Regulations. The Stationery Office, London, UK.
HMG (2011b) Construction Products Regulations. The Stationery Office, London, UK.
HMG (2012) National Planning Policy Framework (updated 2023). The Stationery Office, London, UK.
HMG (2014) Contaminated Land: Statuary Guidance. The Stationery Office, London, UK.
HMG (2015) Construction (Design and Management) Regulations. The Stationery Office, London, UK.
HMG (2016a) Construction Site Waste Regulations. The Stationery Office, London, UK.
HMG (2016b) Environmental Permitting (England and Wales) Regulations. The Stationery Office, London, UK.
HMG (2017) Conservation of Habitats and Species Regulations. The Stationery Office, London, UK.
HMG (2021) Environment Act. The Stationery Office, London, UK.
HMG (2022) Modern Methods of Construction Strategy (MMC). The Stationery Office, London, UK.
ICE (Institution of Civil Engineers) (2022) *Low Carbon Concrete Routemap*. ICE, London, UK.
Pallett PF and Filip RK (2019) *Temporary Works: Principles of Design and Construction*, 2nd edn. ICE Publishing, London, UK.
RICS (Royal Institute of Chartered Surveyors) (2024) *Whole Life Carbon Assessment for the Built Environment*, 3rd edn. RICS, London, UK.
Tayler H (2020) A short guide to reusing foundations. *The Structural Engineer* **98(11)**: 4.
The Concrete Centre (2020) *Specifying Sustainable Concrete*. The Concrete Centre, London, UK.
TWf (Temporary Works Forum) (2019) TWf2019.02: Working Platforms: Design of Granular Working Platforms for Construction: A Guide to Good Practice. https://www.twforum.org.uk/viewdocument/working-platforms-design-of-granu (accessed 10/01/2025).

TWf (2023) TWf2023.01: Embodied Carbon in Temporary Works. https://www.twforum.org.uk/viewdocument/embodied-carbon-in-temporary-works (accessed 10/01/2025).

TWf (2024) TWf2024.04: How to Reduce Carbon in Excavations. https://www.twforum.org.uk/viewdocument/twf2024-04-how-to-reduce-carbon-in (accessed 10/01/2025).

UKGBC (UK Green Building Council) (2021) *Net Zero Whole Life Carbon Roadmap*. UKGBC, London, UK.

USEFUL WEBSITES

Autodesk – www.getconstructioncloud.autodesk.com
Boss Cabins – www.bosscabins.co.uk
BRE (Building Research Establishment) – www.bregroup.com
BSI (British Standards Institution) – www.bsigroup.com
CIOB (Chartered Institute of Building) – www.ciob.org
CIRIA (Construction Industry Research and Information Association) – www.ciria.org
EA (The Environment Agency) – www.gov.uk/government/organisations/environment-agency
IStructE (Institute of Structural Engineers) – www.istructe.org
RICS (Royal Institute of Chartered Surveyors) – www.rics.org
Safe Site Facilities – www.safesitefacilities.co.uk
The Climate Group – www.theclimategroup.org
The Concrete Centre – www.concretecetre.com
The Planning Inspectorate – www.gov.uk/government/organisations/planning-inspectorate
The Plastic Piling Company – www.plasticpiling.co.uk
TWf (Temporary Works Forum) – www.twf.org.uk
UKGBC (UK Green Building Council) – www.ukgbc.org
WRA (Wood Recyclers Association) – www.woodrecyclers.org

FURTHER READING

Autodesk (2022) Green Construction: A Growing Global Trend. https://construction.autodesk.com/resources/guides/green-construction-a-growing-global-trend/ (accessed 10/01/2025).

Autodesk (2023) Green Construction: A Global Transformation. www.construction.autodesk.eu/resources/emea-resources/green-construction-a-global-tranformation (accessed 11/08/2024).

Balson K, Summerson G and Thorne A (2014) *Briefing Paper: Sustainable Refurbishment of Heritage Buildings – How BREEAM Helps to Deliver*. BRE, Watford, UK.

Berryman C, Watkins G, Crozier F and Parsons J (2023) *C809: Suitable Management of Surplus Soil and Aggregates from Construction*. CIRIA, London, UK.

BRE (Building Research Establishment) Sustainability: Practical Guidance and Support for Delivering Genuine Sustainability Benefits. See https://bregroup.com/expertise/sustainability/ (accessed 10/01/2025).

BSI (2013) BS EN 206:2023+A1:2021: Concrete. Specification, performance, production and conformity. BSI, London, UK.

BSI (2023) BS 8500:2023: Concrete. Complementary British Standard to BS EN 206. Method of specifying and guidance for the specifier. BSI, London, UK.

BSI (2024) PD CEN/TS 1090-201:2024: Execution of steel structures and aluminium structures – reuse of structural steel. BSI, London, UK.

CIRIA (Construction Industry Research and Information Association) (2013) *Lean Construction Guides C725, C726, C727, C728, C729, C730 and C769*. CIRIA, London, UK.

Climate Group (2022) Net Zero Concrete Programme. https://www.theclimategroup.org/our-work/press/net-zero-concrete-programme-launched-climate-group-and-global-built-environment (access 10/01/2025).

Environmental Services (2008) Pollution and Noise from Demolition and Construction Sites. www.clec.uk/sites/default/files/esconstructioncop.pdf (accessed 11/08/2024).

Gibbons OP and Orr JJ (2022) *How to Calculate Embodied Carbon*, 2nd edn. IstructE, London, UK.

Gorse C and Dritleff B (2023) *Guide to Sustainability in the Built Environment.* Chartered Institute of Building, Bracknell, UK.

Gowler P, Astle P, Bajars K *et al.* (2023) *Circular Economy and Reuse: Guidance for Designers*. IStructE, London, UK.

Guthrie P and Mallett H (1995) *SP122: Waste Minimisation and Recycling in Construction.* CIRIA, London, UK.

Kwan J, Dickinson C and MacLeod C (2023) *C811: Environmental Good Practice on Site*, 5th edn. CIRIA, London, UK.

Norman J, Ibell I and Broadbent O (2023) Debunking sustainability myths is vital for civil engineers to reduce reuse and repurpose. *Civil Engineering Special Issue Magazine* **176(6)**.

SCI (Steel Construction Institute) (2019) *SCI P427: Structural Steel Reuse*. SCI, Ascot, UK.

UKGBC (UK Green Building Council) (2023) Net Zero Carbon Buildings 101. www.ukgbc.org/wp-content/uploads/2023/04/UKGBC-Net-Zero-Buildings-101-pdf (accessed 11/08/2024).

UKGBC, Chartered Institute of Building, IStructE *et al.* (2024) Policy Position Paper: Embodied Carbon Regulation – Alignment of Industry Policy Recommendations. https://www.istructe.org/getattachment/05855cb7-990a-477e-8672-9329aa3307ca/attachment.aspx (accessed 10/01/2025).

Winch L and Hurst S (2023/2024) Using less stuff. *Temporary Works Forum Yearbook 2023/24*, pp. 25–26. www.library.myebook.com/Hub/temporary-works-forum-2023-24-yearbook/4906 (accessed 11/08/2024).

Winch L and Hurst S (2023/2024) Are clients doing enough to lower embodied carbon in temporary works? *Temporary Works Forum Yearbook 2023/24*, p.59. www.library.myebook.com/Hub/temporary-works-forum-2023-24-yearbook/4906 (accessed 11/08/2024).

Zurich Municipal (2021) Sustainability in a Changing Built Environment. www.zurich.co.uk/news-and-insights/ (accessed 11/08/2024).

Ray Filip
ISBN 978-1-83608-525-6
https://doi.org/10.1108/978-1-83608-524-920251016

Chapter 14
Technology

Dr Matthew Hall
Richter Associates Ltd, UK

14.1. Introduction

Technology (including software) has revolutionised the construction industry and continues to do so; the specific field of temporary works is no different and has seen many changes in the nearly five decades since the Bragg Report (Bragg, 1975). Perhaps only the oldest readers of this chapter will recall the 1970s: using slide rules, logarithmic tables, performing structural calculations by hand using moment distribution, drawing on parallel motion drawings boards with compasses and drawing pens. Mainframe computers were slowly being introduced and could be utilised to perform calculations. Personal calculators became affordable to many in the 1980s and were considered a major advance for engineers. Simple personal computers started to become common on sites and in design offices in the 1990s. Temporary works design engineers wrote analysis programmes in languages such as Fortran and produced Excel spreadsheets to carry out complex calculations for items such as slope stability, earth pressures / stability of sheet pile retaining walls and structural analysis of steelwork frames. Design offices replaced drawing boards with AutoCAD, which made producing and revising drawings a much simpler process. Site teams wrote programmes and spreadsheets to monitor costs and programmes, calculate quantities of materials and document control. The speed of communication changed as email replaced fax machines and letters. More recently hardware, software, the digital world and other technology, such as robots, drones, geospatial modelling / visualisation and artificial intelligence, have developed exponentially. However, there is anxiety and fear of this change among some and there needs to be a certain mindset to fully utilise and integrate new technologies in the temporary works workplace and promote healthy thinking around innovation.

The integration of new technology and innovation can enhance the temporary works process, from site management to design and construction. Everyone needs to try to keep up to date with new technologies and innovations (at least with the core concepts) that could impact their own workplace and embrace the available opportunities. To maximise the benefits of any new system, process or technology, users require training to familiarise themselves with the intricacies.

Perhaps one of the most common mistakes with technology is the coming and going of large concepts / big ideas that are supposed to 'change everything'. These big 'state-of-the-art' ideas can cost a lot of time and money and should only be undertaken because of a specific need and business case with a measurable impact. State-of-the-art for any industry can be difficult to qualify; very often people wish to aim for a complex solution based on the limits of technological advancements. Often the most elegant and effective solutions are the most straightforward. Mini iterative innovations are likely to be far more effective, achievable and perhaps cost-effective than attempting to introduce a complete overhaul.

It is important to measure the actual impact that technology has before examining its uses. New ways of doing things can be exciting but often fall short of making a big enough contribution to cover their exploratory costs. It is important that when a new technology is revealed, its mechanics and business model are understood before specific uses are found for it (unless it has been commissioned for a specific purpose). Often the best solutions come from several ideas merging to solve a specific problem or task. Users should not look for a problem that can fit their new tool, attempting to jam in and justify its use in their work, but first seek out problems that need to be fixed. Often these can be found where people are arguing or under pressure as often inefficiencies and issues can be identified that could be overcome or eliminated altogether.

14.2. Procedural control systems

One of the most effective areas of recent innovations is in managing the temporary works process to BS 5975-1:2024 (BSI, 2024), especially on large and complex projects with multiple temporary works activities. Projects such as High Speed 2 (HS2) have commissioned bespoke management systems which are integrated, paperless, available for use by multiple parties simultaneously, automatically update documents, monitor progress and costs, send reminders and share information. The aim is to allow earlier identification of issues and make the overall process more efficient by simplifying administration.

Some useful features can include the following.

- Competencies of companies and individuals can be stored, with the system then being interrogated to identify the most appropriate parties to carry out specific functions.
- Temporary works registers can be linked to master construction programmes (with dates and parties involved) and enable temporary works coordinators (TWC) and others in the project or design team to forecast and track items, such as progress of multiple designs (and design checking) and onsite inspections carried out by others. Process bottlenecks can be flagged up in advance so timely measures can be introduced.
- Interface issues with external stakeholders such as utility companies can be flagged up.
- Design brief templates for commonly used items, such as hoardings, mobile crane pads and formwork, can be partially automatically populated from a database of common information.
- Site inspection checklists can be generated from design briefs and completed designs and sent to relevant site supervisors. The system will track non-compliances (such as site changes) and update when the process is complete.
- Real-time monitoring of items such as deflection or settlement can be linked to the system.
- Systems can be operated remotely on site using iPads, with the master system being updated automatically.
- Data analysis can be carried out, with reports on performance being produced for auditing and learning lessons.

This area will continue to develop, with tools such as machine learning gathering an increasing dataset of activities to learn from and to understand the results of actions.

14.3. Design software

Software for design, modelling, drawing, detailing, procedural control, project management and fully integrated document control are now essential items in any construction environment. There is still ample room for continued development and innovation in making the temporary works process more accurate and efficient. The available software is now extremely powerful, with user interface paramount.

Not all temporary works designs must be carried out using software. It can sometimes still be quicker and cheaper to design certain simple items by hand or incorporate repetitive calculations into simple spreadsheets. Sometimes the time spent setting up software models can outweigh any potential material savings, especially as often the temporary works designer will have a limited choice – that is, temporary works need to be robust and often utilise what is currently available on site. Also, the turnaround for temporary works designs can be almost instantaneous – that is, a site may contact a designer with an issue that requires immediate resolution. However, a designer's time can often be more effectively spent dealing with overall concept and human relationships, utilising appropriate software for more demanding and complex problems and calculations to avoid human error.

There are concerns regarding the inappropriate use of temporary works design software. All software should be thoroughly checked for potential 'bugs' and the output verified by other methods. It should be used by trained design engineers (may be under supervision) who appreciate the importance of the input parameters and what function the software is performing. Software should be updated regularly with academic advancement and as regulations / standards change. Temporary works designers need a good understanding of concepts, such as types of loads / actions, structural / geotechnical behaviour, load paths, equilibrium, potential modes of failure, sequencing, material behaviour and limitations, appropriate serviceability / deflection limits, buildability, safety and, perhaps most importantly, 'what looks right'. Designers need a 'feel' for the appropriate answer and critical stages / conditions; design software should then be used to verify and quantify specific details. An overreliance on software by unqualified persons can lead to unsafe solutions. Without a basic engineering understanding, those using the software will input inappropriate or incorrect parameters, will then be misled and not appreciate unrealistic answers / solutions. For example, software such as CADS Piled Wall Suite is commonly used and has been available for many years. The likelihood of it making an error within its code is extremely unlikely as it has been checked and verified many times. However, input parameters for soils can be very sensitive, with some degree of interpretation required. The inputting of geotechnical parameters still requires a suitably qualified and experienced engineer to review the appropriately detailed ground investigation and make an engineering judgement call. The engineer must still make the key decisions, with the software being used to speed up the analysis / calculation process and to remove the element of human error in the mathematics. Ultimately, it is still the engineer's responsibility to ensure they deliver a design which is safe and appropriate.

Finite element methods (FEM) are popular for solving complex problems. FEM is a numerical method for solving partial differential equations in two or three dimensions. FEM subdivides a large system into smaller 'finite elements' and resolves the stresses and strains for the structure they represent. As this subject has matured, it is now possible to simulate almost any system encountered in construction. It provides confidence in geometrically complicated structures and how they will react to various load cases. Using modern streamlined FEM software packages, it is possible to simulate large arrangements of temporary works and globally analyse across a whole structure. However, these powerful tools have their issues. Modelling and analysing a structure are time-consuming (therefore expensive) and require a certain amount of expertise. They are not commonly used unless there is a challenging structure which will benefit from the more global analysis. Instead, simple analysis, hand calculations and engineering judgement are regularly used for simple structures (for speed and efficiency) for what is considered as the worst-case scenario / member.

Since the introduction of computing in the modern construction design process, computers have evolved from a single computer which could occupy an entire room and take many hours to resolve a simple beam calculation, to extremely powerful modern computers that are taken for granted. However, engineering culture should continue to push these tools further, with increasingly precise modelling and increasingly complex analysis being carried out with ever-increasing efficiency as computing power continues to advance. While this is an admirable approach, there is also an opportunity for temporary works designers to create their own bespoke tools rather than relying on permanent works tools alone.

To ensure these tools are more efficient than hand calculations, many structures could be described with a lightweight linear static 3D space frame analysis. Using modern computing, the solve times for such items have reduced since their first inception from many hours down to seconds. If temporary works can be generally considered to be lightweight with only hand calculations to match / beat, then extremely streamlined tools can be created for a variety of common temporary works solutions. It is exciting to see what tools will continue to be developed in this space, specifically built around the evolving needs of temporary works design.

14.4. Drones

Drone technology is starting to offer a convenience to construction projects which was not previously considered. They can efficiently and safely survey, monitor, clean and explore tight spaces or difficult to access areas. Without drones, scaffolding, towers, access machinery and other temporary works to facilitate safe access would be necessary. These take time and cost to design, erect and dismantle, specific permissions may be required and their use may be problematic adjacent to or on live infrastructure, such as road bridges or rail. Advanced drones are easily controlled by experienced operatives and can be equipped with high-resolution cameras and sensors, allowing engineers to conduct very detailed tasks on structures, such as bridges, tunnels, chimneys, colling towers, culverts, pipelines and buildings. This technology enhances safety by reducing the need for manual inspections in hazardous environments and can also significantly cut costs and time. Drones can access confined and otherwise inaccessible areas, providing crucial data that aids in maintenance, repairs and overall project management, thus driving innovation and efficiency. The costs of hiring drones and operatives can be small when compared to costs of items such as scaffolding.

Some specific examples for advanced drone technology include the following.

- Whereas asset managers previously had to travel from turbine to turbine, climb the mast and carry out inspections, Alerion's AI-driven wind farm inspection drones, armed with infrared cameras, are now able to autonomously fly to the top of turbines, inspect for overheating of bearings and other key inspection activities, then move on rapidly.

- Drones fitted with high-pressure, small-diameter hoses can be used for in situ cleaning and de-greasing inside difficult to access pipework used in hospitals, petrochemical works and laboratories. Human access into the pipework may be dangerous and problematic and the cost of removing and replacing the pipework after cleaning could be expensive.
- Drones can be used to accurately survey stockpiles and calculate their volumes and monitor temporary cuttings or embankments for potential instability.

These types of innovations will become even more accessible with the continued reduction in the costs of chips such as the Hailo-8L. Soon, more specific start-up ventures will be formed for lightweight intelligent drones and autonomous vehicles.

14.5. Robotics and augmented labour

Robotics and augmented labour are increasingly become more feasible. While there is still a lot of promotional-only robotics use, these are starting to be less science fiction and more commercially viable each day. One area of significant growth for robotics is on Japanese construction projects, where there is a significant shortage of workers. This has led to many innovations around creating bipedal robots that are able to carry out repetitive tasks, or robotic arms which can be mounted and then assist operatives with lifting. There are also examples of workers having exoskeletal suits to reinforce what they can lift in one go and reduce repetitive strain injuries. In the UK, the focus is less on efficiency but more on risk mitigation and safety. Successful runs of using robotics to remove humans from hazardous situations have been completed with effective results. For example, installing sprayed concrete tunnel lining is a dangerous operation – a collapse of the material can result in severe injury or fatality of operators. On London's Crossrail tunnelling projects, a robotic arm with a spray attachment was used and remotely operated to spray tunnel lining in certain key sections. This had the added benefit of accuracy (reducing time and waste), with sensors improving the ability to identify when minimum thicknesses were met. As these subjects advance and trickle-down technology becomes more affordable, there will be a continuing rise in the use of robotics and cybernetic augmentation. These systems will continue to reduce and, in some circumstances, eliminate the need for temporary works and reduce the need for humans to carry out repetitive heavy lifting operations.

One example of robotics replacing repetitive tasks is in the USA, where buildings are increasing in size and there is often the need to lay a huge number of bricks. Human bricklayers have been replaced by bricklaying robotics that can lay long walls accurately, neatly and with no risk for humans to work at height or suffer any repetitive strain. These capabilities are likely to continue to evolve to a point where most (perhaps all) manual labour on site will be replaced by robotics and be able to achieve a near zero harm risk to humans in construction.

14.6. Internet of Things

Internet of Things (IOT) is a general term for a network of sensors, motors or monitoring solutions across assets to gather relevant data and, in some scenarios, remotely operate plant. These networks of sensors allow immediate information gathering and, in some examples, send back commands to react to changing conditions in real time.

Like many new topics in technology, it has cemented itself where useful and been experimented and disregarded elsewhere. Some of the best examples can be seen in live monitoring of loads in props supporting large and deep excavations. The loads in very large props vary in a cyclic manner due to thermal expansion and contraction (day and night cycles). These large-scale temporary works, with repetitive use, have utilised this technology effectively, providing a reassuring output with an ability to trigger warnings so that remedial action can be taken. Sensors can be relatively cheap but the interpretation of their data and modelling this into engineering knowledge can be time-consuming, so single set-up uses are rare.

Sensors are commonly used along rail lines to monitor track movements (perhaps due to nearby excavations) and on slopes (temporary and permanent) to act as trigger warnings of unexpected movement. These types of solutions can be far more reliable than inspection rotations, as they are constantly active and can trigger warnings immediately, rather than letting them deteriorate further. These solutions can appear cost-effective but there are costs associated with setting up the system and interpretation.

Data collection using IOT can provide training data for artificial intelligence (AI) and machine learning, as previously unthought-of connections come to light across a more complex structure. Then by feeding the input controls to large systems to be controlled by algorithms trained by that data, a huge facility can become self-optimised and much more efficient. This is an example of the synergy between raw data and AI, which is discussed in section 14.7.

14.7. Artificial intelligence

Artificial intelligence (AI) is currently a hot topic; it is relatively early days in its development and utilisation in the field of temporary works. This section aims to explain what it is. Those involved in temporary works should then be encouraged to understand the possibilities and explore how it could be utilised for their own specific applications.

In general, AI can be split into two camps:

- language model products / other emerging pre-made tools
- utilisation of machine learning and other numerical techniques.

Both can be powerful, but their limitations must be respected.

14.7.1 AI language models

The utilisation of large natural language models like Open AI's Chat GPT can be very useful for processing large amounts of information, gaining insights into complex data and generating new ideas. It is sufficiently advanced that people are finding out it can do things it was never intended to be able to carry out. These models work with the same algorithms that are common in a mobile phone, such as predictive typing (when sending messages). If this type of algorithm is given an increasing amount of data to cross-reference, it can come up with seemingly intelligent best fits for next words for long periods, given the context of previous words and the prompt that was requested. This kind of intelligence can become more and more advanced as it is trained with more information. Currently, the major technology companies are building vast new data centres, with their own power grids, just to fuel these models, with more information to advance their abilities. This is amazingly powerful but must be utilised correctly. There are two issues to be mindful of.

- The AI model looks at the prompts and tailors the response to what you would like to hear. If your prompt is down a certain avenue with your own predictions, it will find those again and reflect them back, reinforcing your mindset that it is correct. However, if the model has been trained on mass incorrect data (e.g. the content it has read is based on outdated opinions or theory), it will find the most statistically significant and potentially incorrect information and report that back. No actual intelligence has been involved. This can be overcome by specifying: 'go look up the latest information' or base it on references after a certain date.
- In theory, you can ask AI anything and it will attempt its absolute best to fill in a response that looks right, based on what it has read. A specific example could be that you can ask it to 'design some formwork' and it 'will have a go'. The results will just statistically be the best fit, based on what it has seen before, and the actual numbers may make little sense.

Despite these shortcomings, there are many possibilities with these new AI tools, providing the human element has sufficient knowledge and experience to act as an arbiter between it and the real world. AI also has some great pure numerical and coding abilities, which make it very powerful.

14.7.2 Machine learning

This relates to AI machine learning models of raw data. Similar to AI language models, the theories come directly from first principles and can then be applied to specific problems. For this method to be useful, appropriate datasets are required for a complex problem in which the data of various inputs has been collected over a long period, and a way to score this data is required. Machine learning can be extremely powerful at taking large quantities of historical data and seemingly making intelligent decisions for the future, most of the time in greater quantities than a human could digest in a lifetime.

A specific example is from a water treatment plant which had IOT sensors installed to monitor its effectiveness. This problem had a series of input variables and various settings for different hardware around the plant. The output variables were graded as 'efficient' or 'not efficient'. These were repeated over and over in time stamps for thousands of rows of data. This complicated mass of variables was trained into a machine learning model. The output model was given a set of input information and it created the best fit position of all the variables it can control to give the most efficient output. AI could control this massive complex operation by automatically adjusting pumps and valves around the plant far more efficiently than a team of humans adjusting the pumps and valves manually.

The challenge in the field of temporary works is that we would need neat data sets, similar to the example above, to feed into AI. Often in temporary works there is something unique or different in the scenarios faced. We may use hand calculations or analysis software which may have very different approaches between designers, or drawings that are created manually in computer-aided drafting (CAD) software which can differ depending on the drafter's individual style. Training an AI model around most temporary works subjects would be an extremely complex problem to deal with. For AI to become a fully groundbreaking and utilised tool, it is likely the construction industry would need to undergo significant additional digitalisation.

14.7.3. General artificial intelligence

AI is certain to be a huge part of the future. At the time of writing, Microsoft has confirmed the construction of seven nuclear reactors just to power the computational training of these AI tools. It seems to exponentially be growing more advanced by the day. A huge advantage of AI, particularly the large natural language models, is that much like us, what it writes is based on information it has previously come across and its 'experiences'. The collective wisdom of these artificial models is staggering, as they process more written information than humans could possibly process and retain in many lifetimes. It has begun to open the door for coding and to create technical content creatively, without having to fully understand the subject matter. Being able to explore new avenues and combine concepts freely like this is hugely beneficial. Furthermore, it can advise on how to go about implementing new ideas and adapting how it explains them, based on how much experience there is in the topic. This is enabling people to think more freely about any possibilities, without the fear of investing too much time into a new idea which may not yield any measurable impacts.

The construction industry should try to embrace these new tools and be mindful of their development as they become more a part of everyday life. It is not inconceivable that, in the near future, AI temporary works tools and consultants will exist in every site engineer's pocket.

14.8. Future predictions

Temporary works has developed in recent decades, with more specialised tools and ways of working becoming available. Initially permanent works tools were adapted for temporary works purposes but, as the development continues, temporary works will create its own specific tools. An embrace of new emerging technologies will have a positive and uplifting impact on all our lives, as we are able to obtain new levels of both risk mitigation and agility when constructing. It is exciting to look forward into the future possibilities, but it will need the dreamers and reachers to continue to drive forward its adoption. Any 'would-be' technologists should be confident that, even though temporary works can seem rigid and set in its processes, there is always room for innovation and improvement. Forward thinking and a drive to innovate and embrace technology will benefit all those who will interact with the buildings and structures we leave behind for decades to come.

REFERENCES

Bragg SL (1975) *Final Report of the Advisory Committee on Falsework*. Department of Employment and Department of the Environment, The Stationery Office, London, UK.

BSI (2024) BS 5975-1:2024: Management procedures for the control of temporary works. Code of practice. BSI, London, UK.

USEFUL WEBSITES

BSI (British Standards Institution) – www.bsigroup.com
CIRIA (Construction Industry Research and Information Association) – www.ciria.org
Confidential Reporting on Structural Safety (CROSS) – www.cross-safety.org
TWf (Temporary Works Forum) – www.twforum.org.uk

FURTHER READING

Debney P (2020) *Computational Engineering*. Institution of Structural Engineers, London, UK.

DRONES

Al-Dosari K, Hunaiti Z and Balachandran W (2023) Systematic review of civilian drones in safety and security applications. *Drones* **7(3)**: 210. 10.3390/drones7030210 (accessed 05/10/2024).

Choi HW, Kim HJ, Kim SK and Na WS (2023) *An Overview of Drone Applications in the Construction Industry*. Seoul National University of Science and Technology, Seoul, Republic of Korea. www.mdpi.com (accessed 05/10/2024).

Drone Deploy (2024) Exploring Drones in Civil Engineering. www.dronedeploy.com (accessed 05/10/2024).

Drones CE LAB (2023) Drones in Civil Engineering. www.dronescelab.com (accessed 05/10/2024).

GCELAB (2023) Utilising Drone Technology in the Construction Industry. www.gcelab.com (laccessed 05/10/2024).

Low-Macrae A (2017) Engineering with Drones. www.istructe.or/resources/blog/engineering-with-drones (accessed 05/10/2024).

ROBOTICS

Construction Placements (2023) What is Construction Robotics and its Applications in Construction Industry? www.constructionplacements.com (accessed 05/10/2024).

Grigoryan EA and Semenova MD (2020) Automation of the construction process by using a hinged robot with interchangeable nozzles. *Materials Today: Proceedings* **30**: 380–387. 10.1016/j.matpr.2019.12.381 (accessed 05/10/2024).

ISARC (International Association for Automation and Robotics in Construction) (2023) About Us. www.isarc2023.com (accessed 05/10/2024).

McNeil Engineering (2023) The Impact of Automation and Robotics on Civil Engineering Construction Processes. www.mcneilengineering.com (accessed 05/10/2024).

Plan Radar (2023) Smart Construction in 2023: Five Current Trends and Future Predictions. www.planradar.com (accessed 05/10/2024).

INTERNET OF THINGS

EngineeringCivil.org (2023) Civil Engineering and The Internet of Things. https://engineeringcivil.org/articles/news/civil-engineering-the-internet-of-things/ (accessed 05/10/2024).

IoT Marketing (2023) How IoT Is Transforming Civil Engineering. https://iotmktg.com/how-iot-is-transforming-civil-engineering/ (accessed 05/10/2024).

Mosaberpanah MA (2023) An in-depth survey demystifying the Internet of Things (IoT) in the construction industry: unfolding new dimensions. *Sustainability* **15(2)**: 1275.

Vilajosana I (2023) Civil Engineering and the IoT. *AEC Magazine*, 7 August. https://aecmag.com/opinion/civil-engineering-and-the-internet-of-things/ (accessed 05/10/2024).

ARTIFICIAL INTELLIGENCE

BibLus (2023) AI in Civil Engineering: 6 Application Examples, All the Advantages and New Challenges for the Future. https://biblus.accasoftware.com/en/ai-in-civil-engineering-6-application-examples-all-the-advantages-and-new-challenges-for-the-future/ (accessed 05/10/2024).

Chandler DL (2023) Integrating Humans with AI in Structural Design. MIT News. https://news.mit.edu/2023/integrating-humans-ai-structural-design-0302 (accessed 05/10/2024).

CROSS (Collaborative Reporting for Safer Structures) (2024) Report ID 1210: Unqualified Engineer's Unsafe Computer Aided Design of a Retaining Wall. https://www.cross-safety.org/uk/safety-information/cross-safety-report/unqualified-engineers-unsafe-computer-aided-design-1210 (last accessed 05/10/2024).

Kelland O and Wei E (2023) How Civil Engineering Can Use AI to Step up its Digital Transformation. https://www.ice.org.uk/news-views-insights/inside-infrastructure/using-ai-to-step-up-digital-transformation (accessed 05/10/2024).

Pure Future AI (2023) Five Best AI Tools for Civil Engineers (Free and Paid). www.purefuture.net (accessed 05/10/2024).

Springer (2023) *AI in Civil Engineering Journal*. https://link.springer.com/journal/43503 (accessed 05/10/2024).

Zhang C, Kim J, Jeon J *et al.* (2023) What Does Human-Machine Intelligence Mean for Civil Engineers? *Civil Engineering Source Magazine*, 1 September. ASCE (American Society of Civil Engineers) https://www.asce.org/publications-and-news/civil-engineering-source/civil-engineering-magazine/issues/magazine-issue/article/2023/09/what-does-human-machine-intelligence-mean-for-civil-engineers (accessed 05/10/2024).

Ray Filip
ISBN 978-1-83608-525-6
https://doi.org/10.1108/978-1-83608-524-920251017

Chapter 15
Steelwork

Nick Cook
Laing, Laing O'Rourke and Balfour Beatty (retired)

Ray Filip
RKF Consult Ltd, UK

15.1. Introduction

Steel is a strong and versatile metal which is an alloy of iron and a small amount of carbon. It is available in many different forms (currently there are several thousand variations available), which may have additions of other elements, such as manganese, chromium and silicon (plus many others). The carbon influences the steel's strength and durability when compared to iron alone. Cast iron and then wrought iron began to be replaced by steel, which has been commonly used in construction since the late 19th century / early 20th century. It has many uses in temporary works, which could include:

- internal frames to support trenches, excavations and cofferdams
- proprietary equipment for formwork, falsework, scaffolding, access and working platforms
- king posts for retaining walls (post and lagging walls) or to provide support to reinforcement cages
- crane foundation grillage beams and lifting beams
- proprietary and bespoke access bridges for plant and transportation vehicles
- portal frames to support site cabins above roads and pavements
- sheet piles, guide frames and piling gates for installing sheet piles, tubular bearing piles and bored foundation piles
- posts and rails for signboards or hoardings
- façade retention systems and structural propping
- needling beams for forming openings in load-bearing walls
- temporary bracing to provide lateral stability during construction of a steel frame structure
- reinforcement to temporary concrete structures
- support and protection systems for pipes and cables
- proprietary and bespoke guardrail systems
- proprietary tie systems.

Steel has benefits and disadvantages, some of which are summarised in Table 15.1.

Table 15.1 Some advantages and disadvantages of steel

Advantages	Disadvantages
Robust and resistant to onsite use and abuse	Heavier than most common building materials
Can be recycled many times	Energy intensive to manufacture and ore mining can create environmental issues
Properties are well-established and design methodologies have been developed over many decades	Can distort in a fire unless adequate protection is provided (e.g. intumescent paint)
Readily available in numerous different types, shapes, sizes and grades (plus subgrades), with different finishes. Can be manufactured to tight tolerances	Without protection the finishes can be damaged during transportation or lifting
Good strength-to-weight ratio. Good tensile, bending and shear strength and good compressive strength when adequately restrained. Lower grade steel exhibits extensive plastic behaviour	Higher-grade steels become increasingly brittle and susceptible to brittle fracture around welded joints if correct procedures are not followed
Can be easily and relatively quickly joined (bolts, welds, clips etc.), assembled / removed or modified, either in a workshop or on site. Can be prefabricated and transported to site for quick assembly	Maximum transport lengths need to be considered, especially into city centres, and maximum weights to suit available craneage
Skilled and experienced operatives along with specialist equipment and machinery are available	Long unrestrained length can be prone to instability of flanges in compression which can cause buckling. Also, instability of webs in compression
A 'paper trail' can be created for traceability and auditing purposes	Can corrode over time when exposed to oxygen and moisture – often this is not a major concern in most temporary works scenarios. A significant exception is longer-term usage in a marine environment
Steel sheet piles can be hammered, vibrated into most soil conditions and then extracted for reuse	Electrical conductivity could cause issues when close to live electrical supplies

15.2. How to specify steel

As with all items of temporary works, early identification and planning are crucial to success. Section sizes need to be determined for structural performance and robustness; see BS 5975-2:2024 (BSI, 2024a). This should be combined with availability, ease of transportation, lifting, assembly and removal. Figure 15.1 shows some common section shapes.

Figure 15.1 Some common sections (editor's own)

From left to right: universal column, universal beam, square hollow section, circular hollow section, rectangular hollow section, parallel flange channel, equal / unequal leg angle

To fully specify steel, the following information should be provided – for example, using a hot rolled universal beam, as shown in Figure 15.2:

- nominal overall height and width ('*D*' and '*B*' dimensions) – the flange and web thicknesses vary so weight per metre is also specified (e.g. 406 × 140 × 46 kg/m universal beam (UB)). Within any section size, several different weights are available and these are manufactured by varying the roll positions during production to change the thickness of the flange (dimension *B*) and web (dimension *D* is constant thickness, it then increases at the root radius '*r*'). Care is needed when detailing as the overall dimensions *B* and *D* vary slightly with weight per metre
- grade – minimum yield strength (e.g. S275 or S355, formerly called grades 43 and 50, respectively) and subgrade of steel. The 275 represents a tensile yield stress of 275 N/mm^2
- number of sections required and length of each section
- surface finish – for example untreated, powder coated, galvanised, intumescent paint for fire protection and so on
- fabrication details – for example drilled holes for bolts, end plated for splicing or web stiffeners (see section 15.4), plus arrangements for lifting heavy sections / fabrications (e.g. lifting eyes)
- required delivery date and identification marking for installation sequence and traceability.

Figure 15.2 Universal beam dimensions (editor's own)

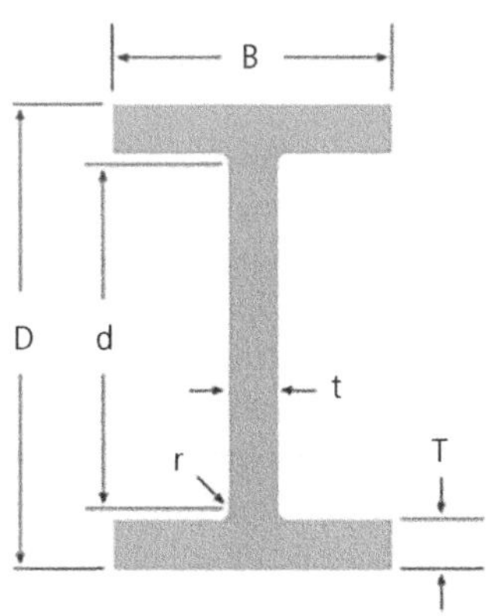

Steel suppliers' quality control procedures should be regularly audited, and they should provide trustworthy 'paperwork' to ensure their products are fully traceable. Many steel section sizes are derived from imperial sizes (e.g. 406 mm is 16 inches); however, metric equivalent sections are also available. Sites should have robust checking procedures to ensure the correct sections with correct details are delivered – this is just as critical for temporary works steel as for permanent works steel.

The maximum universal beam size available 'off the shelf' in the UK is 1016 × 305 × 487 kg/m. Larger bespoke sections can be fabricated in a workshop. All the various section sizes and properties are shown in section tables (e.g. https://britishsteel.co.uk/what-we-do/construction-steel/sections-product-range/).

15.3. Designing steelwork

Steelwork design can be complex and should be undertaken by competent and experienced engineers with a good understanding of the principles involved and potential modes of failure. Temporary works designs should be checked in accordance with BS 5975-1:2024 (BSI, 2024a), table 2 and a design check certificate issued before fabrication begins. There are some concerns regarding designs carried out by inexperienced / not suitably qualified persons, sometimes with an overreliance on design software that may not accurately model actual site conditions, especially regarding combined stress and eccentric loading regimes.

As discussed in chapter 2.6, in the UK virtually all permanent works are designed using limit state methodology, whereas many items of temporary works are still designed using permissible stress methodology. This is particularly true of steelwork design. Eurocode 3 (BSI, 2005) is the current limit state code of practice and has a more refined methodology which can potentially offer overall weight savings when used for designing complex structures which require hundreds / thousands of tonnes of steel. Many temporary works schemes are much simpler and the permissible stress design code BS 449 (BSI, 1969) is still used. It considers allowable stresses for different grades of steel. For example, a needling scheme to form a new 3 m-wide opening in a load-bearing brick wall may only require four steel beams, with a total combined weight probably not exceeding 250 kg. These would generally be designed as a simply supported beam with a point load at midspan; limiting deflection would be the overarching design consideration. BS 449 (BSI, 1969) would provide a very simple, very quick, well-established method which would result in a robust section. Any potential steel weight savings from using Eurocode 3 (BSI, 2005) would be minimal – perhaps no weight saving at all.

As mentioned above, limiting deflection is often a governing criterion in designing temporary steelwork. This can apply to simple items such as needling where excessive deflection of the steel beam would result in the supporting brickwork cracking. It also applies to large, supported excavations where excessive deflection of waler beams leads to ground movement around the excavation, which in turn could cause damage to adjacent roads or services. If large steel props are used within such excavations, then thermal expansion and contraction should be considered. Designers need to be mindful that steelwork will deflect, stretch, compress and so on when loaded and determine realistic values of what degree of movement is acceptable. Temporary steelwork should be designed to be robust so it can accommodate some degree of abuse under normal site conditions. The factor of safety adopted should consider the consequences of structural failure.

Note: some errors have occurred because BS 449 (BSI, 1969) and Eurocode 3 (BSI, 2005) do not use the same axes designation. BS 449 (BSI, 1969) uses x, y and z for major, minor and longitudinal axes, respectively.

Temporary works designers (TWDs) should try to limit the number of different sections specified. There can be mismatches with different sections, jointing details can be standardised, potential fabrication / erection errors can be avoided and there are potential economic savings in bulk-buying fewer sections. If common sections are specified, they can be reused. Many contractors will only keep a limited number of section sizes (tend to be same weight for any specific section size) and use them for multiple common purposes (even if in some cases the section may be substantially over-specified). Commonly used sections could include:

- 152 × 152 × 23 universal column (UC) for needling schemes
- 610 × 229 × 101 UB for large cofferdam frames (especially when 'twinned' with batten plates)
- 305 × 305 × 97 UC for propping small excavations.

For temporary works purposes, using lower grades of steel (in the UK commonly S275 and S355, but note that some UK stockists no longer supply S275) is generally preferred to higher grades for the following reasons. Grade S235 may also be used for smaller sections.

- Often the steel sections being used are second-hand / pre-used and there is no way of recognising steel grade on site (assuming the manufacturer's original identification paperwork cannot be found or relied on). Samples would have to be taken and tested in an accredited facility by tensile testing, chemical analysis or impact testing. If second-hand steel is to be used, designers should be consulted and the sections inspected for damage, corrosion and previous fabrication details before they can be approved for use. The exact section size and weight should be determined by measuring the overall section, plus web and flange thicknesses and weighing the section. For design purposes it is prudent to consider the sections to be S275.
- Welding procedures for lower grades are simpler (especially S275), which is more cost-effective. If the section grade cannot be verified and welded joints are required, the steel should be assumed to be grade S355 (for welding purposes). S355 is susceptible to hydrogen embrittlement unless the correct precautions are taken (see section 15.4). Welding becomes increasingly more complex (hence more costly and time-consuming) as the grade increases.
- Lower grades exhibit longer plastic regions which give a visible warning (deflection) of overloading which allows emergency procedures to be enacted. Higher steel grades will provide a lighter structure but become increasingly brittle and offer much less warning of overloading – that is, they can carry more load but if overloaded they can fail suddenly. The Young's modulus of steel is a constant; hence, higher grades offer no benefit for limiting deflection (i.e. stiffness = Young's modulus × second moment of area). BS 5975-2: 2024 (BSI, 2024b), clause 9.3.2.2 recommends that the Young's modulus for all grades of steel should be taken as $210\,kN/mm^2$.

Note: other lower steel grades are commonly used – for example, some scaffolding tube is S235.

It is generally cost-effective to remove temporary steelwork for storage and subsequent reuse. On occasions this may not be possible – for example, it might not be logistically possible to remove a temporary sheet pile retaining wall. Where possible, steel sheet piles should be incorporated into the permanent works for cost efficiency – for example, the sheet piles could be combined with the permanent reinforced concrete retaining wall, thereby saving on the thickness of concrete required. When temporary steelwork has been removed (for storage and subsequent reuse), it is important that records of section sizes, grades and fabrication details are kept, with identification marks added to the sections.

Steel sections can be subject to several different load combinations, creating different stresses and strains (tension, bending, shear, torsion, compression fatigue, etc.). TWDs should consider the most onerous combinations and ensure the overall utilisation of the section is not exceeded. BS 449 (BSI, 1969), sections 6 and 7 give recommendations on loads and section 14 gives recommendations on combined stresses. TWDs should also refer to BS 5975-2:2024 (BSI, 2024b), section 9 (specifically for falsework), plus annexes H and I.

Steel can corrode over time if exposed to moisture and oxygen. For some temporary works schemes, when the steel is exposed for a prolonged period in an aggressive environment (e.g. in a marine environment), protective measures should be considered. These could include protective coating or designing for a sacrificial loss of section whereby the steel sections are thicker than required (for structural performance) to allow for corrosion. Some caution is necessary when temporary steelwork (especially 'second-hand', which may exhibit some degree of corrosion) is connected to permanent concrete. The corroded steel can cause unsightly staining to the concrete – designers and sites should consider cleaning the steel (grit blasting or similar) or painting.

Certificate of Conformity (Conformité Européenne (CE), which will not be valid in the UK beyond 1 July 2025) or equivalent UK Conformity Assessment (UKCA) quality marking applies to steel used for permanent works but is not required for temporary works purposes (see TWf, 2015) with the following provisos:

- it is removed from the permanent works as part of the construction process or, if they are left in place:
  - it plays no part in the permanent works functions
  - its presence cannot have an effect on the performance of the permanent works.

However, it is also recommended that:

- all temporary works welding materials and welded steels (or other welded materials) are CE / UKCA marked, and that the welding process conforms to CE / UKCA requirements
- wherever practicable, temporary works products are CE / UKCA-marked
- whether or not CE / UKCA-m arked products are used, an appropriate quality management regime is required to ensure the works meet specifications and are generally fit for purpose.

15.4. Some common modes of failure

- Brittle fracture – the sudden or rapid fracture of a component under stress, but without undergoing any appreciable permanent deformation or ductility. The term 'brittle' may be considered as one end of a scale, with 'ductile' at the other end. It can be caused by low temperature (temporary works steel can be used in very cold climates or industrial deep freezers for access or propping), vibration (cyclic loading, fatigue), hydrogen being absorbed around welds, incorrect material being used.
- Web buckling / web crushing – normally occurs at points of high loading, such as at intersections of beams, beams supported by columns, mobile crane outriggers on steel spreader beams or props intersecting with waling beams in large supported excavations. There are several forms (see Figure 15.3) and the issue can generally be resolved by designing and adding web stiffeners, as shown in Figure 15.4. This issue was identified as one of the causes of the London Bridge temporary works collapse in 1972 which was investigated as part of the Bragg Report (Bragg, 1975). BS 5975-2:2024 (BSI, 2024b), annex H highlights this issue and recommends: 'For structural steelwork, web stiffeners should be provided at all loading transfer points, including supports, unless calculations are provided to show that such stiffeners are not required for the actual conditions applying and materials to be used on site'. Designers can avoid the need for web stiffeners in some positions by changing the thickness of the flange and web locally. Additional plates can also be added below a beam at a critical joint position, which increases the stiff bearing length in the web.
- Tension failure – usually occurs in braces or hangers when applied tension exceeds a member's resistance. Failure happens in stages, from yielding to necking to material failure at the smallest cross-section.
- Compression failure – occurs as a result of compression in the axial axis of the member, such as a temporary prop (column) or brace, and can result in buckling. The greater the slenderness ratio of the element, the greater the possibility of buckling (see BS 5975-2:2024, BSI, 2024b, annex I).
- Overloading – any item of temporary works will be designed for a specific load or set of loading conditions (as identified in the design brief). Greatly exceeding the design load could lead to failure and collapse. Changes in the loading conditions should be communicated to the designer for review and comment. Overloading could also be caused by an accidental impact (e.g. from a swinging load on a crane or a falling object during a demolition project) and the potential progressive collapse mechanism should be avoided at design stage. Temporary steel should be designed for an impact load in the most unfavourable direction / location, with a degree of redundancy, unless the likelihood of impact can be eliminated or sacrificial members introduced to absorb impact loading. BS 5975-2:2024 (BSI, 2024b), table 12 provides guidance on dynamic and impact forces.
- Eccentric load – occurs when the applied loading is not applied to the axis of the supporting member and consequently induces a moment. Off-centre connections should be avoided where possible; however, TWDs and those producing design briefs should appreciate and consider construction tolerances to establish a realistic degree of eccentric loading (in any direction), which, if not addressed, can result in overstressing sections in combined loading conditions.

Figure 15.3 Types of web instability (editor's own)

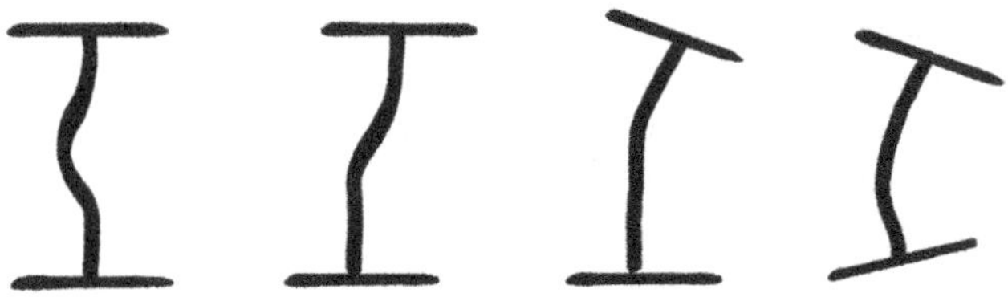

Figure 15.4 Addition of web stiffeners at points of load concentration (editor's own)

Note: this shows a common usage of temporary steelwork for a site cabin set up above a pavement and road.

- Lateral instability – can often be an issue when lifting deep bridge beams into postion by crane, whereby the beams can twist in the centre due to lateral torsional buckling. This is a temporary loading condition and the solution is to lift braced pairs of beams; however, this has an influence on the crane capacity, footprint and foundations required. BS 5975-2:2024 (BSI, 2024b), clause 7.2.2.2 provides guidance on minimum lateral forces.
- Connection failure – joints can often be weak points. Bolt holes should be the correct size for the bolt (drilled not cut with a gas torch), with correct hole spacing and edge distances.
- Rolling laminations or impurities – occur when using thick sections for temporary works purposes. The thicker the plate, the greater the chances of these problems. BS 449 (BSI, 1969) allowable stresses are reduced with increase in matarial thickness. When welding thicker sections and plates, pre-heating is normally required to prevent thermal shock and cracking during the welding process. 'Big welds' create a considerable amount of heat and a welding engineer will be needed to produce a welding procedure to reduce this likelihood.
- Fatigue – this is only rarely considered in temporary works. For example, temporary steel bridges subject to several years of intensive traffic movement, steel grillages to support tower cranes, cofferdams exposed to a tidal / wave environment for a prolonged period and structures exposed to intensive and extreme prolonged wind loading may be subject to fatigue. It can affect bolted and welded joints which, wherever possible, should be positioned at points of low stress and access should be provided for regular inspections. Mitigation measures should be provided where necessary, such as ensuring that welds have smooth shapes and transition to the parent metal. See BS 5975-2:2024 (BSI, 2024b), clauses 5.1.5, 5.3.5, 11.12 and 6.4.2.23, which in particular highlight the issue of vortex excitation as a result of an issue on a high-profile project which resulted in litigation.
- Fire – steel is susceptible to distortion and loss of strength with intense heat generated in a fire. Protection is required in the form of encasing in concrete, fireproof boarding or intumescent paint.

15.5. Cutting and joining steel

Steel can be cut by saw or gas torch (oxygen-butane burning gear), bent into a radius (by hydraulics), plated and joined. The most common methods for joining steel are bolts and welding, although proprietary fixings such as 'Lindapter' clips can also be used. Joint positioning requires careful thought and design. Where possible, it is generally advantageous to avoid placing joints at positions of maximum bending (often midspan or support positions) or maximum shear (often at support or load positions); however, sometimes this may be unavoidable for transporting, lifting or handling reasons.

15.5.1 Bolts

Bolted connections are often used in temporary works and offer benefits but also some challenges, such as the following.

- Prefabricated (manufactured in a workshop) temporary works assemblies can be bolted together on site quickly, especially important when time is a premium – for example, during rail possessions or road closures.
- Assemblies can then be removed quickly and easily when no longer required.
- If site welding is not required, then 'hot works' procedures and permits are not required.

- Additional bolt holes can be drilled on site if necessary to allow for errors or unforeseen circumstances.
- By providing slotted holes, a degree of construction tolerance can be accommodated.
- Bolt positions can be positioned on flanges or webs or both. If end plates or splice plates are added, then sections can be joined for ease of installation.
- TWDs should ensure correct hole sizes are provided and consider minimum spacing between bolts and minimum edge distances.
- When subject to vibration, nuts can work loose. This should be avoided by using 'spring washers', 'double nuts', 'lock nuts', nyloc, split pins in the bolt shank or wrapping steel-fixers tying wire beneath the nut. Installing bolts with the nut uppermost should be avoided where possible to prevent bolts falling out if the nut works loose.

Bolts can be used to resist tension and shear (or both). A bolt in single shear is shown in Figure 15.5(a) and double shear in Figure 15.5(b). Double shear has two effective cross-sections resisting the load and will have double the shear strength of a bolt in single shear.

- High-strength friction grip (HSFG) bolts are tightened to a defined pre-load to clamp the connecting plates together, ensuring the load transfer occurs through the developed friction between mating surfaces. This type of joint relies entirely on friction. It is imperative that there is nothing on the joint (e.g. grease or paint) as it is assembled which would lubricate and reduce the friction. All tightening methods need the components to be brought together to a snug-tight condition before commencement of pre-loading. Whichever method of bolt tightening is chosen, bolts and nuts should always be tightened in a staggered pattern. When a bolt group comprises more than four bolts, tightening should be from the middle of the joint outwards, ensuring that all the piles are properly pulled together in full contact. For very large HSFG bolt groups, a bolting specialist must be involved to ensure the correct procedure is used. Load indicator washers have raised pips which compress when the torque is applied. Feeler gauges should be used to determine whether the correct torque has been applied.
- Torque control bolts (TCB) are a form of HSFG which are tightened by a special tool which performs all the tightening from one side of a joint and then breaks off a spline when the required torque has been achieved. These bolts make inspection easier.

Different grades of bolt are available. Metric bolts have the grade formed on the head except for grade 4.6 (400 MPa nominal tensile strength with nominal yield strength of 60%), which are now only occasionally used for temporary works purposes. The most common is grade 8.8 (800 MPa nominal tensile strength with nominal yield strength of 80%), which is shown

Figure 15.5 (a) Single shear joint; (b) double shear joint (editor's own)

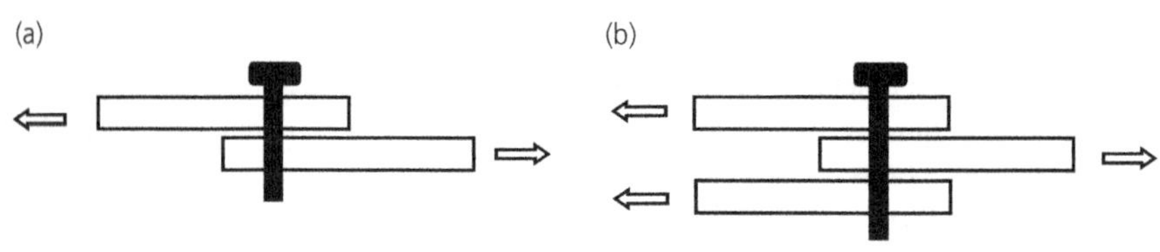

Figure 15.6 (a) Grade 8.8 bolt; (b) recognising nut grades (reproduced with kind permission of N Cook)

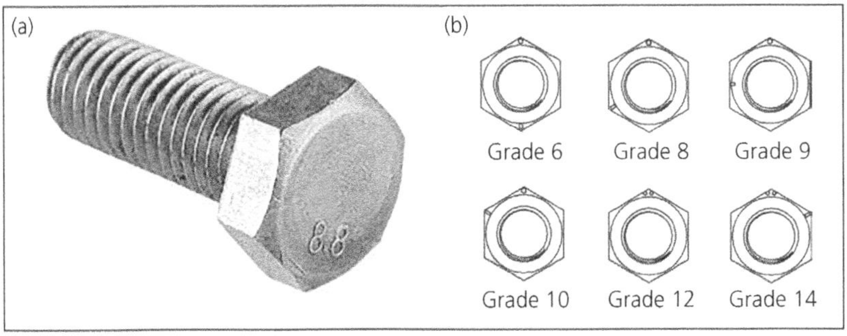

in Figure 15.6(a). Higher grades are available, such as grade 10.9 which is used in HSFG. Nuts should be matched with the grade of bolt and can be identified by a 'clock face' system, as shown in Figure 15.6(b).

If bolt holes need to be created on site, they should not be cut with a gas torch otherwise any shear load will not be transferred from the sides of the hole to the bolt shank. Drill bits should be used, but for larger diameters a magnetic drill (which clamps magnetically to the work surface) should be used for safety and accuracy.

Figures 15.7 and 15.8 show where construction tolerance should be considered for the bolted assembly to ensure safe and efficient assembly (also see section 15.6). Where tolerance has

Figure 15.7 Construction tolerance needs to be considered for this bolted connection (editor's own)

Note: One option is to drill the holes in the square plate on site to suit the final bolt positions.

Figure 15.8 Tolerance has not been adequately considered for temporary tension bracing (reproduced with kind permission of N Cook)

Note: The flat plate bracing is bolted to the steelwork and is intended to work in tension but because of lack of tolerance it has had to be forced into position, resulting in its bowed appearance; the structure would need to sway significantly before the bracing took any load.

not been adequately considered, the site team will have to 'force' steelwork (possibly using tirfors or 'podger bars') to fit or will have to fabricate on site. The issue would be the same for temporary and permanent steelwork.

15.5.2 Welding

Welding can be carried out under controlled conditions in a workshop or on site where it is more difficult and consequently different techniques are used. Welding has the benefit that steel can be cut and / or joined together on site to allow for tolerance, unforeseen circumstances or errors. All welding should be carried out by a competent welder (coded welders who provide proof of their skills set through testing) and welds should be inspected and tested for quality as required (see section 15.5.3). 'Hot works' permits and screens (to protect the public and other operatives) are required if welding is to be carried out on site (also for gas cutting, grinding etc.). The Health and Safety Executive (HSE) has extensive guidance on safety precautions, appropriate protective equipment, storage of equipment (especially gas cylinders) and dealing with welding fumes on their website (see Useful websites section at the end of this chapter). Some common types of welding are described:

- Manual metal arc (MMA) – the most appropriate method for site welding, especially for temporary works purposes. It uses a transformer and welding rods (known as sticks, which are coated in flux to shield the area being welded from contaminants in the atmosphere such as dust or moisture) of a defined length. The rod diameter and specification depend on the type of weld and material being welded. The flux-coated welding rods produce a slag on the weld surface, which must be removed before the next run of weld is added. Slag remaining within the welding will result in imperfections which may result in weld failure.
- Metal inert gas (MIG) – an arc welding process that uses a continuous solid wire electrode, heated and fed into the weld pool from a welding gun which also feeds a shielding gas (inert gas such as argon). The shielding gas can be blown away, exposing the welding area to contaminants in the atmosphere; hence, this technique is less common on site. This technique is more difficult for site work as the machinery tends to be larger and a greater degree of protection may be necessary.
- Tungsten inert gas (TIG) – a combination of gas and MIG welding. The welder holds a gun similar to MIG but the electrode is fixed and is purely used to strike an arc which produces a pool of molten metal. A filler wire is then introduced manually to top up the molten pool. An inert gas is fed around the electrode to protect the molten pool. This techique is generally not used on site.
- Submerged arc welding (SAW) – an arc welding process that involves the formation of an arc between a continuously fed electrode and the workpiece. A blanket of powdered flux generates a protective gas shield and a slag which protects the weld zone. This techique is generally not used on site.
- Fillet welding – the process of joining two pieces of steel together perpendicular or at an angle and commonly used for temporary works processes. The weld forms a triangular shape with a 'throat' and 'legs'. The leg length is easiest to measure but the minimum thickness is at the throat (this is what is generally specified). In general, fillet welds with a leg length over 8 mm are built up with multiple passes (Figure 15.9(a) shows three passes).
- Butt welding (can be single- or double-sided) – formed by placing two pieces of steel end to end (ends are prepared beforehand) and welding along the joint. Quality control procedures are essential, otherwise without adequate preparation a full-strength weld cannot be formed. Inspection and testing (see section 15.5.3) should also be carried out to ensure quality of the finished weld. Designers should specify the root gap, root face and preparation angle (see Figure 15.9(b)).

Weld gauges are available to make measurement simpler and more accurate.

To avoid hydrogen embrittlement when welding grade S355, the correct welding rod or filler must be used to reduce risk. The correct procedures must be used, such as keeping the welding rods dry and warming the weld. The flux coating on a welding rod (used to protect the area being welded from the atmosphere) can be hygroscopic (will absorb water), resulting in hydrogen from

Figure 15.9 (a) Fillet weld; (b) single-sided butt weld (editor's own)

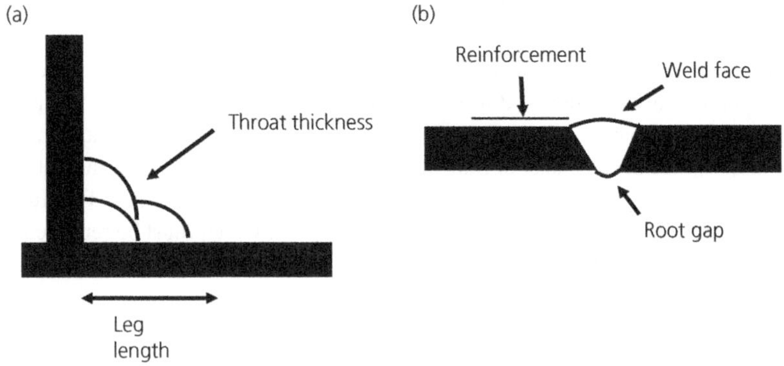

Figure 15.10 Site cut and welded shear stops for vertical reaction from inclined excavation props (editor's own)

the absorbed water being introduced to the molten weld metal. Welding generates a great deal of heat which causes steel to expand and then shrink when it cools. Distortion can occur which may cause problems with 'fit' for complex shapes. Thermal shock may also occur when welding thick sections so pre-heating and insulation to control cooling can be used. A welding procedure should be produced by a qualified welding engineer. Figure 15.10 shows a common welded site detail.

15.5.3 Weld inspections and testing

All welds used for temporary works purposes should be visually inspected for cracking, impurities, lack of continuity and correct size. If a visual inspection identifies potential issues or the welds are deemed to be safety critical, then there are several other methods which can provide greater clarity / certainty.

- Dye penetrant testing or dye penetrant inspection – a widely used, low-cost inspection method used to check surface-breaking defects in all non-porous materials. A penetrant is applied to the material by dipping, spraying or brushing. Capillary action allows this low-surface-tension fluid to penetrate into clean and dry surface defects. The excess penetrant is removed and a developer is applied which draws the penetrant out of the defect, making it visible to inspection.
- Magnetic particle inspection (MPI) – used to detect surface defects in ferromagnetic (or magnetic) materials. A magnetic field is generated inside the component and a magnetic particle powder is applied. The magnetic flux leaks through any defects and attracts the particles, highlighting any defects.
- Ultrasonic inspection (UT) – uses ultrasonic waves to measure internal defects in a variety of different materials. UT can be used to detect a variety of issues such as corrosion, thinning and internal planar defects. It also identifies cracks / voids and is useful for testing pressure vessels, storage tanks and welds used in a high-pressure environment. The method is also useful for inspecting forgings, castings and structural steel welds.

15.6. Site work

Workshops generally have a better controlled environment and a greater choice of more sophisticated equipment to produce better quality work. However, some degree of site work and fabrication is required. Site fabrications will generally require additional inspections to ensure quality and correctness. Welding will often be MMA, which is generally slower and the welds are not as 'neat', but for temporary works purposes this is rarely a concern as long as they perform structurally as intended. Aesthetics of temporary steelwork is rarely considered, so occasionally it may have a coat of primer applied. Exceptions could be where the steel has a public interface (such as steel gantries for cabins above pavements). Where corporate image is important, the steel will be painted and regularly repainted on longer-term projects.

Efficiency and safety of assembly / removal and maintaining stability of steel structure (temporary works or permanent works structure) in the temporary / not complete condition are primary considerations. Steelwork cannot be simply erected in a haphazard manner and needs careful planning. To maximise efficiency, ideally steelwork should be fabricated in a workshop, but delivery logistics (maximum prefabrication lengths / weights that can be transported and site crane limitations for offloading and positioning) may limit the amount of offsite work that can be realistically achieved. Steelwork should be stored securely on site with finished surfaces protected; it should be numbered and stacked in sequence to allow for efficient assembly. Prior to offloading and assembly, checks should be carried out to verify the correctness and accuracy of the elements being delivered. Dismantling sequences should not merely be 'reverse of installation'. As much planning and attention is required for a removal sequence (including maintaining stability to prevent an unplanned collapse) as for an assembly sequence.

All designers need to follow the 'principles of prevention' and need to consider an assembly sequence, with particular attention to temporary lateral stability. Steel frames should be assembled from a braced bay (additional temporary bracing may need to be provided) for stability. Alternatively, the frame can be designed as a portal frame with robust moment connections. Portal frames are particularly useful for site cabin set-ups above roads or pavements where bracing would obstruct access beneath. The indicative assembly sequence should be agreed and communicated, with residual risks being highlighted by the designer and addressed by the site team. Any proposed changes to the agreed sequence (or specific details) should be communicated to the designer for review and approval.

There are numerous proprietary pieces of equipment for lifting and positioning steelwork, such as lifting frames and remote release shackles that allow the crane to be released from the load without requiring operatives to be exposed to work at height.

Efficient and safe assembly / removal of temporary steelwork needs to be considered, especially if the assembly must be carried out within a limited time – for example, during rail possessions or road closures. Detailed planning is required for logistics, access and lifting and careful attention should be given to construction tolerances. Often trial assemblies are time and money well spent. They can be used to determine whether the structure can be assembled within the required timescale and can be used to check that assembly tolerances are acceptable. On site, the steel assembly may have to interface with items previously constructed (e.g. foundations) or may connect to an existing structure. Setting out should be checked and further tolerances need to be considered – for example, movement cones around holding down bolts and allowing for grouting beneath base plates.

If tie rods are used (e.g. for temporary cross-bracing) then slack must be taken out of the system for the ties to work effectively. Turnbuckles are commonly used for this purpose.

The *National Structural Steelwork Specification (Black Book)* is produced by the British Constructional Steelwork Association (BCSA, 2023) and contains guidance on acceptable tolerances in both production and fabrication of steel components and structures. It states what materials and products should be used and how work (fabrication and erection) should be carried out in order to ensure that the completed structure meets the designer's assumptions and the client's needs. Design standards, such as Eurocode 3 (BSI, 2005), give design rules that are only valid if strength, dimensional accuracy and freedom from defects are within presumed limits. The performance of a structure required by regulations will also only be achieved when construction is within presumed limits. This document is normally assumed to be the minimum specification for a steel structure and designers may introduce further specifications and tighten tolerances. Various tolerances (due to rolling tolerances, position, length, verticality, in the horizontal plane) should be considered by designers and site teams and these will have a direct impact on connections.

Steelwork assemblies require inspection to ensure correct sections and bolts are used, they are tightened correctly, weld size, type and quality is as required and overall construction meets the design requirements. Steelwork assembly and subsequent inspection often involve working at height (i.e. this risk cannot be eliminated) and appropriate safety measures need to be considered. Items such as bolts and hand tools may also be inadvertently dropped, so exclusion zones or other safety measures should be provided. Chapters 5 and 6 discuss these issues in detail.

REFERENCES

BCSA (British Constructional Steelwork Association) (2023) *National Structural Steelwork Specification for Building Construction*, 7th edn, 1st revision. BCSA Publications, London, UK.

Bragg SL (1975) *Final Report of the Advisory Committee on Falsework.* Department of Employment and Department of the Environment, The Stationery Office, London, UK.

BSI (1969) BS 449: Specification for the use of structural steel in building. BSI, London, UK.

BSI (2005) BS EN 1993-1-8: Design of steel structures. BSI, London, UK.

BSI (2024a) BS 5975-1:2024: Management procedures for the control of temporary works. Code of practice. BSI, London, UK.

BSI (2024b) BS 5975-2:2024: Temporary works – Falsework: design and implementation. Code of practice. BSI, London, UK.

TWf (Temporary Works Forum) (2015) TW15.051: Construction Products Regulations (EU) CE Marking and Temporary Works. https://www.twforum.org.uk/viewdocument/twf-information-sheet-no-1-const (accessed 10/01/2025).

USEFUL WEBSITES

AJN Steelstock – www.ajnsteelstock.co.uk
ArcelorMittal – www.arcelormittal.com
British Steel – www.britishsteel.co.uk
BSCA (British Constructional Steelwork Association) – www.bcsa.org.uk
BSI (British Standards Institution) – www.bsigroup.com
Fischer Fixings – www.fischer.co.uk
Hilti Fixings – www.hilti.co.uk
Metro Fixings – www.metrofixings.co.uk
Parker Steel – www.parkersteel.co.uk
Steel Construction – www.steelconstruction.info
Steel Construction Institute – www.steel-sci.com
Tata Steel – www.tatasteeleurope.com
TWf (Temporary Works Forum) – www.twforum.org.uk

FURTHER READING

BCSA (British Constructional Steelwork Association) (2023) *Handbook of Structural Steel.* BCSA, London, UK.

Hauke B (2024) *Steel Construction.* Ernst & Sohn, London, UK.

O'Driscoll R (2022) Construction Engineering: A Guide to Temporary Works in Construction and BS 5975:2019. www.constructionengineering.uk (accessed 07/01/2025).

SCI (Steel Construction Institute) (2016) *Steel Designers' Manual*, 7th edn. Wiley Blackwell, London, UK.

SCI (2019) *SCI P427: Structural Steel Reuse*, 2nd edn. SCI, Ascot, UK.

Vayas I, Ermopoulos J and Ioannidis G (2018) *Design of Steel Structures to Eurocodes.* Springer Nature, Cham, Switzerland.

Ray Filip
ISBN 978-1-83608-525-6
https://doi.org/10.1108/978-1-83608-524-920251018

Chapter 16
Temporary and demountable structures

Malachy Ryan
Alan White Design Ltd, UK

16.1. Introduction

Temporary and demountable structures are typically required for a 'short' time and the design and management of these structures is less onerous than that of a permanent structure or building. They can be utilised during the construction of major civil engineering projects, construction of permanent buildings, at festivals, music or sporting events and as short-term storage. Temporary structures and demountable structures are essential for the successful completion of these projects and events, as they provide additional operative accommodation, stages and storage spaces. Temporary structures fall within the definition of 'construction work' in Construction (Design and Management) Regulations (CDM) (HMG, 2015).

This chapter will focus on the use of temporary demountable structures on construction sites and at festivals or events. It will not consider temporary scaffold roofs or tensile structures as these are considered elsewhere.

The main guidance documents for temporary structures are:

- TG20:21: *Good Practice Guidance for Tube and Fitting Scaffolding* (NASC, 2021)
- BS EN 12811-1 (BSI, 2003)
- *Temporary Demountable Structures: Guidance on Procurement, Design and Use* (IStructE, 2017)
- BS EN 1991-1-4 (BSI, 2005).

16.2. Public events temporary structures

Demountable structures are temporary buildings which are designed for short-term, repeatable use. The detailing and design of the structural elements and connections allow for easy assembly and disassembly to move the structures from event to event. This type of structure is typically used for stages, exhibitions, concerts and sporting events. Reference should be made to *Temporary Demountable Structures: Guidance on Procurement, Design and Use* (IStructE, 2017) for specific guidance on procurement, design, erection, use and dismantling.

The Health and Safety Executive (HSE) has produced guidance on temporary demountable structures which is available on their website. The structural principles are similar to those for construction site temporary structures but there are a few additional points to note.

- Public safety is critical. An emergency plan (including a fire safety plan) with clear evacuation routes for large crowds is essential. High-profile events can be targets for extremists; security considerations and provisions are also essential.
- Maximum occupancy limits must be established and implementing effective crowd management strategies is essential to prevent overcrowding and ensure the safety of occupants.
- These structures are used for a very short time (could be a few days) at any location and need to be assembled and removed easily and quickly. They can be made from proprietary modular components which can be clipped, bolted or screwed together (avoiding onsite cutting and welding) and designed to be easily transportable from venue to venue. Standard designs can be developed for typical locations, with specific consideration given to foundations and any unusual wind effects (or other environmental effects) at each location.
- The overall layout should ensure the paying public has a degree of comfort and a clear view of the event. Specialist advice should be sought regarding acoustics and lighting.
- For music and sporting events, the design should specifically consider high dynamic and cyclic crowd loading.
- If used at the same location for several events, then thorough examination and any necessary maintenance should be carried out between events.
- The Events Structure Industry Training Scheme (ESITS) is a nationally recognised training and accreditation programme for those involved in the design and build of proprietary demountable structures in the events, media and entertainment sectors.

Recent notable UK examples include temporary spectator facilities for the 2012 London Olympic Games and the 'Pyramid Stage' / audio towers / lighting towers at Glastonbury music festival.

16.3. Construction site temporary structures

Temporary construction site structures are erected to meet a temporary need that could last from several months to several years. They are typically required for uses including site accommodation, welfare facilities, workshops, stores, materials or soils testing laboratories and temporary weatherproofing. These structures are not permanent, so their design reflects this short-term requirement. Corrosion, aesthetics, concrete cover, fatigue and exposure to extremes of weather are less of a consideration. The design and management of these structures should be governed by the guidance of BS 5975-1:2024 (BSI, 2024a) and BS 5975-2:2024 (BSI, 2024b) and reference should be made to BS EN 1991 (BSI, 2005) for environmental loads.

Scaffold roofs are commonly used to provide weatherproofing during new build or refurbishment projects. The guidance on the design, installation and maintenance of scaffold temporary roofs is covered in NASC TG20:21 and TG9:18 (NASC, 2018, 2021) as the primary design guides for temporary roofs. This chapter does not cover the design of scaffold roofs. Some important considerations are how the roof will be installed / removed safely, snow loading and uplift due to wind; all these issues should be addressed in the design. If a scaffold temporary roof relies on another structure for support or stability (e.g. an existing building), then the building requires assessment for the applied roof loads.

Figures 16.1, 16.2(a) and 16.2(b) show some examples of temporary structures.

Figure 16.1 Temporary site workshop made from timber frame and plastic sheeting (editor's own)

Figure 16.2 (a) External and (b) internal view of temporary domed 'bunker' structure at Hinkley Point C (HPC) (reproduced with kind permission of M Parker and A Dixon, both from HPC)

Note: Figure 16.2(a) shows the outside of the two temporary domed 'bunker' structures to provide scale and Figure 16.2(b) shows the inside of one. The structure comprises steel plate welded to a steel frame with a geodesic roof. It allows for the fabrication on the permanent steel liner and steel rings which provide shielding to the nuclear reactor. The permanent works being constructed are safety-critical items; hence, the welding and inspection is to be carried out in a controlled environment to ensure the highest standards and quality of work. This structure is used on site for around 3 years.

(a)

(b)

16.4. Managing demountable structures

The procedure for managing demountable structures involves several key steps to ensure a safe event. Their procurement, design, risk management, end-user experience and safety are detailed in *Temporary Demountable Structures: Guidance on Procurement, Design and Use* (IStructE, 2017).

It is likely that these structures may be subject to approval from the local council authority. In some scenarios, planning and building warrant consent may be required for a temporary structure. Contact should always be made with the local building warrant officer to assess their requirements and understand any specific imposed restrictions (such as relating to local residents' concerns regarding security, traffic, noise, light pollution etc.) to achieve the building warrant. If planning permission is required, then this officer should advise on that aspect also. The requirements will vary from local authority to local authority so it cannot be assumed that approval from a past event will be similar or remain valid for further events. If there is a failure by any stakeholder to comply with the requirements of the Health and Safety regulations or building control regarding temporary structures, this could potentially lead to enforcement action, including prosecution. Seeking guidance from design professionals and regulatory authorities can help prevent potential prosecution and ensure full compliance with the law.

Construction sites are generally fully aware of the management procedures of BS 5975-1:2024 (BSI, 2024b) and have well-established and robust procedures. BS 5975-1:2024 (BSI, 2024b) provides clear guidance for clients, designers, contractors and equipment suppliers. However, those procuring festivals, music or sporting events may not be fully aware of the risks, their legal duties / responsibilities and how to adequately manage these temporary structures. Often, significant reliance is placed on advice from health and safety professionals who may not have the necessary awareness, training or experience of the risks involved with these temporary structures. This could also apply to local authority building warrant officers.

16.5. Specific design considerations

16.5.1 Wind loading on roofs

Wind pressures can affect both the walls and roof of the structure / building. Temporary structures are typically lightweight, often made of scaffolding, timber or aluminium and clad in sheeting or a membrane. Uplift from wind pressure can be one of the primary design considerations. Often these roofs do not have sufficient self-weight to resist uplift (see section 16.5.5) and kentledge or anchorage will be required. Temporary buildings can come in a variety of shapes; some of the more common geometries are detailed below (Figure 16.3(a)–16.3(c)). BS EN 1991-1-4 (BSI, 2005) provides guidance on structural forms for wind load calculations.

- A mono pitch roof (see Figure 16.3(a)) typically has walls at both eaves that are of differing heights. The roof spans between these walls to form a single sloping roof plane.
- A duo pitch roof (see Figure 16.3(b)) typically has walls at the eaves of similar height and an A-shaped roof of two slopes between them.
- A barrel-vaulted roof (see Figure 16.3(c) and Figure 16.4) is a series of arch-shaped structures arranged in a row to form a half-cylinder shape. These roof types are very common for temporary demountable structures.

Figure 16.3 (a)–(c) Common types of roof (author's own)

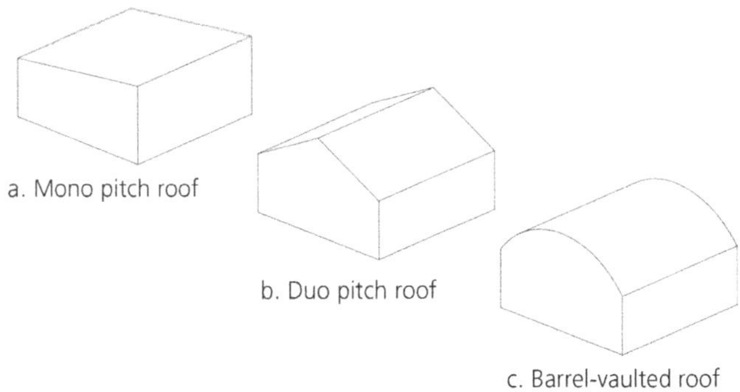

Figure 16.4 Example of barrel-vaulted roof structure (reproduced with kind permission of Trans-Sport TV)

Note – vertical and horizontal reactions from roofs need to be considered by designers, especially duo pitch and barrel-vaulted roofs, which will 'splay' outwards under their self-weight and any other applied vertical load.

16.5.2 Wind loading on walls

The wind pressures applied to the exposed structure walls are a critical consideration in the design and analysis of temporary structures, influencing both structural stability and kentledge requirements. Detailed guidelines for wall loads are provided in BS EN 1991-1-4 and BS 5975-2:2024 (BSI, 2005, 2024b) standards and in TG20:21 (NASC, 2021), offering guidance on pressure and shape coefficients. When determining wall loads, designers should refer to these standards and select the most applicable scenario, based on the site location, structure geometry, duration of the installation and structure configurations. In cases of conflicting information, defaulting to BS EN 1991-1-4 (BSI, 2005) ensures compliance with governing industry standards.

Figure 16.5 Layout of wind orientations (author's own)

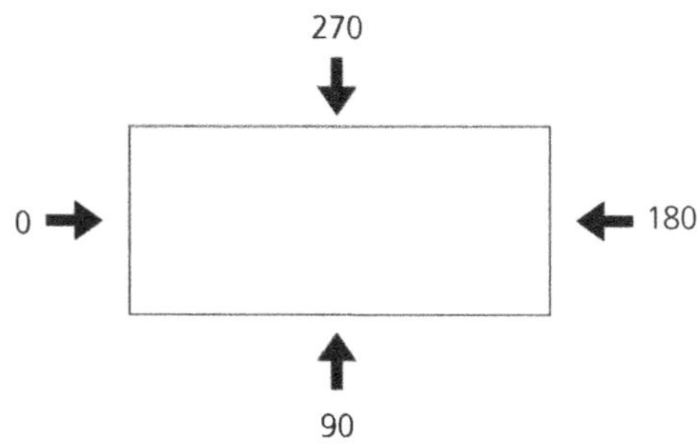

To accurately evaluate wind forces, designers should assess them in at least two separate orthogonal orientations, typically with wind applied at 0° and 90° to the structure. For structures with asymmetric geometries or complex configurations, additional orientations may be necessary to capture the full range of wind loading scenarios. Consideration of orientations at 180° and 270° ensures thorough assessment and enhances the accuracy of wind force calculations (see Figure 16.5).

16.5.3 Internal wind pressure

Figure 16.6 shows that internal wind pressures should be considered in the design and analysis, particularly in scenarios where structures feature dominant openings, such as stages with no front wall cladding. These openings can create unique challenges by allowing wind to penetrate the structure, resulting in internal pressures that exert additional forces on the framework. To ensure structural integrity and safety, designers must assess internal wind pressures in structures with dominant openings. It should also be noted that in a structure with a dominant opening, internal pressures are typically added to external pressures, resulting in a combined net pressure acting on the structure. This net pressure can be substantial, particularly in cases of high wind speeds or structures with large openings. If an internal wind coefficient is +0.5, the positive value indicates that the wind is applied normal 'to' the internal surface. If the external pressure at this location is –0.5, the negative value indicates that the external wind is applied normal 'from' the external surface. These coefficients do not summate to zero; in fact the net pressure coefficient for the wall panel is 1.0. Consideration

Figure 16.6 Explanation of net pressure coefficients (author's own)

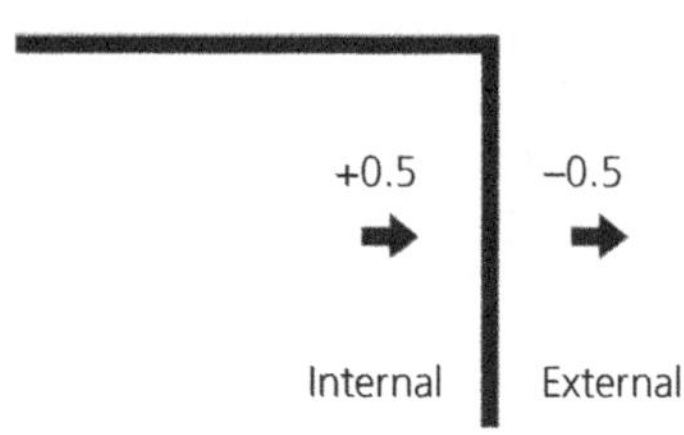

of the pressure both externally and internally and the correct derivation of the net wall panel coefficient are critical design considerations.

16.5.4 Downforce

Downforce on roof panels is an often-overlooked aspect of wind loading. While roof wind pressure coefficients typically account for both uplift and downforce loading, the significance of downforce cannot be understated, particularly in certain scenarios where it can significantly influence the structural capacity of roof members. These roof coefficients encompass both uplift forces, which exert an upward pressure on the roof, and downforce forces, which exert a downward pressure. Understanding the interface between uplift and downforce is critical in accurately assessing the structural response to wind loading. In specific scenarios, such as structures with large roof areas or unique geometries, downforce load combinations may dictate structural capacity and influence the design. Thorough analysis and consideration of downforce loading are essential to ensure the structural integrity and stability of the temporary structure under varying wind conditions. BS EN 1991-1-4 (BSI, 2005) provides guidelines and procedures for accurately determining downforce loads based on site location, roof geometry and structural configurations.

16.5.5 Uplift

Roof uplift should be assessed, as it often determines the most critical kentledge condition for structural stability. A thorough assessment of the uplift forces is vital to ensure a safe temporary structure. In the analysis of uplift, self-weight and wind loads are typically factored at 1.0 and 1.5, respectively, in accordance with BS EN 1990 (BSI, 2023). These factors are provided to ensure conservative estimates and account for potential variations in material weights. Beneficial forces, such as the self-weight of the frame or the base of the stage, contribute to the overall stability of the structure and help counteract uplift forces. Snow loading should not be relied on to resist uplift but will be a major design consideration, especially for large-span roofs (see section 16.11.5). Conversely, adverse forces, such as wind, exert upward pressure on the structure and can lead to instability if not adequately addressed. It is important to note that certain live loads, such as floor loading, lighting and public address systems, should not be included in uplift assessments. These loads may not be present in the structure during storm conditions or wind loading events and including them in uplift calculations will lead to inaccurate kentledge requirements.

16.6. Analysis

Various methods are available for analysing temporary structures, including the use of analysis software or hand calculations, both of which are valid approaches. Hand calculations remain widely used due to their practicality. The choice between software analysis and hand calculations depends on factors such as the complexity of the structure, available resources and engineering preferences. Regardless of the chosen method, the analysis process should encompass several essential considerations to ensure the structural integrity and stability of the temporary structure. These considerations include the following.

- Gravity loads – the calculation of gravity loads is fundamental to assessing the overall structural stability and load-bearing capacity of the temporary structure.

- Imposed and environmental loads – imposed and environmental loads, such as floor loads, wind loads and snow loads, must be accurately determined based on site-specific conditions and relevant design codes.
- Analysis of members – each structural member must be evaluated to determine its capacity to resist the applied loads. This includes assessing factors such as bending moment, shear, torsion and axial forces. Predicted deflections should be within agreed limits. Industry standard deflection limits can be relaxed for temporary structures due to their flexible materials and temporary nature.
- Analysis of connections – connections are essential to ensure transfer of loads between structural members and maintain overall structural stability. Connection details should be carefully examined to prevent potential failure points. Designers should carefully consider how a roof is connected to the remainder of the structure to fully mobilise self-weight and any additional kentledge or anchorage. Structural adequacy of all connections should be ensured. Connection details between sheeting / cladding and the structural frame should be sufficiently robust to prevent the sheeting / cladding from being blown off the main structure.
- Bracing – internal bracing can be provided or, if space allows, external buttressing or guy ropes used. The design of bracing systems plays a crucial role in preventing lateral instability and ensuring the overall rigidity of the temporary structure. Analysis should verify the effectiveness of bracing elements in resisting applied loads.
- Overturning and sliding – checks for overturning and sliding are necessary to assess the stability of the temporary structure under lateral forces. Adequate measures should be implemented to prevent potential overturning or sliding failures (this could include additional kentledge or anchorage).
- Kentledge / anchors – verification of kentledge or anchor systems is required to ensure their ability to resist uplift and provide sufficient stability.

By addressing the above in the analysis, designers can evaluate the structural performance of temporary structures and implement necessary design modifications or reinforcements to ensure their safety and stability throughout their lifecycle.

16.7. Stability

Bracing should be assessed to ensure the lateral stability and safety of temporary structures. While often overlooked, the lateral stability of the framework is essential to prevent structural failure, particularly in adverse weather conditions or during installation / removal. Bracing serves to resist lateral loads and prevent excessive deformation of the structure. It reinforces the framework, enhancing its ability to withstand external forces, such as wind, or dynamic loads from lifting operations. The installation of bracing to full wall elevations can present challenges, especially concerning access and egress points. Designing effective bracing solutions while maintaining accessibility for personnel and equipment requires careful consideration and coordination. Figure 16.7 shows an idealised layout of vertical bracing (VB) in a temporary structure. The bracing should be installed on two orthogonal elevations and ideally in pairs. This layout is not always achievable; however, the designer should ensure sufficient bracing is provided for lateral stability.

Several factors must be considered when designing bracing systems for temporary structures; Figure 16.8 shows some typical configurations.

Figure 16.7 Plan view of temporary structure with idealised vertical bracing (VB) (author's own)

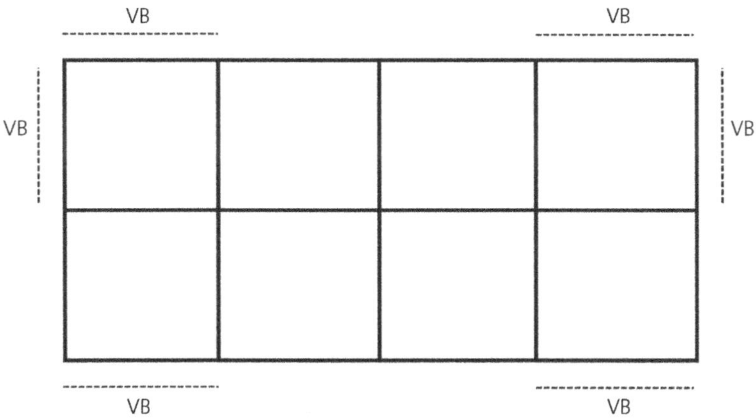

Figure 16.8 Typical bracing configurations: A: cross-bracing, B: compression strut, C: V-bracing, D: knee bracing (author's own)

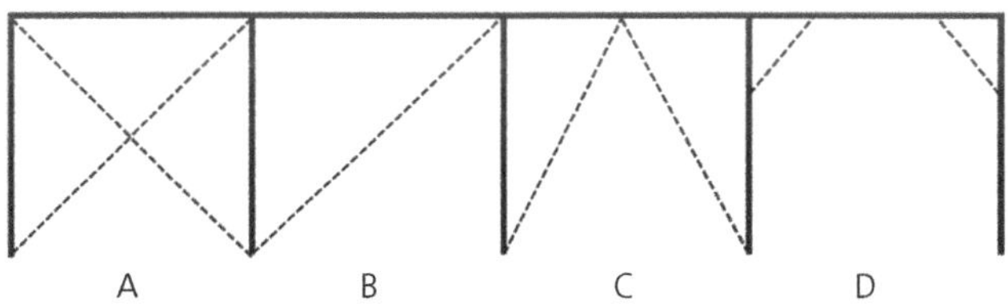

- Structural configuration – the structural layout and geometry of the temporary structure influence the placement and design of bracing elements. Bracing should be installed in pairs and along two orthogonal elevations where possible.
- Access and egress points – bracing arrangements should accommodate access and egress points without compromising safety or obstructing movement within the structure. K-bracing or V-bracing should be considered when access to an elevation is required.
- Flexibility and adaptability – bracing systems should be flexible and adaptable to accommodate changes in site conditions, load requirements or structural modifications during the lifespan of the temporary structure.
- Integration – bracing should be integrated seamlessly with other structural components and systems, such as foundations, kentledge and primary member connections.
- Type of brace – strut member constructed from tubular (or other suitable) structural steelwork sections or a tension only member (i.e. tie rods, wire rope or ratchet straps). Designers should allow for construction tolerance by allowing for slotted holes for bolting strut members to the frame and 'turnbuckles' for adjusting the length (and removing slack) from tie rods.

- The positioning of the bracing should reflect the anticipated construction sequence. As with any braced frame, a 'braced bay / corner' is erected and the sequence should continue from this initially stable point.

16.8. Ground conditions

Designers often encounter challenges related to information available on the existing ground conditions beneath temporary structures. Due to the temporary nature of these structures, comprehensive site investigation data may be unavailable or deemed economically impractical for short-term events. The costs associated with conducting boreholes or trial pits, coupled with laboratory testing, are often commercially prohibitive within the constraints of temporary projects. Considering these constraints, designers are tasked with taking reasonable steps to address uncertainties in ground conditions and limit applied bearing pressures from their structures. *The Structural Engineer's Pocket Book* (Cobb, 2014) provides valuable guidance for estimating allowable ground conditions based on various soil parameters. By utilising this information, designers can make informed decisions regarding the suitability of ground conditions for supporting temporary structures. Additionally, the use of timber or steel spreader pads can be considered to distribute loadings and reduce bearing pressures to acceptable values.

If a temporary structure is founded at a location where there are buried services, it is feasible that there may be failure of these service structures due to downforce overloading. A survey of existing utilities within the proposed site is advisable to ensure the ground is sufficient to provide support to the temporary structure. BS 5975-2:2024 (BSI, 2024b), section 10 provides additional information on the design of foundations and assessment of ground conditions.

16.9. Kentledge

Kentledge is critical in ensuring the stability and safety of temporary structures, providing counterbalancing forces to resist uplift and lateral loads. Designers should carefully consider what to use for kentledge. They should avoid using items that may be easily removed or stolen, such as bags of cement or bricks. Common kentledge solutions for temporary structures include large steel mass weights, large concrete blocks, drums filled with soil or rubble and water tanks. Steel weights offer compact size, robustness and durability, making them suitable for various applications. Concrete blocks provide substantial mass and stability, particularly for larger structures requiring significant counterbalancing forces. Water tanks, such as intermediate bulk containers (IBC), are a popular and common choice for their versatility and ease of transport to site compared to traditional kentledge. However, when utilising water tanks as kentledge, it is essential to implement an inspection process to ensure the integrity of the tanks. Regular inspections help detect any leaks, evaporation or vandalism of the tanks, preventing potential failures and maintaining structural stability throughout the duration of the event. Water tanks should be avoided for long-term usage, as there will be holiday periods (e.g. Christmas shutdown) where regular inspections could become unfeasible, plus the water will stagnate over time and could become a biohazard.

16.10. Ground anchors

Ground anchors offer a versatile and effective means of securing temporary structures, providing essential stability against uplift and lateral forces. These anchors come in various forms, ranging from simple steel pins to more complex screw anchors. Screw anchors have gained popularity for their ease of installation and robust performance. When installing screw anchors, adhere to the manufacturer's guidance to ensure performance and structural integrity. Driving conditions of the anchors can pose challenges, particularly if stiff ground is encountered. Designers must carefully assess ground conditions and provide alternative anchor grouping solutions if driving of the anchors becomes unfeasible. To validate the capacity of ground anchors, proof testing is recommended to verify their pullout capacity. The proof test involves applying a load to the anchor, typically 1.25 times the unfactored uplift it will experience during service. This ensures that the anchor can withstand anticipated loads and provides confidence in its performance under real-world conditions. In instances where an anchor fails a proof test, measures should be taken to address the loss of uplift capacity. This may involve noting the failure value and installing additional anchors at the location until the required global pullout capacity is achieved. By conducting thorough proof testing and implementing corrective actions as required, engineers can mitigate risks associated with anchor failure and ensure the stability of their temporary structure. Proof testing can be conveniently conducted on site using equipment such as excavators. The excavator can be rigged to the ground anchor with the lifting equipment and a load cell. For items such as screw anchors, an excavator boom-mounted torque applicator can be used and the measured applied torque can be related to the capacity of the anchor.

16.11. Site phase

Significant consideration and planning must be given to the installation process and subsequent utilisation of the structure by the end user. It is imperative that installers adhere to the design specifications and criteria to ensure the structural safety and stability of the temporary structure. Site installers have a responsibility to ensure that the structure is erected as per the design (following the agreed assembly sequence), with careful attention to detail throughout the installation process. This includes installing the specified materials, configurations and proposed construction sequence outlined in the design documentation. If the site installation deviates from the design, the designer should be notified of these alterations, which must be confirmed acceptable prior to operation of the structure.

16.11.1 Planning and installation

Risk assessments are essential for the management of temporary structures, encompassing both operational and design risks. Throughout the lifecycle, it is imperative that these risks are identified, evaluated, mitigated where possible or managed by the installer or user.

Example design risks

- Overloading from wind – design considerations should account for site-specific wind conditions, ensuring the structure is robust enough to withstand expected wind loads.
- Insufficient kentledge – design parameters should align with site wind conditions and comply with Eurocode load combinations to prevent structural instability.
- Structural member or connection failure – thorough structural assessments must be conducted to evaluate the structure's ability to withstand various loading scenarios, with all member capacities verified and ratified.
- Ground failure, excessive settlement – designers should take appropriate values for the allowable bearing pressure of the ground and spread the temporary structure's base loads accordingly.

Example operational risks

- Incorrect installation – installation procedures must adhere to design drawings and specifications to minimise the risk of structural failure.
- Access and egress – adequate provisions for clear and visible access and egress routes should be established to ensure the safety of end users.
- Adverse weather conditions – implementation of a comprehensive wind management plan is essential to mitigate risks associated with adverse weather, such as high winds. Extreme snowfall could also cause a structural failure of any roof (see section 16.11.5).
- Ground anchors – onsite proof testing of ground anchors is necessary to verify their effectiveness and ensure structural stability.

16.11.2 Inspection and checklists

Before use, a comprehensive inspection must be conducted by a competent person to ensure compliance with design specifications and drawings. A checklist can serve as a valuable tool during this inspection process, ensuring that no critical aspects are overlooked. Example key points to consider in the inspection checklist include the following.

- Assess the suitability of the design for the site-specific conditions and topography.
- Verify whether the designer has accurately interpreted the project brief and incorporated all major design capacities.
- Confirm that the constructed structure matches with the approved design, including dimensions, configurations and materials.
- Ensure the correct installation of kentledge or anchors as per the design requirements to provide structural stability.
- Check for the installation of spreader plates if specified in the design to distribute loads effectively.
- Highlight any residual risks identified during the design phase and verify that appropriate onsite measures have been implemented to mitigate these risks effectively.
- Assess the need for bracing to enhance structural stability and confirm its correct installation.
- Verify the encapsulation of the structure at the correct elevations to provide adequate protection against environmental elements and ensure safety.

16.11.3 Site sign-off

The competent person overseeing the site sign-off process should ensure the following key points.

- The temporary structure has been installed precisely in accordance with the design drawings and calculations provided.
- Any residual risks identified during the design phase have been appropriately addressed by the site team, and necessary precautions have been taken to mitigate these risks.
- In cases where deviations from the original design are necessary due to site-specific conditions or unforeseen circumstances, these deviations must be carefully reviewed and approved by the designated designer or structural engineer.

By adhering to these guidelines, the site sign-off process ensures that temporary structures are erected safely and in accordance with established design criteria. There is the potential that a representative from a local building control department may want to view the structure prior to approving use. They should be given access to the site to view the completed structure.

16.11.4 Wind management

A wind management plan, with method statements, should be adopted to address extreme wind events. This plan serves as a proactive measure, requiring the site team to enact procedures in response to adverse weather conditions, thus safeguarding both the public and the safety of the structure. The wind management plan should encompass several crucial elements, including a systematic process for monitoring wind conditions, identification of wind speed limitations pertinent to the structure, establishment of wind mobilisation speeds, and detailed corrective actions in response to increased wind activity. These actions may involve the removal of encapsulation or wall panels, dropping of cladding panels, opening of blow-out panels or the installation of storm rigging, as deemed necessary. Action levels (see Table 16.1) serve as a pivotal component of the wind management plan, providing clear guidelines for escalating responses based on wind speed thresholds. Table 16.1 provides an example for a temporary structure; the wind speeds and actions required will vary from structure to structure.

Following any event triggering a level 2 or level 3 response, the structure should undergo thorough re-inspection by the site team to verify the continued integrity of all rigging and connections, ensuring it is safe for future operations.

16.11.5 Snow management

Snow clearance management is essential to ensure the structural integrity of temporary structures, in accordance with BS 16508 (BSI, 2015). The accumulation of snow on temporary roofs poses a significant risk, as excessive or continuous snowfall can lead to overloading and subsequent structural failures. To mitigate this risk, effective snow management strategies must be implemented. One approach involves the physical removal of snow by trained operatives using appropriate safety equipment and techniques. Regular monitoring of snow accumulation levels and prompt clearance as necessary are essential to prevent overloading and maintain structural stability. Alternatively, industrial heating systems can be installed within the temporary structure to elevate its internal temperature and facilitate the melting of snow. This method helps to prevent snow accumulation on the roof, thereby reducing the risk of overloading and structural failures.

Table 16.1 Example of action plan for wind management of a 40 mph limited structure

Action level	Wind speed: mph	Actions required
0	<29	Monitor wind periodically; no immediate action
1	30	Mobilise site operatives in preparation
2	35	Site team to commence pre-determined storm actions
3	40	Complete storm actions prior to this windspeed

Note: when monitoring wind speeds on site, it is important to understand that all design values are measured as a 10 minute mean value rather than a gust speed.

16.12. Conclusions

This chapter highlights the essential role of temporary structures in various settings, from construction sites to festivals and events. By following industry standards, such as BS 5975 (BSI, 2024a, 2024b), BS EN 1991 (BSI, 2005) and *Temporary Demountable Structures* (IStructE, 2017), stakeholders can ensure the safe design, installation and management of temporary structures. It emphasises the importance of thorough inspections, proactive wind and snow management strategies and competent oversight during site sign-off processes. This chapter serves as a resource for professionals involved in temporary structure projects. By prioritising structural integrity, safety and compliance with industry standards, stakeholders can mitigate risks and ensure the successful use of temporary structures for their intended purposes.

REFERENCES

BSI (2003) BS EN 12811-1: Temporary works equipment – Scaffolds. Performance requirements and general design. BSI, London, UK.

BSI (2005) BS EN 1991-1-4: 2005+A1:2010: Eurocode 1: Actions on structures. General actions – Wind actions. BSI, London, UK.

BSI (2015) BS 16508:2015: Temporary works equipment. Encapsulation constructions. Performance requirements and general design. BSI, London, UK.

BSI (2023) BS EN 1990:2023: Eurocode. Basis of structural and geotechnical design. BSI, London, UK.

BSI (2024a) BS 5975-1:2024: Management procedures for the control of temporary works. Code of practice. BSI, London, UK.

BSI (2024b) BS 5975-2:2024: Temporary works – Falsework: design and implementation. Code of practice. BSI, London, UK.

Cobb F (2014) *Structural Engineer's Pocket Book*, 3rd edn. Taylor and Francis, Abingdon, UK.

HMG (Her Majesty's Government) (2015) (Construction (Design and Management) Regulations. The Stationery Office, London, UK. www.legislation.gov.uk/uksi/2015/51/contents/made (accessed 1/11/2024).

IStructE (Institution of Structural Engineers) (2017) *Temporary Demountable Structures: Guidance on Procurement, Design and Use*, 4th edn. IStructE, London, UK.

NASC (National Access and Scaffolding Confederation) (2018) *TG9:18: Guide to Design and Construction of Temporary Roofs and Buildings*. NASC, London, UK.

NASC (2021) *TG20:21: Good Practice Guidance for Tube and Fitting Scaffolding*. NASC, London, UK.

USEFUL WEBSITES

BSI (British Standards Institution) – www.bsigroup.com
ESITS (Events Structure Industry Training Scheme) – www.esits.co.uk
HSE (Health and Safety Executive) – www.hse.gov.uk
IStructE (The Institution of Structural Engineers) – www.istructe.org
MUTA (formerly Made-Up Textiles Association) – www.muta.org.uk
TWf (Temporary Works Forum) – www.twforum.org.uk

FURTHER READING

HSE (Health and Safety Executive) (n.d.) Temporary Demountable Structures. https://www.hse.gov.uk/event-safety/temporary-demountable-structures.htm (accessed 19/08/2024).

MUTA (Made-Up Textiles Association) (2018) *Best Practice Guide – Safe Use and Operation of Temporary Demountable Fabric Structures*. MUTA, Suffolk, UK.

X-Venture – Safety Principles of Temporary and Demountable Structures. One day training course; details available at www.x-venture.co.uk.

Ray Filip
ISBN 978-1-83608-525-6
https://doi.org/10.1108/978-1-83608-524-920251019

Chapter 17
Engineering testing and monitoring

Ray Filip
RKF Consult Ltd, UK

Simon Smith
Contractors Design (CDS) Ltd, UK

17.1. Introduction

Engineering design is rarely a precise science; there is uncertainty, so assumptions must be made. Applied loads are not precise, the construction process has tolerances, material properties can vary and ground conditions can vary significantly; hence, methods of analysis make simplifications. Testing on site (or in a yard or workshop) and follow-up monitoring may be required to verify the assumptions, check performance under load to improve / optimise design, check workmanship and manage risks, or may be specified in contracts as control measures. This chapter explores the specific reasons behind engineering testing and monitoring for temporary works purposes and the temporary works that may be required for testing permanent works items. This should not be confused with inspections of temporary works prior to loading or during the life of the temporary works, some of which are statutory – for example, regular inspection of scaffolding and excavations, see Construction (Design and Management) Regulations (HMG, 2015).

With all testing and monitoring, the equipment is of vital importance. It should be used by trained personnel, be appropriate for the intended purpose, not damaged, calibrated and provide the required information to the required degree of accuracy.

Testing and monitoring can be used to prolong the life of a structure by justifying whether the actual condition is acceptable or better than anticipated and making repairs as required, rather than demolition and rebuilding.

This chapter will not cover testing for items such as lifting appliances (which is covered by Lifting Operations and Lifting Equipment Regulations (LOLER, HMG, 1998a)), personal protective equipment (which is covered by Provision and Use of Works Equipment Regulations (PUWER, HMG, 1998b)), fire, contamination, hazards such as asbestos, water penetration, insulation, air quality and durability. The testing of non-mechanical temporary works equipment (such as scaffolding, falsework and props) is covered by BS EN 12811-3 (BSI, 2002). Further specific information on investigations is available in chapters 3 and 4.

17.2. Testing

Some common examples of temporary works engineering testing include:

- testing of temporary anchors for items such as scaffolding, fall arrest, edge protection, Davit bases, safety netting, temporary structural fixings, fixings for temporary brackets supporting services (diverting existing services or providing temporary services for tunnelling), conveyors for spoil removal, ties for access stair towers and ties for plant such as hoists or mast-climbers
- temporary beams / frames for load and integrity testing of piles
- load testing floors, slabs and balconies in existing buildings to ascertain capacity
- compressive strength testing of fill material for suitability as working platforms and haul roads
- testing edge protection systems in accordance with BS EN 13374 (BSI, 2013)
- strength testing of concrete for striking of formwork and falsework
- pressure testing pipelines
- testing wire ties for reinforcement cages
- testing quality and strength of onsite welding
- geotechnical testing to ascertain properties of soils and rocks or to check on compaction
- tensile load testing of temporary ground anchors and soil nails
- trial or test jacking / moving heavy items – for example, test bridge slide
- testing reinforcement bars or tie rods which are post drilled and anchored into a base material (such as a concrete slab) – often this may be because of a specific site problem that has to be rectified
- full-scale load testing for innovative temporary works items prior to use on site
- testing historic building materials (or other materials) to determine properties as they may be used in a temporary works scheme. BS 5975-2:2024 (BSI, 2024), clause 5.2 states:

> where doubt exists as to the properties, by composition or condition, of the material or component, it should where necessary be submitted to a test of the load / deflection type, to determine stiffness and strength, wherever possible in accordance with an appropriate British Standard. Where a sample piece is cut from the section and tested to check its strength, the sample should be cut from an area which will not be critical to the resistance. Testing on site of materials or components should only be specified for quality control purposes, as even the simplest tests require disciplined test procedures. If the performance of an item of equipment is not known, or is in doubt, and calculations are impracticable, appropriate tests should be carried out. These tests should be related to likely site conditions to determine necessary values. A single test might give misleading results, and this should be taken into account.

Testing is often necessary because of a specific issue or risk where there is uncertainty (relating to properties, capacities, dimensions, workmanship for safety critical items or condition) and reasonable assumptions cannot be justified. These issues should be identified at an early stage so there is time for adequate planning. Responsibility for assessing the structure (that ties or anchors rely on), selecting, specifying testing regime, carrying out the testing and analysing the results should be clearly identified.

Once the issue / risk has been clarified and testing identified as a solution, specific objectives must be determined before the type and extent of testing is specified. There may be restrictions to consider, such as site access, working space and availability of specialist equipment, which may limit the type of testing that is possible. In all cases sufficient time to carry out the testing, evaluate the results and consider value engineering should be provided.

17.3. Testing strategy

The cost and time taken to carry out testing should be measured against potential benefits in terms of improving safety, design optimisation and improved practicality for the overall construction process. Where possible and practicable, an initial investigation / assessment should be carried out and this may involve some preliminary proof testing to identify the suitability of the methodology or product capacity (e.g. anchors). Adequate planning is required to ensure the testing is carried out safely and in an appropriate manner that can yield meaningful and accurate results.

Any testing should be carried out by suitably qualified personnel, using certified instrumentation which has been correctly set up and which gives the required degree of accuracy (e.g. load application and measurement of movement). The type of testing specified should be appropriate for the purpose, with realistic performance criteria, and should be representative. The magnitude and direction of the loading should be determined in the preliminary assessment. The following items should be considered.

- Objectives of testing – to confirm assumptions, to determine safety criteria or for quality control.
- Type of test to be carried out – can multiple tests be carried out simply, quickly and at low cost or is it only possible to carry out a single test? Are there any restrictions on the type of testing that can be carried out?
- Who will carry out the testing, what equipment will be used and who will interpret the results?
- Number of different scenarios – testing should consider a likely range of varying materials (e.g. soils, construction materials etc.).
- Testing should be carried out at different representative locations which will then not be used again during the project to avoid clashes or detrimental effects.
- Consequences of failure – for safety-critical tests (e.g. for supporting personnel), where there is significant uncertainty (soil properties, properties of base material for anchors) or for a high-risk location (adjacent to a railway line), the number of tests and magnitude of test load is increased to give a greater certainty. Additional supervision and quality control may also be necessary. This should be determined by a job-specific risk assessment.
- Type of load – magnitude and direction of load plus any combination of loads. Load could be tensile, compressive, shear, bending or a combination.
- How many tests are required to provide the required degree of certainty?
- What is being measured and how will the measurements be made and recorded – are the measurements discreet or continuous and to the required degree of accuracy over a sufficient time?

Testing could be used to determine performance or to validate the quality of the installation. The results should be interpreted correctly otherwise this may compromise safety and / or cause significant economic / programme implications. Specific guidance is available for some types of testing, which will be discussed below.

Designers / specifiers of testing should allow for some flexibility in their thinking, as testing may not always provide results as anticipated and further testing or alternative solutions may be necessary. If testing does not yield results as anticipated, potential causes such as human error, equipment error, non-representative test location, lack of supervision or incorrect anchorage should be investigated. If errors or anomalies cannot be eliminated, then alternative solutions should be considered.

17.4. Testing anchors

Testing of anchors is an important safety requirement which is part of the inspection process and applies to anchors which are used for temporary and permanent works purposes. Anchor selection is based on several variables which can include the magnitude and type of load, type of base material to be fixed into and condition of the base material, and installation logistics. The Construction Fixings Association (CFA) has a wealth of freely available information on its website, such as installation videos, method statements and guidance notes. CFA form 8539/04 (CFA, n.d.) is a test request form which highlights the detail required to help the tester meet the test objectives and a procedure for site testing of construction fixings.

BS 8539 (BSI, 2012) covers all aspects (selection, supply, installation, testing and certification) of fixings drilled in concrete and masonry for safety-critical applications and was published because of several failures which caused fatalities. It gives two reasons for load testing anchors:

- proof testing to verify the quality of the installation (by far the most common site test carried out); carried out on a small sample and for each different type of anchor and into each different base material
- allowable load testing to determine the performance / design resistance of the anchor in a material not covered by European Technical Approvals (ETA) to determine safe loading information.

Many anchors have ETAs and CE marking which means they have been rigorously tested so the capacity of the anchor should be well-established, but often the issue may be the structural fabric into which the anchor is fixed.

Many anchors which are light duty or non-structural do not require testing. In addition, testing will not be required if the following criteria are met (the manufacturers will often provide resistance load data):

- approved anchor (with relevant ETA) is specified by competent person
- the base material is sound and is covered by manufacturer's guidance
- anchor is installed by competent person (under supervision) following the manufacturer's instructions (hole diameter, depth of hole, cleaning of hole, mixing

resin if used, inserting fixing etc.). Particular attention should be paid to manufacturer's guidance on minimum spacing between anchors and minimum edge distances (e.g. to edge of a slab).

Exceptions could be for safety-critical fixings, if fixing into a variable (or where properties are not certain) base material or where there is a contractual requirement. BS 8539 (BSI, 2012) defines safety-critical as an application where failure of the anchor could:

- result in collapse or partial collapse of the structure
- cause risk to harm life
- lead to significant financial loss.

Anchors will rely on the quality of the base material and the overall structure that they are fixed to, as the base material or parts of the structure may fail before the anchor. There are numerous types of anchors (expanding, self-tapping, nylon plug and resin, to name a few) serving different purposes from numerous manufacturers. The correct anchor must be specified (by a competent person) and purchased (sometimes cheaper versions are purchased which claim to be similar), and installed and used as per manufacturer's instructions. Section 17.2 highlights some of the various uses for anchors in temporary works. Anchors which need to be tested should be identified early in the design process to allow sufficient time for testing to be carried out. These will generally be for heavy-duty structural applications, into a base material with uncertain or unproven properties, where resistance values are not provided or safety-critical, as described above.

The testing could be:

- non-destructive proof load testing, where the test load is selected to provide adequate assurance of correct installation or to determine design resistance with an appropriate factor of safety for the risk involved and the base material is not damaged
- destructive pull-out testing (this should be done in areas that are not visible or can be repaired), where the load is sufficiently high to cause failure of the anchor, bond between resin (if used) and base material or base material.

A variety of purpose-made 'pull testers' for tensile testing are available, such as from Hilti, Hydrajaws or Staht (see Useful websites section at the end of this chapter). A common method for testing anchors in shear is shown in Figure 17.1.

Figure 17.1 Shear test for anchors (editor's own)

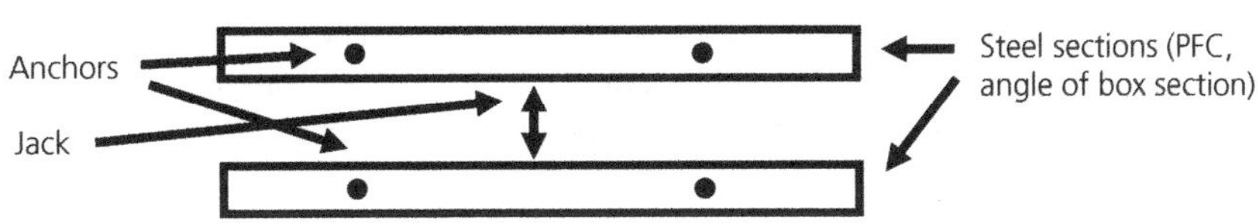

The maximum test load and number of tests to be carried out should be determined in advance. They should be representative and appropriate to the risk and uncertainty involved. Initial testing should consider each different type of anchor, each different base material, different elevations, as well as different states of weathering or repair of the base material. BS 8539 (BSI, 2012) outlines the number of tests, test load and assessment of results. CFA guidance note (CFA, 2023) provides a procedure for site testing and Hilti has also produced a guide (Hilti, 2024). In circumstances where safety of personnel relies on the anchor, high-risk areas or where there is a contractual requirement, all anchors may have to be tested.

17.4.1 Testing scaffold anchors

There are several different ways to provide stability to a scaffolding, especially if it is sheeted or has netting. Perhaps the most common is to anchor the scaffold into the base material of the structure (existing structure for refurbishment and demolition projects or into the new structure as it is being constructed). The testing of scaffold anchors is one of the most common site tests and generally uses purpose-made pull testers. The National Access and Scaffolding Association (NASC) has produced *TG4:19: Anchorage Systems* (NASC, 2023) which provides guidance on which type of anchor is suitable for different base materials.

Responsibility for gathering information on the structure and the base material that the scaffold will be anchored into should be identified early in the design process. In many circumstances this will be the permanent works designer (structural engineer). The designer should be satisfied that the structure itself will not fail when loaded by the scaffold. For example, the joints in a timber frame structure may not be designed for the tensile loads from scaffold ties.

The type of anchor should then be selected to suit the base material; TG4:19 (NASC, 2023), sections 2.2 and 2.3 discuss the different types of ties and the suitability of certain anchors into different base materials. The tie pattern (locations) and working load are then determined. TG4:19 (NASC, 2023), section 3.1 also discusses loads arising from fall arrest used during the construction or removal of the scaffold.

TG4:19 (NASC, 2023), section 5 discusses installing anchors and there are a few points of caution to note.

- Drilled holes should be the correct size for the anchor. This is particularly important for self-tapping anchors whereby a hole which is too large will affect the capacity of the fixing and a hole which is too small will make the anchor difficult to install and it may break during installation.
- Manufacturers will generally give load recommendations into sound concrete.
- Expanding anchors should not be used in hollow bricks or perforated bricks (only solid bricks) as they struggle to carry load and can work loose. Brick parapets can be problematic due to weathering and lack of weight of brickwork above. TG4:19 (NASC, 2023), section 5 has guidance on fixings into brickwork.

A series of preliminary tests is carried out to check the suitability of a particular fixing and base material to determine the load capacity where there is no data from the manufacturer for the base material. The tests should be spaced out and the number of tests should be representative (TG4:19, NASC, 2023 recommends a series of five tests) for different base materials, different fixings, different elevations or states of weathering / repair. The location of the tests should be selected such that they do not affect the appearance of the structure or if any damage occurs it can be repaired. Additional tests may be carried out if there are problems or failures and the designer consulted. The recommended test load is 2 × the working load and creep effects may need to be considered. For safety-critical fixings, this load may be increased. The testing should also be representative of the type of applied load. The results of these tests are then used for design briefs.

Proof testing is then used to ensure the quality of the installation. TG4:19 (NASC, 2023), section 6.2 recommends a test load of at least 1.25 × the working load and 5% of all anchors (with a minimum of three), spread evenly throughout the project, to be tested. If failures occur, then the frequency of testing should be increased. The test load and number of tests can also be increased if a risk assessment demands this – for example, for high-risk areas or where the consequences of failure are severe (failure could lead to casualties).

17.5. Load testing slabs and floors

The capacity of an existing slab or floor often needs to be determined for temporary loading during construction or demolition (live loading due to operatives with tools, material storage, plant loading, back-propping or accumulation of demolition rubble). Often this can be done by calculation, by ensuring any temporary induced stresses and distribution of stresses do not exceed the stresses in the permanent condition (bending, shear or punching). There are occasions when the exact properties of the slab or floor cannot be fully determined, or the condition may be questionable, and a load test is specified to verify assumptions. Care should be taken to ensure the whole floor area is covered. Serious issues have occurred when previous alterations or repairs have not been identified – for example, filling in previous openings / voids.

Before any testing is specified and carried out, as much information as possible should be gathered, such as type of construction, spans, thicknesses and materials used. Preliminary assessments and calculations should be carried out to provide an initial estimate of capacity and anticipated deflections. The magnitude, type and position of the temporary loads should be considered:

- distributed load from operatives with tools, material storage or accumulation of demolition rubble
- track or wheel loading from machinery
- point loading from props or machinery outriggers.

The testing regime should consider the likely position of these loads and consider the most onerous locations. The type and magnitude of test loading should be based on the preliminary estimate and on a risk assessment of the consequences of failure. Generally, the magnitude of the test load should be greater than the applied temporary loads but more likely 1.5 × to provide additional confidence and a suitable factor of safety as load testing does not guarantee capacity of the overall structure.

There are inherent risks associated with loading the slab with kentledge (such as water-filled containers or concrete blocks), as operatives may be required to enter a hazardous area where it could be difficult to manage and rectify a potential runaway failure, which could lead to disproportionate collapse. Where possible, a temporary testing arrangement should be devised, which utilises a remotely controlled jacking system for load application and a remote system for measuring deflections. Props with spreader beams can be used at different locations on the slab or floor. The size and stiffness of the spreader beams and spacing of props can replicate different types of temporary loads. The load is applied in increments by hydraulic jacks, with each load increment maintained and deflections continuously recorded. An allowance should be made for any joints in the props and elastic shortening of the props under load. Displacement monitoring during loading and unloading is essential to confirm elastic response and recovery. The jacking method requires the structure to be exposed beyond the finishes. An alternative method is to install jacks with ties above the slab to be tested and anchor these ties to slabs beneath (to utilise the self-weight of slabs beneath), then jack the slab being tested downwards.

The deflection should be proportional to the applied load and should be broadly in line with preliminary predictions, but if the deflections increase unexpectedly, the load should be removed. Jacks have the advantage (over kentledge) because the load can be quickly and safely released from a remote position. Excessive deflection could signify that the slab or floor is unable to carry the loading at certain positions, leading to superficial damage to finishes or structural damage, such as cracking of slabs. Solutions could include back-propping, spreader beams or different machinery, or areas cordoned off to prevent loading.

17.6. Load testing piles

Load-bearing piles are often tested to confirm the installation and performance under loading. This testing can be used to verify design assumptions and to determine the load against settlement behaviour, which could then be used to optimise (value engineering) the design for projects where there are numerous piles (could result in significant overall cost and time savings). This section discusses static load testing – the most common is a maintained load test (less common is a constant rate of penetration test) which requires a spreader beam / frame with anchor piles or kentledge for reaction. Piling subcontractors will have well-developed procedures and standard designs for a series of fabricated beams / frames for testing up to a maximum load. This beam / frame is designed to span between anchor piles or to allow significant kentledge to be loaded. The frame will be anchored to the reaction piles (e.g. using Macalloy bars). The load is applied by a hydraulic jack and the beam / frame should be sufficiently stiff for test loads which can exceed 100 MN to 200 MN (megaNewtons) without excessive deflection and the anchor piles should be designed for the applied tensile reaction from the frame. The jack, beam / frame (and kentledge, if used) are removed after the test, the reaction piles are left and may be part of the permanent works.

The maximum test load to be applied should be agreed in advance to ensure an appropriate testing system is provided. The position of the test should be planned well in advance and the principal contractor should ensure there is adequate access to the test pile (or pile group and anchor piles), adequate working space and a safe working area for the piling subcontractor to carry out the testing. Figure 17.2 shows a test frame with anchor piles. The reader's attention

Figure 17.2 Pile test (note the working space requirements) (editor's own)

is drawn to the access and working space requirements for lorries to deliver / remove the frame and for a crane to lift the frame plus any kentledge blocks (unless anchor piles are used).

Two publications provide further information:

- Federation of Piling Specialists (FPS) *Handbook on Pile Load Testing* (FPS, 2006)
- Institution of Civil Engineers (ICE) *Specification for Piling and Embedded Retaining Walls* (ICE, 2016).

Piles may also be tested for lateral loading to determine the piles' ability to resist lateral loads, such as wind and seismic loads. Horizontal jacking forces can be applied using other piles or kentledge blocks for reaction.

The integrity of piles (installation workmanship) may also be tested using sonic echo and transient dynamic response methods to ensure the shaft is free from major cracks and voids, using highly sensitive accelerometers coupled to the pile head.

17.7. Bespoke systems for testing innovation

The construction industry tries to innovate with new ideas and techniques to make projects safer, more efficient, more sustainable, quicker to construct or to solve a specific site problem. Innovative ideas will require careful design based on information available at the time but to provide additional assurance, testing (and perhaps follow-up monitoring) may be necessary. This additional work could be a contractual requirement to satisfy clients or approving authorities that the innovative design will behave as anticipated. Below are two specific examples from the editor's experience.

Figure 17.3 shows a specifically designed and constructed bespoke temporary testing frame in a plant yard to test the performance of an innovation for motorway gantries. The testing regime was required as part of the Highways Agency (now Highways England) approval process, as the intention was to utilise this innovation at numerous locations, where logistics and ground conditions allowed. The temporary frame was loaded (a series of concrete blocks were suspended) to replicate various design loads / load combinations and the performance of the holding down detail, load transfer system from mast to base and foundations was measured and assessed.

Figure 17.4 shows the abutment piles and piers for a unique bridge in Milton Keynes. The specific site issues were a high and fluctuating water table (with artesian pressure) and building a bridge within an existing housing estate without excessively disturbing the residents. A design was developed to utilise steel bearing piles which could also be left protruding above ground level to form the piers and then be cast directly into the bridge deck (an integral bridge). The design was further developed to form the pile / piers from four connected sheet piles to form a box pile which could then be jacked into the ground (silent piling). The pile / piers were designed in accordance with recognised standards, but site testing was specified to

Figure 17.3 Motorway MS3 gantry foundation test frame (editor's own)

Figure 17.4 Westcroft bridge foundation and piers load testing (editor's own)

confirm actual site performance against calculated predictions. Two vertical maintained load tests and a lateral load test were carried out. Figure 17.4 shows the temporary works jacking and testing frame arrangement (see Filip, 2005).

In both these examples, the design of the temporary testing frame had to be sufficiently stiff and robust to ensure there was negligible deflection under loading and the loading was effectively transferred to the item of permanent works being tested. Appropriate instrumentation was required to record the results to an appropriate degree of accuracy.

17.8. Other common testing

- Early age strength testing of concrete – generally carried out to allow striking (removal) of wall / column formwork, soffit formwork (and supporting falsework), as well as to determine strength of permanent works columns that are required to provide support to top-restrained falsework systems. For striking soffit formwork and supporting falsework, the concrete slab should have achieved sufficient compressive strength to support itself, plus any additional temporary applied loads to the slab, such as live loading, construction plant and material storage. Traditionally the crushing of concrete cubes / cylinders was used to determine the rate of strength gain. Several other methods offer more accurate results, such as LOK-tests and maturity testing as the in situ concrete is being tested. *ICE Temporary Works: Principles of Design and Construction* (Pallett and Filip, 2019, chapter 24) has further information.
- Pressure testing of pipelines – pipelines will be tested prior to operating to ensure the system can operate safely at the test pressure (i.e. checking the integrity of joints, fittings and other components). A temporary restraint system is often required which may comprise steelwork and a concrete thrust block to prevent movement during testing. Construction Industry Research and Information Association report R128D

(CIRIA, 1994) describes the design of concrete thrust blocks. This testing and associated safety precautions are covered in Pallett and Filip (2019, chapter 29).
- Testing reinforcement ties – Pallett and Filip (2019, chapter 32) and Temporary Works Forum (TWf, 2020, 2021) provide an explanation as to the importance of reinforcement ties in preventing rebar cages collapsing in the temporary condition (before they are encased in concrete). Ties are also critical to ensure large cages can be safely lifted, rotated and positioned. TWf (2021), appendices A and B provide a specification and details of a temporary frame arrangement for testing ties to determine a safe working load for ties on that site, with adequate consideration given to limiting displacement and site quality control measures.
- BS EN 13374 (BSI, 2013) classifies temporary edge protection systems into three classes (A, B and C), based on potential falling height and inclination of the surface the operatives will be working on. The standard specifies the loading and the maximum permitted deflection to replicate persons falling against the systems. Class B systems require static and low dynamic testing (swing bag test) and class C systems require a high dynamic load test (rolling cylinder test). Often the issue is how the systems are connected / restrained. The capacity of any fixings and the structure which the edge protection system is connected to should be determined by calculation or testing if necessary.
- Testing site-won materials – there can be significant financial and environmental benefits in utilising materials that exist on the site rather than importing. Often sites would like to utilise demolition rubble for hardstanding, haul roads and piling mats. The rubble should be sorted to remove timber, plastic, metal, glass and so on. Concrete can have a range of compressive strengths and bricks can have an even wider range. Testing would be necessary to determine the properties for suitability and design purposes.
- Soil and groundwater testing – ground and groundwater conditions, at the locations for items of temporary works, often need to be confirmed on site. For a description of California bearing ratio test, plate testing, shear vane and pocket penetrometer, see chapter 4. Groundwater will often be tested for contamination.
- Testing onsite welding – the quality of onsite welding can be checked by visual inspection, non-destructive testing and larger-scale load testing (see chapter 15).
- Bolted connections in steelwork – for example, torque tests for high strength friction grip (HSFG) bolts.

17.9. Monitoring

The purpose of monitoring (and evaluation of the results) is to measure actual performance against predictions, to identify any variations from the predictions (to ensure safety), to identify any long-term effects / behaviour / trends and to provide feedback on the predictions to improve effectiveness in the future. In some cases, monitoring may be a contractual requirement (party wall agreements) to provide information for any insurance claims or legal cases.

Some examples of where monitoring is used in temporary works include:

- measuring settlement of temporary foundations after loading (e.g. foundations for tower cranes or propping)
- measuring variations in groundwater and water pressure over time during excavation

- measuring earth pressure and movement of retaining walls supporting excavations and measuring movement of adjacent structures, services, roads or railway tracks that may be affected by the excavation
- measuring heave at the base of a large excavation
- measuring load in large props (e.g. supporting an excavation) for variations due to temperature
- measuring ground movements around tunnelling and measuring movement of structures, services, roads or railway tracks above the tunnel that may be affected by the tunnelling
- stability of temporary slopes and battered excavations
- measuring cracks in the fabric of a structure during refurbishment, excavation, underpinning of foundations or demolition
- measuring movement of a retained façade or party wall over time to prevent damage
- measuring ground movement and settlement of adjacent structures during dewatering; the groundwater quality is also likely to be tested during dewatering before it can be discharged
- measuring noise, vibration and exhaust fumes from heavy machinery, especially during demolition works
- measuring applied preloading and stressing loads (e.g. for post-tensioned structures).

The purpose and scope of any monitoring should be clearly identified and responsibility apportioned; the various parties involved need to reach an agreement. The following questions should be asked.

- Why do we need to monitor?
- What will be the cost of monitoring measured against benefits? Are the risks understood?
- What do we need to monitor?
- Where will we be monitoring – are there any logistical restrictions?
- Who will do the monitoring?
- How will the monitoring be done, to what degree of accuracy, how often and for how long?
- Who will interpret the results? How do we know whether unusual measurements are real or anomalies due to errors in measurements?
- What do we do if the monitoring identifies problems? Are there agreed limits and an action plan for problems?

Before the work starts, a precise three-dimensional reference frame (a baseline / starting point of precise measurements and observations) of the existing situation must be formed. Ideally this should be done well in advance of the works to allow for any possible seasonal effects (particularly for groundwater monitoring). Any subsequent monitoring should be based on this starting point and trigger levels for action should be established, as described in sections 3.9 and 4.3.2. Critical stages in the construction process should be identified and the monitoring regime may be enhanced (e.g. increased frequency) at these stages.

There is a wide variety of specialist equipment available, ranging from simple measurements with a tape and photographs to remotely controlled drone photogrammetry, scanning and 3D/4D digital surveys. Specialist advice should be sought to select the most appropriate equipment / instrumentation which can provide relevant information to the required degree of accuracy, over the required time.

Readings can be affected by the weather and human error. If readings are taken regularly or continuously, then these effects can be identified. There may also be occasions where 'unusual' readings are recorded. These should be investigated and compared against other readings to identify whether something is 'going wrong' or whether it is a rogue reading (there are examples of these being caused by vandalism, birds sitting on target points or tampering by site operatives).

As mentioned in section 2.11, there will always be some degree of deflection / movement / settlement which monitoring will identify. What needs to be agreed is the extent that is deemed acceptable, and any agreed limits should be realistic (zero should not be an acceptable option).

17.9.1 Some specific considerations

- All buildings and structures are subject to natural movement due to changes in items such as temperature and wind. It is important to differentiate between these natural movements and movements caused by construction activities.
- For structural cracks, photographs and measurements should be taken prior to work commencing and subsequently simple techniques such as 'tell-tales' or 'Demec studs' can identify whether cracks are opening. Regular photographs will form a permanent record.
- Verticality, cracks and differential movement / settlement should be monitored for retained façades and party walls. This will be required during partial demolition or refurbishment, during underpinning of foundations or during an adjacent excavation (e.g. basement) causing undermining. Party wall agreements often drive the need for monitoring.
- During excavation, the verticality of the retaining wall, earth pressures, prop loads, water levels and movement behind the wall can be continuously monitored. If there are railway tracks nearby, there will also be a contractual requirement to monitor track settlement.
- For tunnelling projects in urban areas, settlement and movement (especially differential settlement / movement) above the line of the tunnel are continuously monitored. Measures such as grouting (could be carried out ahead of the tunnelling or during tunnelling) can be used to limit the risk.
- If multiple points are being monitored simultaneously and continuously, then some degree of synchronisation is necessary.

REFERENCES

BSI (2013) BS EN 13374:2013+A1:2018 (corrigenda January 2019 and August 2021): Temporary edge protection systems. Product specification. Test methods. BSI, London, UK.

BSI (2002) BS EN 12811-3:2003: Temporary works equipment – Load testing. BSI, London, UK.

BSI (2012) BS 8539:2012+A1:2021: Code of practice for selection and installation of post-installed anchors in concrete and masonry. BSI, London, UK.

BSI (2024) BS 5975-2:2024: Temporary works – Falsework: design and implementation. Code of practice. BSI, London, UK.

CFA (Construction Fixings Association) (2023) *Procedure for Site Testing Construction Fixings*. CFA, Leicester, UK (available as free download from their website).

CIRIA (Construction Industry Research and Information Association) (1994) *Report R128D: Guide to the Design of Thrust Blocks for Buried Pressure Pipelines*. CIRIA, London, UK.

Filip R (2005) Recent advances in quiet and vibration-less steel pile installation. *DFI (Deep Foundations Institute), 30th Annual Conference, Chicago, IL, USA*.

FPS (Federation of Piling Specialists) (2006) *Handbook on Pile Load Testing*. FPS, Bromley, UK.

HMG (Her Majesty's Government) (1998a) Lifting Operations and Lifting Equipment Regulations. The Stationery Office, London, UK.

HMG (1998b) Provision and Use of Work Equipment Regulations. The Stationery Office, London, UK.

HMG (2015) Construction (Design and Management) Regulations. The Stationery Office, London, UK.

Hilti (2024) The Possibilities and Limitations of Anchor Site Testing. www.hilti.co.uk/content/documents/e1/OSTExpertArticle.pdf (accessed 28/10/2024).

ICE (Institution of Civil Engineers) (2016) *Specification for Piling and Embedded Retaining Walls*, 3rd edn. ICE Publishing, London, UK.

NASC (National Access and Scaffolding Confederation) (2023) *TG4:19: Anchorage Systems*. NASC, London, UK.

Pallett PF and Filip RK (2019) *Temporary Works: Principles of Design and Construction*, 2nd edn. ICE Publishing, London, UK.

TWf (Temporary Works Forum) (2020) TWf2020:03: Temporary Condition of Reinforcement Cages Prior to Concreting: Part 1 – Management Guidance. https://www.twforum.org.uk/viewdocument/rebar-guide-part-one-management (accessed 10/01/2025).

TWf (2021) TWf2021:01: Temporary Condition of Reinforcement Cages Prior to Concreting: Part 2 – Technical Guidance. https://www.twforum.org.uk/viewdocument/rebar-guide-part-two-technical (accessed 10/01/2025).

USEFUL WEBSITES

BSI (British Standards Institution) – www.bsigroup.com
CFA (Construction Fixings Association) – www.the-cfa.co.uk
CIRIA (Construction Industry Research and Information Association) – www.ciria.org
ETA (European Technical Approvals) – www.eta.de
FPS (Federation of Piling Specialists) – www.fps.org.uk
Hilti – www.hilti.co.uk
Hydrajaws – www.hydrajaws.co.uk
Keltbray Technical Services – www.keltbray.com
NASC (National Access and Scaffolding Confederation) – www.nasc.org.uk
Staht – www.staht.com
TWf (Temporary Works Forum) – www.twforum.org.uk
Vibration Monitoring Services Ltd – www.vmservices.co.uk
Wentworth House Technical Services – www.wentworth-house.co.uk

FURTHER READING

BSI (2012) BS EN 795:2012: Personal fall protection equipment – Anchor devices. BSI, London, UK.

BSI (2005) BS 7883:2005 (amended 2007): Code of practice for design, installation, use and maintenance of anchor devices conforming to BS EN 795. BSI, London, UK.

Buenfield NR, Davies RD, Karimi A and Gilbertson AL (2008) *C661: General Fixings – Intelligent Monitoring of Concrete Structures*. CIRIA, London, UK.

Bungey JH and Grantham MG (2018) *Testing of Concrete in Structures*, 4th edn. CRC Press, London, UK.

Cashman PM and Preene M (2020) *Groundwater Lowering in Construction*, 3rd edn. CRC Press, London, UK.

CFA (Construction Fixings Association) (2023) *Guidance Note – Procedure for Site Testing Construction Fixings*. CFA, Leicester, UK.

CFA (n.d.) Form 8539/04: Test Request. https://www.the-cfa.co.uk/Support/Forms/ (accessed 10/01/2025).

Gaba A, Kelly L, Tully K *et al.* (2017) *C760: Guide on Embedded Retaining Wall Design*. CIRIA, London, UK.

Gilbertson AL (2019) *C777: General Fixings – Selection and Whole Life Management*. CIRIA, London, UK.

Head JM and Jardine FM (1992) *TN142: Ground Borne Vibrations Caused by Mechanical Construction Works*. CIRIA, London, UK.

IStructE (Institution of Structural Engineers) (2023) *Subsidence*. IStructE, London, UK.

Nicholson D, Tse CM and Penny C (1999) *R185: The Observational Method in Ground Engineering*. CIRIA, London, UK.

Stacy M, Denton S and Potle S (2019) *C778F: Management and Design of Safety Critical Fixings*. CIRIA, London, UK.

TWf (Temporary Works Forum) (2019) TWf19.101: Static Load Pile Tests. https://www.twforum.org.uk/viewdocument/twf-information-sheet-no-7-stati (accessed 10/01/2025).

Weltman AJ (1980) *PG7: Pile Load Testing Procedures*. CIRIA, London, UK.

There are numerous videos on YouTube relating to anchor installation and testing (too many to list individually).

Ray Filip
ISBN 978-1-83608-525-6
https://doi.org/10.1108/978-1-83608-524-920251020

Chapter 18
Why do temporary works fail?

Ray Filip
RKF Consult Ltd, UK

Martyn Ostcliffe
Health and Safety Executive, UK

18.1. Introduction

Temporary works failures and 'near misses' still occur on a regular basis. Site workers and members of the public have been injured and killed as a result. Two of the aims of BS 5975-1:2024 (BSI, 2024a) and BS 5975-2:2024 (BSI, 2024b) (see BS 5975-1, clause 4.1.1.5) are to:

- minimise the chances that errors will be made
- maximise the chances that errors will be found.

Well-established, robust procedures for the design, design checking, construction, inspection, loading, use, maintenance and dismantling of temporary works have been in place for many years; however, things still occasionally 'go wrong'. Often, it is only operative experience and vigilance (and occasionally some good luck) that have prevented the occurrence of incidents with severe consequences.

The *Bragg Report* (Bragg, 1975, p. 7) highlighted the main reasons for temporary works failures as:

> each failure has two elements; the technical cause which led to the collapse; and the procedural errors which allowed the faults to occur and to go undetected and uncorrected … none of the improvements considered necessary can be achieved unless a determined effort is made to improve the training of all concerned.

The *Bragg Report* (Bragg, 1975, p. 35) also highlighted

> evidence on medium and small building operations, of deplorable standards.

The Bragg Report divided technical failures into three categories:

- where the loads on which the design was based were different from those applied
- where the design itself was inadequate for the specified loads
- where the works were not constructed according to the design.

The *Bragg Report* (Bragg, 1975, part 2, pp. 20–35) highlighted the following reasons for technical failures:

- the neglect of horizontal forces and a failure to make provision for adequate lateral and longitudinal stability
- progressive collapse
- overloading – falsework inadequate for the design load, genuine mistakes in calculations, the way the load was applied not properly allowed for in the design or load not applied in the way the designer specified
- inadequate foundations
- support of beams – inadequate bearing (beam not long enough), disregard of the effects of deflection (load redistribution), failure to restrain beams against twist
- beam web overstressing due to lack of technical knowledge to consider buckling and / or bearing failure
- dimensional and verticality tolerances
- eccentricities of applied load to the axis of supporting members
- setting out errors
- defective or inadequate materials
- inadequate design consideration given to dismantling procedures.

The *Bragg Report* (Bragg, 1975, part 3, pp. 38–44) highlighted the following procedural inadequacies:

- no design brief or inadequate design brief
- design not followed on site and site alterations made without reference back to the designer
- design errors not detected
- confusion over specific responsibilities.

This chapter is based on extensive research and analysis of temporary works-associated incidents (failures and near misses) from several different sources:

- Health and Safety Executive records
- questionnaire answers received from 30 experienced temporary works professionals
- the editor's and close colleagues Peter Pallett's and David Thomas's accumulated collection of records
- published information from Standing Committee on Structural Safety (SCOSS) (Note: now known as Collaborative Reporting for Safer Structures)
- Collaborative Reporting for Safer Structures (CROSS)
- recent published information and specialist websites such as *Construction Enquirer*, *Contract Journal*, *LinkedIn*
- Institution of Structural Engineers (*Temporary Works Toolkits*: IStructE, 2017a and IStructE, 2017b)
- Institution of Civil Engineers (*New Civil Engineer*)
- *Ground Engineering* magazine
- Temporary Works Forum website resources – *Causes for Concern*.

18.2. Analysis

Six hundred 'incidents' (associated with temporary works) were identified and analysed. Incidents included:

- collapses of temporary works
- unsafe use / actions relating to temporary works resulting in an unsafe situation / condition / near miss
- collapses or damage of permanent works because of a temporary works failure or unsafe use / actions relating to temporary works
- collapse of machinery due to misuse, equipment failure or foundation failure.

Where it was possible to extract the relevant information from the reports / articles (some of the information was vague), the causes that most directly led to the 'incidents and near misses' were identified and classified. It is worth noting that many incidents and near misses had several causes. Most of these incidents and near misses have occurred in the last 10 years (although several high-profile incidents from the last 25 years were also reviewed), mainly in the UK; however, a few international incidents were also identified.

It must be stressed that this research is a 'snapshot' in time, from limited sources and not all reports provided all the relevant information. However, it provides an indication and paints a picture of the main issues. It is also interesting to compare these findings with those of the *Bragg Report* (Bragg, 1975).

18.3. Summary of causes

It is fortunate that these incidents and near misses have caused relatively few fatalities and serious injuries. This is perhaps due to good fortune and actions taken by those on site, but all incidents had significant financial and programme implications, with some examples where specific companies became insolvent as a result. However, there have been some very high-profile failures with multiple fatalities and injuries – for example, in 2007 in Can Tho, Vietnam, bridge falsework failure resulted in 50 fatalities and 82 injuries.

18.3.1 Incidents by sector

Where information was available, the 600 incidents were classified by industry sector, as shown in Figure 18.1.

18.3.2 Type of failure and near miss

The 600 incidents were analysed and, where information was available, the type of incident or near miss was identified, as shown in Figure 18.2.

18.3.3 Procedural issues

The 600 incidents were analysed and, where information was available, procedural issues were identified and classified. 63% had procedural issues, as shown in Figure 18.3. Some incidents could be attributed to more than a single cause.

Figure 18.1 Temporary works incidents by industry sector (author's own)

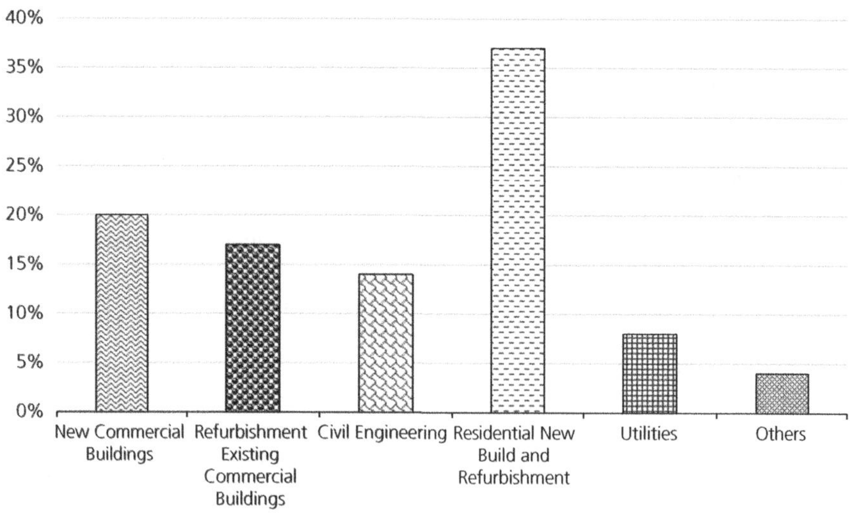

Notes on Figure 18.1:

- Commercial buildings include offices, retail, warehouses, hospital, schools.
- It was not possible to easily split residential new build and refurbishment, as many projects involve some aspect of both and many reports did not clearly specify the exact nature of the project.
- Others include mining, marine, rail, demolition, events/entertainment.

18.3.4 Design issues

The 600 incidents were analysed and, where information was available, design issues were identified and classified. 48% had design issues, as shown in Figure 18.4.

Where a design had been carried out, some of the more concerning issues were as follows.

- Design calculations provided were 'cut and paste' from a previous unrelated project and did not reflect the actual site conditions.
- There was an over-reliance on software answers without checking they were appropriate.
- Design calculations provided for a supported 3 m-deep excavation in a city centre used completely unrealistic ground parameters which were unrelated to the actual location. Also, no allowance was made for any surcharge loading, despite the excavation being adjacent to a busy road which was to remain open to traffic.
- Inappropriate design assumptions were made and subsequently these assumptions were not confirmed on site.
- Structural design calculations were carried out using inappropriate software by a building surveyor with no design experience and no appropriate professional indemnity insurance.
- The 'designer' used calculations from a third party for an unrelated project and 'passed them off' as his own calculations. The designer had no temporary works design experience.

Figure 18.2 Types of failure, near miss or unsafe condition (author's own)

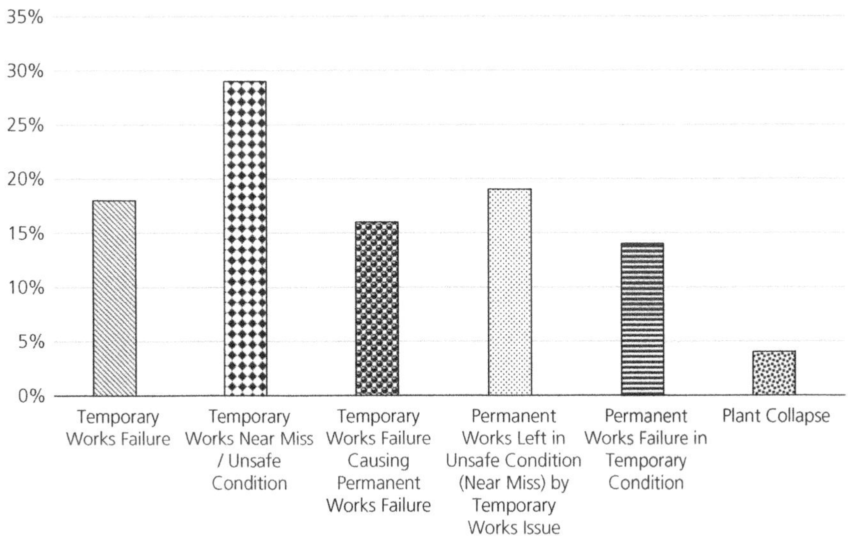

Notes on Figure 18.2:

- Temporary works failure causing permanent works failure includes:
  - unsupported trench undermining gable end of a building and when trench collapsed the wall also partially collapsed into the trench
  - failure of inadequate temporary structural propping (Acrow props) causing structure to partially collapse
  - part of a building was being demolished, but the agreed sequence was not being followed and temporary propping had not been installed, which led to an unplanned collapse of the whole building
  - inadequate temporary works was provided for a basement excavation, which led to significant ground movement and extensive damage to the neighbouring building
  - part of a steel frame warehouse collapsed because a contractor removed bracing, without authorisation, because it was 'in the way'.
- Permanent works failure in the temporary condition includes:
  - slender precast bridge beam toppled over during installation as temporary stability had not been considered
  - a mezzanine floor in a building was overloaded by the storage of construction materials on it
  - a brick wall around a garden collapsed due to a significant amount of excavated spoil being piled against it
  - a 5 m-high brick wall was designed to be restrained by a floor which sat on the wall; the wall blew over in high winds before it was completed (it was freestanding) as no consideration had been given to sequencing and temporary stability
  - a thin reinforced concrete slab in a building failed as a large air conditioning unit was being wheeled across it on pallet trucks; no consideration had been given to the substantial point loads being imposed on the slab
  - metal decking flooring system deflected excessively during the concrete pour. The site team were using lasers to pour the floor to a level rather than a thickness and had not considered temporary deflection. As the metal decking deflected, the laser system was instructing more concrete to be poured to achieve level.
- Plant collapse includes:
  - mobile crane foundation failure
  - tower crane base failure
  - failure of working platform leading to collapse of tracked machines (crawler cranes, excavators, piling rigs)
  - failure of foundations beneath smaller machines, such as MEWPs and lorry loaders
  - failure of lifting equipment, lifting eyes or structural failure of the item being lifted.

Figure 18.3 Procedural issues (author's own)

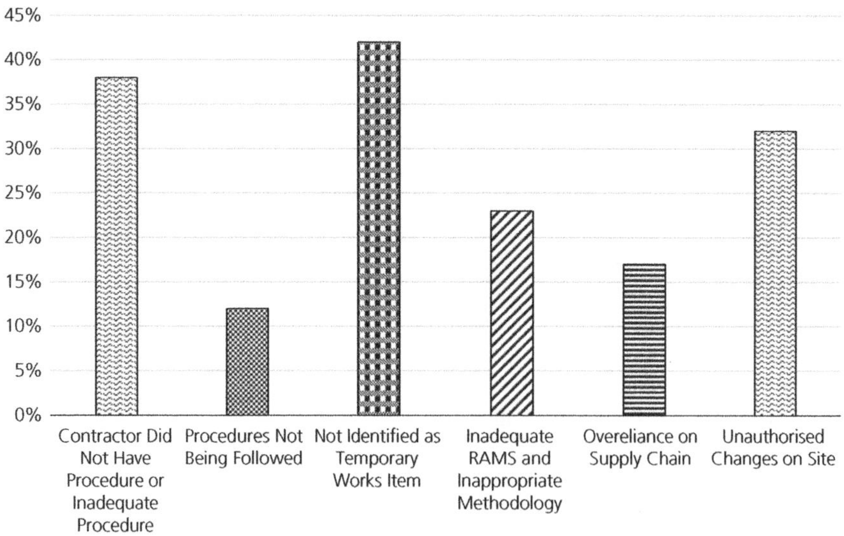

Notes on Figure 18.3:

- Contractor did not have an adequate procedure includes:
  - contractor did not have any temporary works procedure
  - contractor had copied a temporary works procedure from another contractor, but it was not appropriate for the type of works they were involved in.
- Procedures not being followed includes:
  - contractor had a temporary work procedure, but it was not being used on site and not being audited to ensure compliance
  - site team working outside of the procedure
  - not carrying out inspections or testing as per requirements.
- Not identified as temporary works item includes:
  - site team did not realise the issue was temporary works and therefore did not have any procedures or design.
- Inadequate RAMS (risk assessment/method statement) and inappropriate methodology includes:
  - RAMS was generic (cut and paste from a previous project) and did not reflect the actual work being carried out on the project. Also, the site team were not briefed in the methodology or specific risks involved
  - site team storing excess surface run-off water at the base of the sheet pile wall, which caused the ground to soften (reduced passive resistance) and led to excessive deflection
  - access scaffolding was being erected on a timber-joisted flat roof. No RAMS or survey had been carried out to identify the capacity of the flat roof (below the felt it was in poor condition) and the roof partially collapsed under the weight of the scaffolding.
- Overreliance on supply chain includes:
  - principal contractor had no in-house temporary works expertise and was completely reliant on advice from supply chain
  - principal contractor appointed an external temporary works coordinator without checking they were competent to carry out the role and provided no training or briefing on their procedures.
- Unauthorised changes on site (lack of change management) includes:
  - substituting components
  - not following agreed sequence
  - changing layout of temporary works scheme to suit unforeseen site conditions – for example, to avoid services, to accommodate setting out errors or procurement errors.

Figure 18.4 Temporary works design issues (author's own)

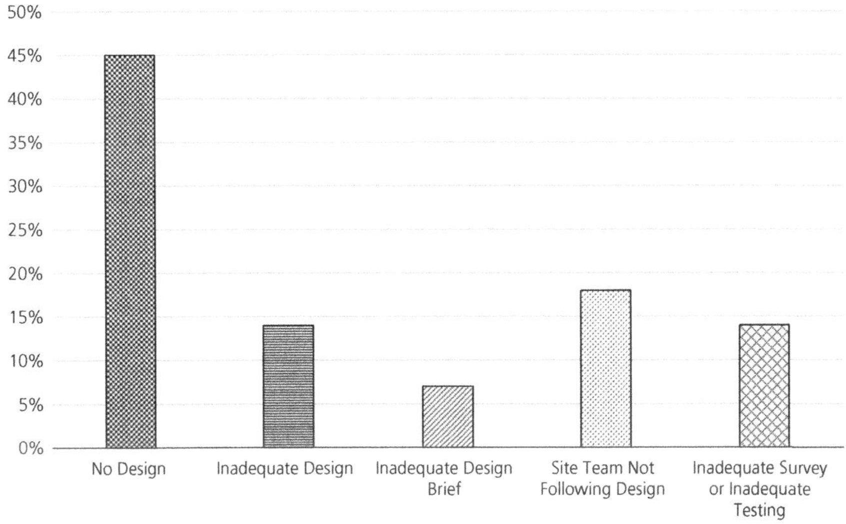

Notes on Figure 18.4:

- Inadequate design issues includes:
  - design errors, including not identifying critical failure mechanisms
  - inadequate factors of safety
  - misuse of software
  - incompetent/not trained/not experienced designer
  - inappropriate design assumptions
  - inadequate checking of design
  - lack of consideration of construction sequence and the temporary condition of the permanent works, which often could be the most onerous condition.
- Site team not following design includes:
  - substituting or omitting critical items without consultation
  - not following the agreed sequences
  - lack of communication between site team/temporary works coordinator (TWC) and designer.
- Inadequate survey/testing includes:
  - structural or geotechnical survey not carried out
  - too little information gathered
  - inappropriate surveys (wrong location or not gathering relevant information)
  - assumptions relating to uniformity/repetition/condition being made but not verified
  - inadequate testing of scaffolding anchors installed into a poorly maintained brick wall.

- The design failed to consider the construction sequence and did not identify critical stages and most onerous condition / loads for the design of individual components.
- There was an inappropriate use of factors (factors of safety and partial factors).

18.3.5 Human resource issues

The 600 incidents were analysed and, where information was available, human resource issues were identified and classified. 42% had human resource issues, as shown in Figure 18.5. Some incidents could be attributed to more than a single cause.

Figure 18.5 Human resource issues (author's own)

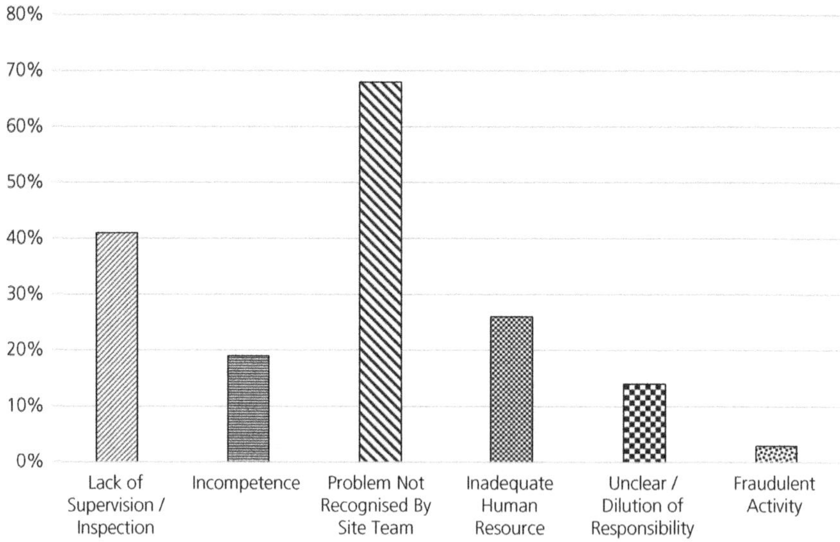

Notes on Figure 18.5:

- Incompetence includes:
  - not trained for the role and not understanding responsibilities
  - lack of experience.
- Inadequate human resource includes:
  - temporary works coordinator was too busy to carry out their role effectively as they had numerous other roles on site (e.g. site manager, appointed person for craneage, first aider etc.)
  - temporary works coordinator was based on another project, visiting site regularly
  - lack of resource to supervise or inspect temporary works before loading.
- Unclear / dilution of responsibility includes:
  - temporary works coordinator not formally appointed, not given any training or job description and did not realise they were supposed to be the temporary works coordinator.
- Fraudulent activity includes:
  - forging of testing certification
  - forging qualifications – for example, CSCS or CPCS cards
  - deliberately not following agreed sequences for financial gain
  - substitution of components with 'cheaper'/'inferior' components (e.g. anchors or other fixings) without consultation or permission.

18.4. Principal findings and commentary

- The findings of the *Bragg Report* (Bragg, 1975) into temporary works failures 50 years ago have a striking resemblance to the findings of the study carried out for this chapter. As an industry we have made significant progress, but we still seem to have many similar issues to 50 years ago.
- Figure 18.1 shows that residential projects seem to have a disproportionate number of incidents, especially when compared to civil engineering projects. Civil engineering projects tend to be technically more complex, with greater risk profiles, but are

generally managed by experienced and technically competent professionals with well-established and robust temporary works procedures.

- There are still many smaller and medium-sized principal contractors who do not recognise the risks associated with temporary works and do not have robust procedures to manage these risks.
- Some 'building contractors' involved in residential work (new build housing, housing refurbishment and extensions) have a 'basic' / 'simplified' temporary works procedure. This is a reasonable approach (see BS 5975-1, BSI, 2024a, clause 1), as the procedure should be proportionate to the risks involved and complexity of the work. However, some play little more than 'lip service' to implementing this procedural control on site, with time and financial constraints often being quoted in mitigation. Inadequate (often generic) risk assessments are carried out and a safe sequence of works for installation and removal of temporary works is often not considered.
- Figure 18.2 shows that there are a significant number of incidents where temporary works issues result in failure or unsafe situations for associated / adjacent permanent works. This should be an area of particular concern.
- Figure 18.3 shows that lack of procedure, items not being identified as temporary works and unauthorised changes on site are major issues.
- Figure 18.4 shows that there are still many incidents relating to a lack of design solutions. Often these incidents occurred on smaller projects, with less management and supervision, where there may be less awareness of temporary works risks and issues. Poor designs were often associated with a lack of information. Consequently, assumptions were made that subsequently proved to be inappropriate. However, there were a significant number of incidents related to poor design practice (see section 18.3.4). Where an adequate brief was provided and an adequate design carried out, the number of incidents was very rare.
- Designers should ensure they are not over-reliant on software. They need to understand the problem and issues and ensure the input and output information is appropriate.
- Figure 18.5 shows that human resources are a significant issue. This includes a lack of temporary works management expertise, a lack of competence among operatives and a lack of site supervision. There are also many examples where those responsible for managing temporary works (e.g. TWC) were not sufficiently competent, were not formally appointed and did not fully understand their responsibilities. Further, they had to juggle several roles on site (this is almost always the case on small and medium-sized projects), were based on another project, or had inadequate support / resource to fulfil their TWC role.
- The construction industry is fragmented, often with long and convoluted supply chains and a drive to pass on as much risk and responsibility as possible.
- Many principal contractors and most smaller contractors / local builders have little or no in-house temporary works specialist knowledge (including in-house temporary works design capability). They rely entirely on their supply chain to design, procure, install, inspect, load and remove temporary works. This can prove to be problematic as often there is no oversight or 'sanity check' by a knowledgeable and experienced individual to question and challenge what is being suggested / implemented. Although not specifically mentioned in many of the original reports, there may also be a lack of interface management between various subcontractors working in proximity and with

the surrounding area. Often their only focus is on their own section / item of work – that is, what they are being paid to do under their contract. There are various industry good practice examples where contractors have individuals / teams of temporary works specialists who have a roving remit to advise and help site teams as necessary.
- The industry continues to work on low margins and tight (often unrealistic) programmes, so cutting corners remains an issue.
- Failure of temporary works equipment is rare as generally most reputable suppliers will provide quality equipment. Some issues were identified where equipment was not being used in an appropriate manner, alternative (often cheaper) equipment / components were substituted without consent and the equipment became overloaded. This indicates a lack of understanding and a lack of control on site.
- A significant number of issues were identified where a temporary works design was carried out by a permanent works designer who had no / little previous temporary works design experience. It should be more widely recognised that temporary works design is a specialist activity that requires specialist skills and experience (see chapter 2). Often the most onerous temporary condition was not identified. All designers need to be aware of their responsibilities under the Construction (Design and Management) Regulations (HMG, 2015) to follow the general principles of prevention, consider the need for temporary works, provide robust schemes and demonstrate how their designs can be built (and subsequently removed) safely.
- In recent years there have been significant strides in sharing information (good and bad practice) for the benefit of the wider industry. This should be further encouraged, especially when digital collaboration can be utilised.
- BS 5975-1:2024 (BSI, 2024a), clause 4.2 comments that universities and colleges should include temporary works in their syllabus for construction courses. Some progress has been made but, overall, the take up of this is relatively slow and sporadic. This can result in a knowledge gap, with those entering the industry not being able to challenge questionable design and / or site practices. Recent years have seen advances in temporary works training but there are concerns regarding the relevance and quality of some of the training being provided, especially when delivered by a tutor with questionable / limited temporary works experience.

18.5. Main causes of incidents and failures for some common items of temporary works

This section provides some more specific and technical details for some common items of temporary works. For each main category, some common issues are highlighted. There are many generic issues which can apply to almost all temporary works items; these have been highlighted in sections 18.3 and 18.4.

18.5.1 Formwork

- Formwork is assumed to be a common and repetitive low-risk site operation and not designed.
- Concrete mix is changed from that assumed in the formwork design and / or the concrete is poured too fast (rate of rise not controlled by site team), leading to increased pressure on formwork.

- Poor quality components (excessively worn or damaged) and poor workmanship are often associated with inadequate supervision / inspection of formwork prior to pouring concrete.
- Components from different suppliers are mixed and components are substituted without referring to suppliers / designers. These components may not be compatible.
- Uplift forces during concrete pouring (particularly for single-sided formwork) are not adequately considered.

18.5.2 Falsework

- Inadequate consideration / design is given to lateral loading and stability, especially for top-restrained falsework. This includes the unauthorised removal of bracing ('it got in the way' is a common reason given) by those unaware of its importance.
- Poor quality components (excessively worn or damaged) are used or components from different suppliers are mixed (these components may not be compatible).
- Poor workmanship leads to falsework being erected 'out of plumb' and eccentric loading, often associated with inadequate supervision / inspection of falsework prior to pouring concrete.
- There is local overloading of the falsework system by placing large point loads – for example, loading out with reinforcing bars by crane and landing these loads onto falsework bearers.
- Inadequate foundations are laid.

18.5.3 Trenches and excavations

Most fatalities and injuries occur in relatively shallow trenches (less than 2 m deep) due to collapse. Deeper trenches and excavations are more likely to have a full bespoke temporary works design and robust management procedures to ensure safety.

- Inappropriate assumptions are made that the ground is stable and does not require temporary support. The ground is often cut too steep to be stable.
- No initial ground investigation is made and there is no temporary works design, leading to inappropriate or ineffective temporary supports being provided.
- There is no consideration given to changing ground or groundwater conditions (e.g. for long trenches).
- There is no consideration given to time-related effects and changing weather conditions.
- There is excessive surcharge adjacent to the trench or excavation, leading to excessive deflection of the temporary supports or groundworks collapse.
- The trench or excavation undermines existing structures.

18.5.4 Fencing and hoarding

- Items are used in an inappropriate location without adequate consideration of the associated risks.
- There is no design or an inadequate design which does not fully consider wind, crowd, accidental impact or other loading.
- Items are not assembled or used on site as per design or manufacturer's instructions (for proprietary items).

- Components deteriorate over a prolonged period of use. This is most likely because of a lack of appropriate assessment, inspection and maintenance or the designer not considering longevity.
- Items are added, such as netting and signage to mesh fencing, without adequate consideration of increased wind loading.

18.5.5 Scaffolding

Scaffolding for access and working platforms is perhaps the most common item of temporary works across the whole of the construction industry, from major civil engineering projects to domestic extensions / repairs.

- Scaffolding is assembled unsafely by incompetent operatives. Many scaffolds can be 'recognised standard assemblies' (bespoke design is not required) but rely on the skill of the scaffolding team to assemble correctly.
- Components are of poor quality (damaged or badly worn). There are also some examples of 'copycat' components being used.
- Unauthorised onsite modifications are made – often items such as ties and bracing are removed because 'they are in the way'. This is mainly due to a lack of planning and coordination from an early stage.
- Sheeting, netting or large signage are added without adequate consideration of increased wind loading.
- Ties fail in high winds. Too few or inappropriate ties were provided, there was inadequate or unrepresentative testing of ties and the structure to which the ties were fixed failed (e.g. failure of brickwork to which the scaffold was tied).

18.5.6 Mobile cranes and other plant

It is often assumed that the manufacturer or supplier of machinery (such as mobile cranes, tower cranes and piling rigs) is responsible for the foundations beneath their machinery. This is not the case – the manufacturer / supplier should provide loading information, but it is the hirer / user that should ensure this loading is adequately transferred into the ground or structure supporting the machinery. Ultimately, the principal contractor should ensure this is done.

- There is inadequate investigation of the ground or structure beneath the machine.
- Inadequate spreader mats are provided beneath mobile cranes (and other machines such as lorry loaders and scissor lifts). This may be because the ground beneath the crane is not as expected, assumptions are made that mats suggested by the crane supplier are appropriate (bespoke design or design check not carried out) or a different crane (different footprint and outrigger load) or other machinery is provided by the supplier.
- Machinery fails due to worn components and lack of maintenance / inspection.
- There are soft spots beneath or within a working platform beneath a tracked machine. An example could be due to excavations being carried out within the working platform (to remove an obstruction) and not being adequately reinstated.
- Working platform degrades over time due to heavy tracking of machinery, poor weather conditions (e.g. heavy rain) or spoil not being removed. This is associated with a lack of inspection and maintenance.

18.5.7 Structural alterations

- There is no survey or an inadequate / inappropriate survey. The structural nature and condition of a structure must be understood but often no survey is carried out or an inadequate / inappropriate survey is carried out.
- An appropriate risk assessment is not carried out and a safe sequence of work is not determined.
- There is no design, leading to inappropriate / insufficient temporary works solutions being installed.
- Structural components are overloaded by the accumulation of rubble, storage of materials or machinery. Often inappropriate equipment is used for the magnitude of applied loading.
- Structural instability is caused by the removal of load-bearing items or items that provide lateral stability (shear walls or bracing). This is often associated with a lack of survey information and not following a safe sequence of work (occasionally due to cost-cutting).

18.5.8 Stability of reinforcement cages

A reinforcement cage, especially for slender walls and deep base slabs, should be considered as temporary works until such time as the surrounding concrete has achieved sufficient strength to be self-supporting.

- There is a lack of awareness of the potential for collapse and hence it is not considered as a temporary works issue by permanent works designers and the site team.
- There is excessive cage deformation or failure of ties during lifting operations. It is often assumed that a cage is much stiffer than it is.
- Slender vertical bars are unstable due to 'out of plumb' assembly, self-weight, eccentricity and wind loading.
- There is failure at discontinuities / splices (scissoring failure).
- There is accidental impact due to machinery, swinging load on a crane or due to installation of formwork (if formwork is installed after the cage is constructed).

18.6. Case study

One of the authors was involved in the investigation of an incident on a major civil engineering project to construct a new power station.

A rectangular box culvert (each precast concrete segment was approximately 3 m high × 4 m wide × 3 m long and the culvert was around 500 m in total length) was being installed in an open-cut, battered excavation in silty sand. Appropriate RAMS for the excavation and installation of the culvert segments had been carried out. The slopes of the batter had been designed and were being used correctly on site. The excavation and installation of the culvert had to be halted, as it had to cross beneath a site access road that was being used at the same time. The intention was to pause installation of the culvert until the road could be diverted, a process that could take several weeks. The site team took the decision to place a piece of formwork over the end of the culvert and then backfill to make the area safe. The formwork had previously

Figure 18.6 Culvert sections with formwork positioned to retain backfill (author's own)

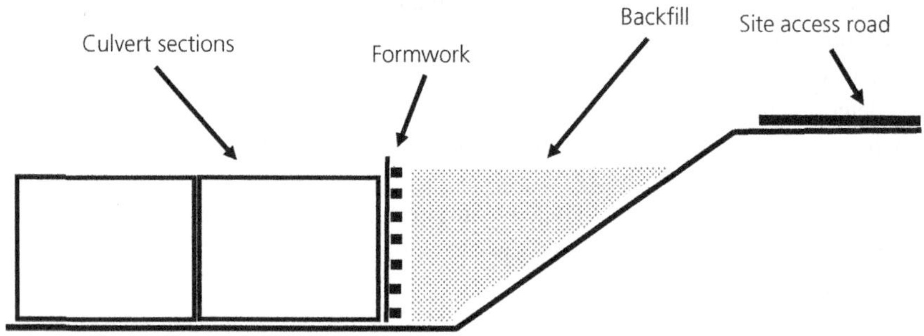

been used for a concrete pour and was positioned to prevent the backfill material from entering the culvert (see Figure 18.6). The formwork was approximately 3.5 m high × 6 m wide and weighed around 1.0 t. The formwork was lifted into position by mobile crane, which provided support to the formwork long enough for excavators to place the backfill against the formwork. The formwork was not tied / fixed to the culvert and relied on the backfill for stability. RAMS was not carried out for this operation, the site team did not consult with site management or designers, did not make any notes of their actions or communicate any information to others.

Several weeks later, the site access road had been diverted and a different team were to excavate the backfill and remove the formwork to allow the final sections of culvert to be installed. Their intention was to remove the backfill and then connect the formwork to a mobile crane for lifting out. The new team did not realise the formwork was not fixed to the concrete culvert and relied solely on the backfill for stability. They were following the original RAMS for the excavation and culvert installation, which did not mention the formwork. During excavation of the backfill, an operative entered the excavation to remove some backfill material from the formwork with a shovel, prior to it being lifted out by mobile crane. The excavator continued to remove backfill rapidly until the formwork became unstable and fell onto the operative in the excavation, who sustained significant injuries.

REFERENCES

Bragg SL (1975) *Final Report of the Advisory Committee on Falsework. Department of Employment and Department of the Environment.* The Stationery Office, London, UK.

BSI (2024a) BS 5975-1:2024: Management procedures for the control of temporary works. Code of practice. BSI, London, UK.

BSI (2024b) BS 5975-2:2024: Temporary Works – Falsework: design and implementation. Code of practice. BSI, London, UK.

CROSS (Collaborative Reporting for Safer Structures) (2023) Report 1133: Misuse of Standard Details and Notes on Structural Drawings. https://www.cross-safety.org/uk/safety-information/cross-safety-report/misuse-standard-details-and-notes-structural-1133 (accessed 10/01/2025).

HMG (Her Majesty's Government) (2015) Construction (Design and Management) Regulations. The Stationery Office, London, UK.
IStructE (Institution of Structural Engineers) (2017a) *Temporary Works Toolkit Part 5: Temporary Works Failures – What are the Common Causes?* IStuctE, London, UK.
IStructE (2017b) *Temporary Works Toolkit Part 7: Temporary Works Failures – What can we Learn?* IStructE, London, UK.

FURTHER READING

Hewlett B, Jones A, Marchand S and Bell B (2014) Re-visiting Bragg to keep UK's temporary works safe under Euronorms. *Forensic Engineering* **167(2)**: 58–68.
HSE (Health and Safety Executive) (2011) *Report RR834: Preventing Catastrophic Events in Construction*. HSE, London, UK.
Pallett PF, Burrow M, Clark L and Ward RT (2001) Contract Research Report 394/2001: *Investigation into Aspects of Falsework.* HSE with Birmingham Centre for Railway Research and Education, HSE, London, UK.

Ray Filip
ISBN 978-1-83608-525-6
https://doi.org/10.1108/978-1-83608-524-920251021

Index

Printed and bound by CPI Group (UK) Ltd, Croydon, CR0 4YY

10/12/2025

14789260-0001